Earth to Heaven

Earth to Heaven

The Royal
Animal-Shaped Weights of Burma

DONALD AND JOAN GEAR

SILKWORM BOOKS

This book is dedicated to all those who, over the past many centuries, provided the material which enabled the authors to enjoy for so long the many kinds of pleasure they obtained from the study of the Burmese weights.

ISBN 974-7551-20-9

First published in 1992
This edition is published in 2000 by
Silkworm Books
6 Sukkasem Road, T. Suthep
Chiang Mai 50200, Thailand
E-mail: silkworm@loxinfo.co.th
Website: www.silkwormbooks.info

Cover photograph and design by Umaphon Soetphannuk
Printed by O. S. Printing House, Bangkok

5 4 3 2

Contents

List of Maps

List of Tables

List of Plates

Foreword

THE ROYAL ANIMAL WEIGHTS are almost the only remaining portable metallic evidences of past Burmese activities which are, or were, relatively common. Partly this is because of the destruction and looting of temples and pagodas by Kublai Khan's men for fortifications in 1284 AD, by the Burmese in the 11th and 18th centuries AD and by the Mons in the 15th and 18th centuries AD. Partly it is because of the continual looting of these structures for which royal approval was given in 1526 AD. Partly it is because any metal object, especially of precious metal, was re-worked repeatedly into different forms either for re-sale or re-use. The study of the weights has provided us with many pleasures. They formed a foundation for an appreciation of the romance of Burmese and Asian history and myth. It was, and still remains, a pleasure to recall the visits in search of knowledge to villages and monasteries, sometimes remote, and the discussions with these mainly delightful people in a usually beautiful land. There was the beauty of the shapes of the larger weights and, occasionally, their patination. There was the satisfaction of searching for, and sometimes discovering, a new specimen. There was the joy of delving ever further back in time, seeking for the symbolic meanings. From the nature of our usual work and all the foregoing, there was a feeling for the long-gone time when the wild animal formed an essential part of western man's physical, emotional and rational life. Though we did not have the advantages of the resources of a wealthy university of which to make use, also we did not have the disadvantages. We had the pleasures of finding out for ourselves, of admiring the work of the 19th century 'giants' and searching in the modern literature for matters which might be relevant. Of course, there are unwitting faults and errors in our work. We should be glad to learn of them.

The authors must express their thanks to the many interested Burmese monks, professional men, academics, artisans and friends who went out of their way to help in this investigation during the late fifties and early seventies. In particular, we would like to thank U Lu Pe Win, U Aung Thaw, U Maung Mya and U Hla Thamein, formerly of the Archaeological Survey, Rangoon; Daw Ama and U Hla Myint of the Ludu Press, Mandalay; Shwegaingtha U Thaw Bita of the Anuyokda Chaung Monastery, Sagaing; U Myaing of the old weight village at Kyauksittaung, Mandalay; our old friends U Tun Tin, former Postmaster-General and U San Dun, former magistrate, who directed us to so many people and Richard Bone. Our thanks to the many other people who are not mentioned by name here or in the notes, who also helped.

We would also like to express our thanks to Avery Research Administration of Birmingham, whose assistance from 1972 to 1975 in obtaining copies of publications and sending them to us in various countries was a kindness and of great help. A number of correspondents in several countries have assisted with particular matters. Our thanks to them for their time and knowledge. Dr. G. Houben of Zwolle, Holland and Dr. G. Prunner of the Hamburgische Museum fur Volkerkunde were kind enough to suggest improvements to the 1975 draft. In particular, Dr. Prunner's knowledgeable comments made us realise that we needed to delve much deeper than we had done. More recently we have profited from the expertise of Dr. Frances Wood of the Chinese section of the British Library and of Dr. Patricia Herbert of the Burmese section of the British Library. We are grateful to them and to the staff of the library of the Percival David Foundation. Dr. D.A. Swallow of the Victoria and Albert Museum was exceptionally kind in adding to her many commitments by reading and commenting on the whole of the typescript. Mrs. Ruth Willard of San Francisco was so kind as to give her valuable comments on Appendix 3.

The authors' acknowledgements are due to the following institutions for permission to reproduce the illustrations, Plates 1, 2, 6, 7, 9 and 56: Pitt-Rivers Museum; Plate 4, Bodleian Library.

Writing Conventions

THE SPELLING OF non-English words is that employed by the older British writers, especially Harvey. Diacritical marks usually have been omitted.

The older country names have been employed usually, e.g. Persia for Iran, Siam instead of Thailand, Burma instead of Myanmar, but Vietnam instead of Annam. Chinese province names are usually given in their older form or both older and newer are shown together.

Indo-Aryan has been used for the greater part instead of Indo-Iranian or even Indo-European.

Christian era indications of BC and AD are used instead of Common era.

Usually the word 'duck' has been used without a sexual connotation. Where masculinity is to be emphasized, the words 'drake' or gander' are employed. Where the nature of the bird, duck, goose or swan is not evident, the word 'anserine' has been found convenient. Where the evidence is even less clear, the word 'avian' has been employed to indicate an unspecified kind of bird.

In general the intention has been to standardise, simplify, reduce consideration of possible errors in matters not essential to the work and to use terms and expressions familiar to the non-specialist.

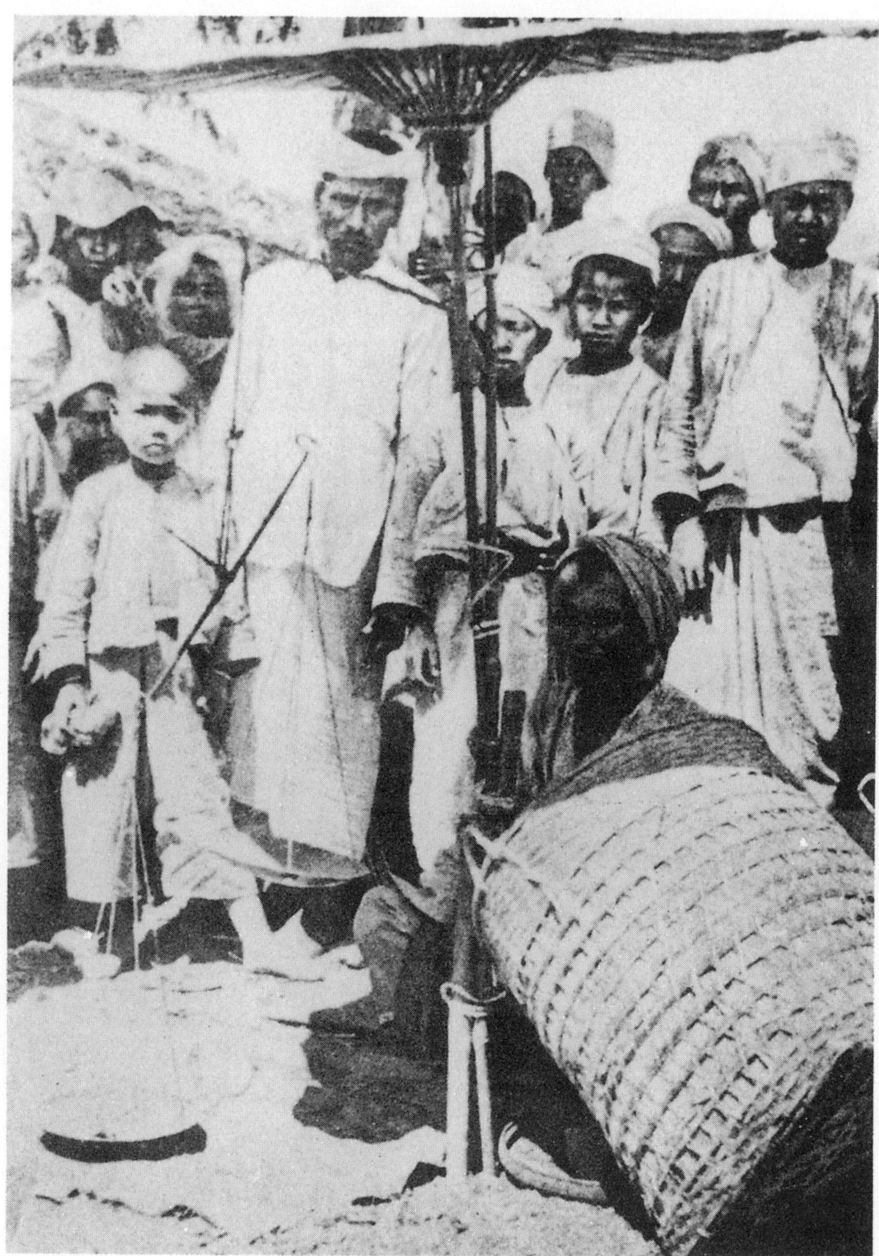

Plate 1: Burmese scales in use
Fort Stedman bazaar, 1917

Introduction

From about the 15 th century A.D. the animal-shaped weights of the Burmese empires were made in different shapes of three (possibly four) different kinds. One is a somewhat feline-like beast. Another is elephant-like and used only in north Siam. A third is a bird, usually clearly duck-like. They all stand on plinths or bases of different shapes. In many cases the base-sides bear an impressed small sign of a geometric or animal-shape. The weights are made of a cupriferous alloy. They range in mass from about 2 to 4000 grams, rarely more, on an essentially decimal scale. Originally they were used to weigh relatively high-valued products such as the silver ingots which were used as currencies. Dome-shaped marble weights were used for the commoner products. The weights were intended not only for trading purposes but also for religious and political purposes conveying as part of their message the powers of the divine spirit in heaven and the earthly god-king. They were also intended to imply the Buddhist disapprobation of stealing through the use of incorrect weights. Because of the monarchical symbolism and for other reasons mentioned elsewhere in the text, the 'official' weights are 'royal' weights.

Although production of the royal weights ceased over 100 years ago, they are used occasionally to this day. In 1917 people from Yunnan and Kiangsu in China and other distant regions visited the markets on the Inle Lake in eastern Burma (see Plate 1) where sets of animal weights were still in use.[1] In 1970 some conservative traders and monks practising traditional medicine still used sets of animal weights in Upper Burma. Elsewhere in Burma the occasional animal weight may be found in use along with car-piston heads and stones in village market places. Some villagers even in Thailand and Laos keep them as mementoes of the past. Copies of ancient weights are still made in the Shan states. Otherwise they are to be found on the traders' stalls of those towns commonly visited by tourists. There are also small private collections, at a guess about 50 in Burma and some in Thailand. Outside of Burma the weights were still in use (1970) along the more remote routes that pass from north-east Burma into Yunnan, north Thailand, north Laos and the Thai cantons of Vietnam.[2] The trade in the weights has resulted in the export of relatively considerable numbers from Burma and northern Thailand mainly to Bangkok and Singapore.

The study of the weights was made firstly to ascertain their physical characteristics and provide a rational classification. Thus this forms the first part of the work and deals with the mass units, mass scales, dimensions, frequencies, materials and shapes. Included are comments concerning the usage and manufacture of the weights. The second objective was to discover the symbolic meanings. So the identification, origins and symbolic meanings of the creatures, their parts and the chronological sequence of the weights form the second part. The third objective was to ascertain why creatures which are found on the ancient Egyptian, Mesopotamian and Persian weights should reappear on the Burmese weights perhaps two thousand years later. So the last part of the work deals with the transference of the lion-duck motif to Burma. It considers the population centres, the means of transport and travel times, the movements of people, trade and routes. Appendices refer to the dome-shaped marble weights of Burma and animal-shaped weight-like objects from Burma and neighbouring countries. Chronologically, the study goes back to long before the present peoples entered Burma. Geographically it ranges to India, the rest of south-east Asia, China, Central Asia, Persia and Mesopotamia. Therefore, where relevant, information concerning the weights, the religions, the associated animal representations and symbolisms of these regions is included.

The field work upon which the study is based was done during the writers' residence in Burma during 1957 to 1960 and 1970 to 1972. It was recorded but not published in 1976. Further studies were made subsequently by the writers and are incorporated here.

Concerning the previous information assembled and published, special acknowledgment of the value of the works of Temple and Decourdemanche is due. Any serious study of the Burmese mass scales and currencies must start with the work of the former. Those works and Temple's studies of the Indian mass scales and their relation to those of China and South-east Asia are outstanding. Just as admirable in its field is Decourdemanche's work on tracing the connections between the ancient mass scales of the Mespotamian region and eastern Asia. In addition there are brief articles by Noettling, 1982: Sale, 1910; Gardner, 1962 and 1968; Fraser-Lu, 1982 and a few articles in Burmese newspapers and magazines. The Brauns published a comprehensive series of photographs and masses in 1983, rather more of figurines than weights, and Mollat made a style and chronological classification in 1984. Some writers concluded that the different styles of the three animal shapes represented several different kinds of animals and related them to one or another of the various associations of oriental mythological animals. Their problems were compounded by the existence, mostly in Siam, of sets of animal figurines used by astrologers and for household shrines. Probably these were used as weights when nothing else was available, a common practice with many weighty objects, including pebbles. Other short references to the physical properties of the weights lie scattered mostly in the older European literature. In Burma itself the few references obtained from the chronicles and manuscripts proved of uncertain value. Extensive inquiry of otherwise knowledgeable people yielded little factual information but much valuable help and direction.

The royal weights are defined here as those which bear genuine signs and which, by their frequency of occurrence in Burma, may reasonably be assumed to be Burmese. They are also those which, though not bearing signs, are of a shape-style[3] and mass very similar to those that do. This definition results in the exclusion of most unofficial copies with the exception of some excellent ones made perhaps 150 years ago (see Appendix 2, 'Unofficial Weights and Weight-like Figurines').

Notes

1 Annandale, p. 195. Temple, 1913, p. 118

2 U Thaw Bita, a monk of Sagaing interested in the animal weights.

3 The weights occur in different styles separable into classes and systems which can be referred to the periods during which they were made. These terms are explained in the introduction to Chapter 5, 'Dimensions, Shapes and Decoration'. Qualifying letters sometimes accompany the references to Group 2. U and V refer to the shapes of the tail cross-section of the Group 2 bird weights. 'r' refers to the mouth appendage shape of one of the Group 2 beast weights.

1
Mass Units and Standardisation

Mass unit

A MASS UNIT is a mass chosen as a base, of which the multiples or fractions form its accompanying scale. For this investigation, commonly used mass units of between 8 and 16 grams have been chosen for the greater part. The Burmese mass unit is the *kyat.*

Style group unit mass data

1078 specimens of weights believed to be genuine animal weights in good condition (incised decoration not worn away) and without addition of material, were weighed in August 1972 in Rangoon, Pegu, Mandalay and Magwe. Since the weights were classified by group and style, the results of the survey form the foundation for a number of conclusions utilised subsequently. Additions of material were rarely observed and subtractions, never.

Variability of unit mass

Those weights selected for mass determination were chosen for their lack of visible wear, corrosion, added material and for their wholeness. About 300 were washed and 700 were unwashed. When a weight's gram mass had been determined it was divided by its *kyat* multiple (obvious from the linear dimensions) to obtain the unit mass of its *kyat.*

Burmese weights were not made as accurately as in Europe and hence vary to a greater extent from a mean. Pre-19th century AD Indian and south-east Asian coins had even less accurate masses while inscribed specimens of the Babylonian *mina* vary between 640 and 918 grams. There are many similar examples. Because of varying chemical composition there was little need in south-east Asia for great accuracy when weighing, for example, silver alloy ingots. Moreover, the problems of maintaining a constant alloy composition, moulding and casting the weights created difficulties in obtaining high accuracy. Also important in this connection is that in India and Asia, precious metal mass scales were based on the numbers of *rati* (*abrus precatorius*) seeds and not multiples of a unit mass. Since the mass of the *rati* seed varied with the climate and hence the location, both the unit mass of the weights and the masses of the weights could vary if seeds were used for calibration though the scale itself remained constant.[1]

Weights in Europe have long been made to precise standards. Hence because of corrosion and wear it would seem that among the Burmese weights, the highest recorded unit mass must be the correct one. Concerning the former, uniform oxidation of 5% of the original material of a Burmese weight would increase its mass by about 2% and its volume by a little less than 2%. That is, a weight originally of 50 mm diameter would increase this dimension by almost 1 mm and so probably obscure the incised surface decoration. Uniform wear of 5% of the original mass of a Burmese weight would result in the uniform loss of about 1 mm, again obscuring the incised surface decoration. If the loss is confined to the two main sides of the weight, about 2 mm would be lost from each side, so removing some of the moulded features.

The differences in the precision of the weighing and the mass units used for comparison also add complications. At least one traveller's weighing was made at a time when his home unit mass had changed and in his case, some uncertainty exists as to which unit mass was used. Steelyards and *bismars* were, and are, in common use in the East especially, but not only, for the commercial products. Both are less accurate than equal-armed balances and the latter is more easily falsified by moving the fulcrum. The Dutch and the Portuguese usually neglected small fractions in their weighings. In parts of India at least, though bearing the same names, retail weights weighed somewhat less than wholesale weights.[2] It cannot now be known with certainty what were the origins and ages (and hence masses) of the weights measured. There are other possible sources of error.

The percentage deviation from the mean mass (accuracy of mass) was calculated for each style period for 596 weights. The distributions were normal, sometimes skewed towards the lower end. The results are shown below.

Table 1: The Accuracies of the Weight Masses

Calculated percentage deviations	GROUP			
	All	1, 2, 3	4	5, 6, 7
	Percentages of weights deviating from means by less than calculated percentage deviations			
±2	51,2	48,5	56,8	48,3
±4	79,5	78,5	81,8	81,3
±6	90,3	91,9	89,7	83,7
±8	95,5	96,8	94,6	93.0
±10	97,8	99,1	96,5	97,6
±12	99.2	100,0	98,5	100,0
±14	99,7	—	100,0	—
No. of masses	596	334	204	48

Variations in unit mass with location

The table 'Variations in the Burmese *Kyat* Mass with Time' suggests that differences in the unit mass of the *kyat* occurred between different locations. Though there is this possibility, it cannot be proved or disproved by these data. However, the table makes it evident that the differences between contemporary *kyat* masses determined at different locations lie within the range of variation of the mean *kyat* mass, i.e. within +/- 10%.

It was stated in the middle of the last century that, from observation, the standard of weight appeared to have been the same all over the country. Since it was in the financial interest of the Burmese monarch to ensure that his masses were identical, so far as was practical, throughout his country, it seems unlikely that the *kyat* mass would have varied from one locality to another, certainly after Burma was united in the mid-16th century.

Variations in unit mass with time

The *kyat* has varied in mass with time. It is evident that the *kyat* mass has increased since before the 16th century from about 11,2 grams or less to 16,3 grams or more in recent times. In Siam (Thailand) the increase in the *baht* mass over a similar period was from 9 or 12 grams to 15 grams.[3,4,5] In India it estimated that the mass of the *rati* rose about 28% from about 1100 to 1600 AD and then remained more or less constant.[6,7] It has been found that over much the same period of time the coin masses rose only by about 9%.[8] In China the more generally used *liang* has remained at about 37,3 grams +/- 5% since about 618 AD,[9] having risen from about 13,9 grams.

The *kyat* and *baht* masses of about the 14th century or before suggest a possible linkage between them and the Indian *tola* of those times. The *tola* weighed between 11,6 and 12,0 grams. The subsequent constancy of the Indian Mogul weight masses and the Chinese *liang* compared with the continuous parallel rise in the *kyat* and *baht* masses until the mid-18th century suggest a discontinuation of this linkage from 1600 AD or before.

The following note provides a possible explanation for the increases in *kyat* mass with time. In India changes in the coin mass scale were frequently introduced either to create an increased income for the ruler and his government or by conquerors as an administrative convenience. In the case of rulers issuing coins they ensured an increased income by decreasing the mass or lowering the precious metal content of the coins. In some cases these changes were made at regular intervals.[10] When weighed metal ingots were used instead of coins and the king was the chief, and perhaps the sole, trader,[11] the obvious way to increase his income from revenue paid in silver was to increase the unit mass. Rulers needed money to pay not only for their personal expenses but for their officers, their religious buildings, their irrigation systems, their wars, the change of capital with each new reign and for other reasons. Personal character, the duration of reign, the need to acquire merit and the ability of his subjects to pay, e.g. after famines, depopulations, civil and foreign wars, could be important considerations in deciding upon an increase in the *kyat* mass. These matters were studied and a list was prepared of those kings of Burma and Pegu who were most likely to have increased the *kyat* mass (see table 'Appropriate Dates for Increases in the *Kyat* Mass'). Also shown on the table are the dates when weight standardisations and revenue inquests are recorded to have taken place.

Table 2: Variations in the Burmese *Kyat* Mass with Time

Reference	Year of Publication	Location	Equivalent Masses		Kyat Mass grams	Style Group	
			European	"Burmese"		Bird	Beast
70	1515	Pegu	3 arratels, 5 ozs Port	1 viss	11,4	—	—
12	—	—	—	—	11,2	6	—
13	1554	Pegu	40 oz Port	1 viss	12,6	—	—
13	1554	Pegu	3,2 oitavas	1 tical	11,5	—	—
13	1554	Dala	41,25 ozs Port	1 viss	11,8	—	—
13	1554	Cosmin (Bassein)	—	—	10,2 or 11,4	—	—
12, 13	1554	Martaban	47,25 ozs	1 viss	13,5	5,7	5
14	1585	Pegu	40 ozs Ven	1 bize	12,6	—	—
70	1693	Pegu	3 lbs Dutch	1 bits	14,8	—	—
15	1727	Pegu	39 ozs troy	1 viece	12,1	—	—
12	—	—	—	—	14,5	4	4
16	1739	Burma	—	—	14,6	—	—
12	—	—	—	—	14,7	—	4H
17	1752	Pegu	3 lbs 5 ozs 5dr	1 viss	15,1	—	—
18	1775	Pegu	3 lbs 5 ozs 5dr	—	15,1	—	—
19	1783	Pegu	1,25 rupees	1 tical	14,1 14,6	—	—
20	1786	Pegu	—	1 viss	15,3	—	—
21	1821	Rangoon	3 lbs 5 ozs 5.33 dram	1 viss	15,1	—	—
12	—	—	—	—	15,3	3	—
12, 22	1821	Pegu	1,138 tolas	1 tical	15,4	—	2t
12	—	—	—	—	15,7	2	2v
12	—	—	—	—	15,8	—	2r
12	—	—	—	—	15,9	2,4E	—
23	1835	Pegu	—	1 tical	15,4	—	—
23	1835	Rangoon	—	1 tical	16,2	—	—
23	1835	Rangoon	—	1 tical	15,1	—	—
24	1852 to 1888	U. Burma	3,56 lbs (average)	1 viss	16,6	—	—
24	—	—	252 grains	1 kyat	16,3	1	—
25	1830?	Burma	—	1 kyat	16,33	—	—

Notes to Table 2

(a) The year shown is that of the publication of the book in which the mass was recorded, not when the mass was determined. The latter would be earlier.

(b) Entries 17 and 18 appear to be repetitions of the information first provided by Brooks, entry 16.

(c) In the Style Group columns, E and H refer to the period. Mouth appendages are referred to by t (trefoil), v (vee), r (rod).

(d) Port. = Portuguese. Ven. = Venetian. lb = pound. oz = ounce. dr = dram.
"Burmese" = Name recorded in Burma by Europeans.

(e) ⅕ shown as 0,2; ¼ as 0,25: ⅓ as 0,33.

Table 3: Approximate Dates for Increases in the *Kyat* Mass

Date	King	Weight Standardisation (WS) or Revenue Inquest (RI)	References Notes	References Source
1853 – 1873	Mindon		a	
1819 – 1837	Bagyidaw		b	
1782 – 1819	Bodawpaya	RI 1803, RI 1784	c	28
1763 – 1776	Hsinbyushin			
1752 – 1760	**Alaungpaya**			27
1758	◊		d	28
1629 – 1648	Thalun	RI 1638		
1605 – 1628	Anaukpetlun			
End 16th Century	◊		e	29
1551 – 1581	Bayinnaung	WS 1563 – 1575	f	30, 31, 32
1531 – 1550	**Tabinshwehti**			
1472 – 1492	**Dammazedi**			
1453 – 1460	Queen Shinsawbu			
1391	◊		g	33
1385 – 1423	Razadarit			
1331 – 1353	Binnya U			
1287	**Wareru**			
1259 – 1317	◊		h	33
1112 – 1167	Alaungsitthu	WS		34
1044 – 1077	**Anawrahta**			

Column 2, ◊ = Siam. Bold names = first king of a new dynasty.

Notes to Table 3

(a) Mindon may have been more concerned with his coin issue and the British Indian weight standards than with the Burmese weights.

(b) Bagyidaw became insane.
(c) Bodawpaya made unsuccessful coin issues and abandoned them.
(d) King Ekat'at introduced a law standardising the weights and measures of Ayudhya.
(e) At the end of the 16th century the King of Ayudhya required foreign merchants to ensure that their weight corresponded with the royal weights.
(f) Chiengmai in Siam was included in King Bayinnaung's standardisation after the indigenous weight standards had been destroyed.
(g) In 1391 Siam and Cambodia obtained the standards of Chinese weights and measures.
(h) Weight standards were established by King Mengrai of Lan-Na.

Pre-Pagan mass unit:
In the south of Burma, the Pyu and Mon (Dvaravati) coins indicate a precious metal mass unit of between 8,5 and 11,0 grams. However, coin and weight mass units do not necessarily agree even though they bear the same name, e.g. the British pound.

Pagan mass unit:
Though no weights from Pagan, the 11th to 13th century AD capital of Burma (see Plate 2) have yet been recognised, information from inscriptions[35] there enables a reasonable estimate to be hazarded. Copper was weighed by the *pisa*, the *buih* and the *klyap*. The *klyap* (equivalent to the *tola*) mass of about 11,8 grams is similar to one of the ancient Indian masses of the *karsha*, the closest being about 11,2 grams.[36]

In neighbouring Lan Na and Sukhothai, the unit masses in the mid-14th century were about 12,2 grams. This mass persisted in north Siam but not in the south, until the 17th century[37] and probably the 19th century, though both states had long disappeared. There is in use in north Thailand and Laos a weight known as the *moeun* or *mu'n*. It has a mass of about 12 000 grams in Laos[38] and 12 700 grams (28 lbs.) in north Siam.[39] In Laos it is part of a decimal scale 1 *mu'n* = 10 *changs* = 100 *damling*. The word *mu'n* also means 10 000 in Siam. *Mu'n* was in use in Sukhotai about 1361 AD also with the meaning of 10 000 in connection with quantities of silver and gold. *Tomaun* is both a Mongol and Persian word meaning 10 000.[40] The word *mu'n* is probably a variant of an ancient and widespread weight name known as *man*, *maund* or similar. The mass unit *man* (*maund*) may have reached India as late as the beginning of the Mahometan invasions in the 8th century AD. However *maund* or *man* in its Babylonian and Accadian form of *mina* is a word some 5 000 years or more in age and, like the Babylonian mass systems, spread over much of Central and Western Asia[41] and so to India. *Mans* with the approximate weight of 12 000 to 13 000 grams like those on north Siam and Laos appear to be almost entirely confined to the southern coastline of India from north of Goa on the west to Orissa on the east.[42] Referring to pre-Pagan times, the Pyu of the Prome region were forced to flee this southern

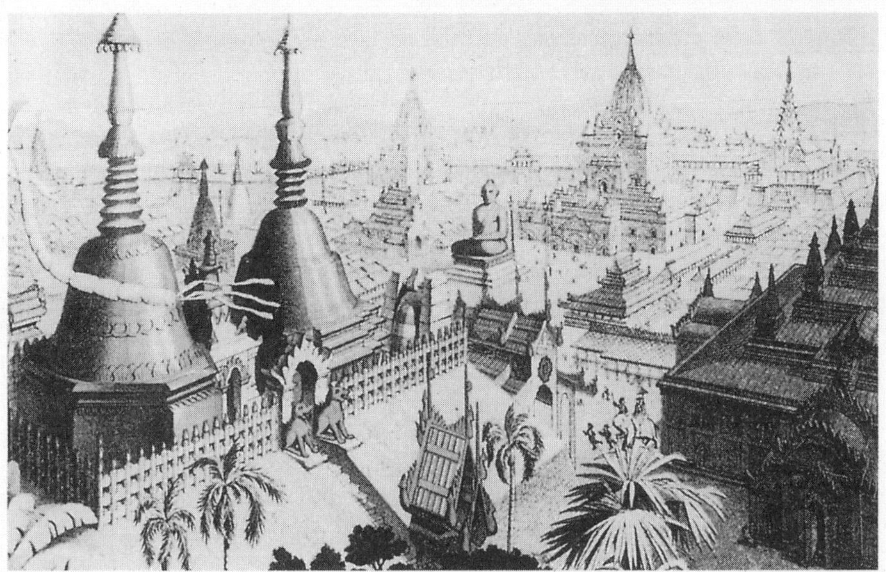

Plate 2: Pagan , 1287 AD

This representation was prepared from 19c. AD views of Pagan and Marco Polo's account of Pagan according to the descriptions given him by Kublai Khan's officers. Pagan was reported to be a very noble city with towers (pagodas) of stone covered with gold and silver (lead?) having bells at their tops tinkling in the wind. When viewed from afar it was "one of the finest sights in the world with its splendidly finished pagodas glistening in the sun." (Yule, ed. H. Cordier, 1903)

settlement of theirs just after 800 AD and joined their kin in central and northern Burma among the Shans and Burmese under the domination of the Nanchao (Yunnan) kings.[43] Here then was the possibility of a combination of the Pyu pre-Pagan unit and the Sukhotai (and Shan?) multiple of 10 000. Under these combined conditions a Pagan mass multiple might have been 10 000 units of about 0,85 to 1,27 grams equivalent to 1 000 units of about 8,5 to 12,7 grams, like the Indian *karsha* or the Pagan *klyap* and also equivalent to the north Siamese *mu'n* and the south Indian *man.* 0,85 grams was the mass unit of the 2000 B.C. Indus Valley civilisations.

Group 6 mass unit:

Within this group the average of all the *kyat* masses computed from the weight masses is 11,2 grams. However, there might be two unit masses, one of 10,2 grams and one of 11,4 grams.[44] Referring to the section 'Group 6 mass scale', it has been shown that the pre-530 AD mass of the *liang,* 13,9 grams is identical with that of one of the members of the Group 6 mass scale. The 2,5 or 3 mass unit member of the scale computed from the few specimens available averages 40,2 grams, a value which lies close to that of the post 618 AD Chinese *liang*[45] and to several of the pre-9th century AD Indian *palas*, i.e. 37,3 grams in both cases.[46] The close approach of the mass of the Indian *tola* to that of the pre-618 AD Chinese *liang* may have led to the alternative name of *tael* (perhaps indirectly through the Malayan weight *tahil*) being given to the pre-530 AD *liang* of 13,9 grams, a name which the subsequent *liang* of 37,3 grams retained. It was at about this time that Indian and Chinese Buddhist contacts were at their peak. The average mass of the greatest member of the scale is 1112 grams which is very close to the pre-993 A.D. Chinese unit, the *huan* and also to the Siamese *chang* of pre-1688 A.D.

It may be relevant here to refer to a considerable change in mass systems which took place in India after the Mahometan conquests of about 1200 AD.[47, 48] This may have had some influence on the Burmese coin (and weight?) masses of the subsequent period. For example, up to about the 11th century, the unit mass of the Arakanese coins (north-west Burma) was about 7 to 8 grams. When coins were next produced in the 15th century, the unit mass had increased to between 10 and 11 grams.[49]

Group 7 mass unit:

The *kyat* mass of the Group 7 bird weights varies between 14,0 and 14,1 grams from the writers' weighings. Other masses from Siam indicate that possibly similar bird weights may have had a unit mass of 12,0[50]

Groups 5 to 1 mass units:

The unit masses of Groups 5 and 4 are about 14,4 and 14,7 respectively. The latter is close to the mass of the north Indian *karsha*, 14,9 grams, used in the weighing of copper and other metals. The Group 3 unit mass is approximately 15, 3 grams, the Group 2 unit mass varies from 15,4 to 15,9 grams and the Group 1 unit masss is 16,3 grams.

The origins of the mass units:

The mass units of Burma may have had an indigenous origin or have been derived from foreign sources. That the Indians probably initiated the first mass units in Burma will be clear from the preceding account. Later a Chinese influence appears to have had an effect. Subsequently, the later mass units were probably indigenous, an inference derived from the increase in the mass unit over the years.

According to several students of weights[51] the unit masses of the Babylonian *shekel*, *mina* and *talent* from about 2500 BC not only spread across west and central Asia south of the Caucasus mountains and Caspian Sea but migrated to India and China. A similarity has been recognised between the mass of the Babylonian light *talent*, the

Assyrian *daric* (a specific *shekel*, the most closely associated with Assyrian lion and duck weights), the modern Chinese *liang* (*tael*),[52] the ancient Indian *pala* and the Etruscan unit mass, each of which amounts to about 37 grams. A similarity also has been seen between the mass of the Babylonian heavy *talent* and the Chinese *hu*. Clearly the weight systems of Eurasia could have migrated and probably did migrate in modified forms from one side of the land mass to the other, just as the weight name (e.g. *denarius*, *mina*) and ancient artifacts have done. However, the variability with time and place of the unit masses involved, e.g. the *talent*, the *liang* and the *pala*, and the considerable geographical gaps existing between the locations of available information, make the search for proof a difficult one.

On the evidence and inference provided, it is probable that in the Pagan region, though perhaps not everywhere in Burma, the *kyat* mass was about 12, 0 grams up to about 1300 AD.

The similarities to the Burmese *kyat* mass of the Siamese unit masses are compared in the following table 'Comparison of Changes in the Mass of the Siamese *Baht* and the Burmese *Kyat*'. Prior to about 1752 to 1782 (Ayudha, the home of the *baht*, was destroyed in 1767 AD), the masses of the *kyat* and the *baht* (if such was its name in north Siam - see note 4 of the chapter 'Mass Frequencies, etc.') were similar and varied similarly with the exception of the weight masses of Chiengmai. This had been conquered by the Burmese in 1558 AD and the Burmese subsequently were supposed to have standardised its weights. It then suffered decades of oppression and conflict. However, in 1595 Burma was in chaos and Chiengmai fell under Siam's suzerainty from which it was not recovered by Burma until 1614. Even long afterwards revolts occurred in the surrounding region.[53] It may be that this long period of turmoil and administrative change is one of the possible reasons for the Burmese tolerance of the persistence of the 12,2 unit mass in Chiengmai when it had been abandoned by Burma and Siam. In addition to the reason given above, this may have been because the Burmese officials who controlled northern Siam as tributary states for so long were somewhat more concerned with controlling the China trade to southern Siam and Burma using their own weights than they were with the local north Siam trade and its associated mass unit and mass scale. This control may explain the periodic copper shortages in Siam whereas the Burmese appear not to have suffered markedly from such shortages until the end of the 18th century when they lost north Siam and had to use lead instead of *ganza* for low-valued payments. The initial reason however, may have been the respect in which the ruler of Lan-Na in north Siam was held by the Burmese at the time of his defeat. He became one of Burma's 37 *nats* or guardian ancestor spirits.

Referring once again to the table 'Comparison of Changes in the Mass of the Siamese *Baht* and Burmese *Kyat*', it is evident that between the 14th century AD and

1615 AD the *baht* changed from about 12,2 grams to between 14,1 and 14,9 grams. The Group 6 average unit mass of about 1452 AD to 1472 AD appears to include two masses, one of 10,2 grams and the other of 11,4 grams while the Group 5 weights have a unit mass of 14,4 (and possibly 14,0) grams. It appears from these figures that the higher unit mass of over 14 grams may have been introduced either during the reign of the Mon King Razadarit (1385-1423 AD) or at some time near the end of Queen Shisawbu's reign and the beginning of King Dammazedi's, i.e. about 1460 to 1472 AD. The wholly decimal mass scale may have been introduced at the same time. See also the sections on 'Trade' and symbolisms of the bird weights and of the beast weight tails.

So since it seems likely that before 1460 AD the *kyat* mass in Burma was about 11 to 12 grams (see also Pagan mass unit), there must have been good reason to change it. This may have been because the Chinese wanted to replace the mass of their old *liang* of about 14 grams (see note 52), apparently still in use throughout an extensive region (see the tables 'Variations of East Asian Precious Metal Scales with Times and Place' and 'Origins of the Mass Scales'), with their later one of approximately 37 grams. In this latter case two *liang* (a more common Chinese silver ingot *[sycee]* mass than one *liang*) would equal approximately five *kyats* or five *baht* of 14 grams which would fit more readily on to a decimal scale than six *kyats* of about 12 to 12,5 grams.

But why did not the Burmese adopt the Chinese mass unit as did the Cambodians, the Annamese (Vietnamese) and the Siamese, at least in their commercial weights? The answer to this question is not obvious but the reason may be that Chinese cultural influence was greater in Cambodia, Annam and Siam than it was in Burma. Perhaps more importantly, the region of the Malayan peninsula and the countries to the east of it were dominated by the Chinese commercial and industrial entrepreneurs as Burma never was.

Standard weights and standardisation

The table 'Appropriate Dates for Increases in the *Kyat* Mass' shows the years when weight standardisation and revenue inquests were undertaken in Burma.

During King Bodawpaya's reign (1782-1819) and probably before as well, it was a criminal offence to use weights which had not been made 'at the palace'. There is evidence from tradition and historical record that special sets of weights were retained by the authorities and used in cases of dispute. King Bayinnaung (1551-1581) is said to have kept a standard set of bird weights in his palace at Pegu and traders were required to use weights equivalent in their masses to those in the palace.[68] It is conjectured that the rare bird weights which bear a small bird on their chests may have been such standardising weights, especially since they are present in Groups 3, 4, 5, 6 and 7. (See Plate 3)

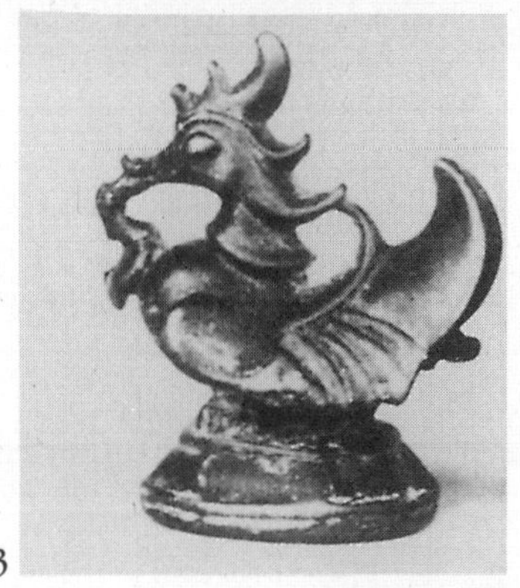
3

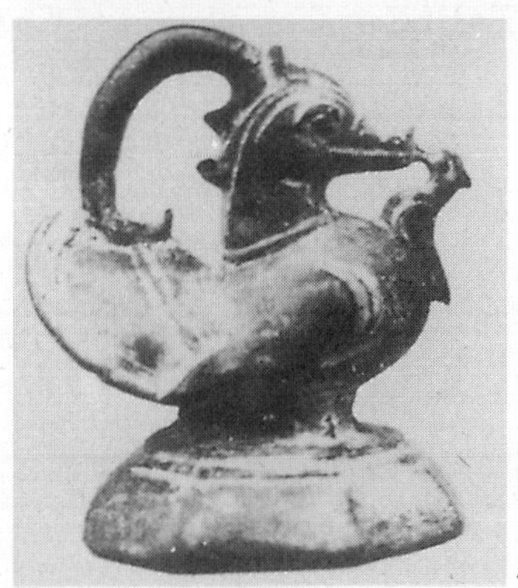
4

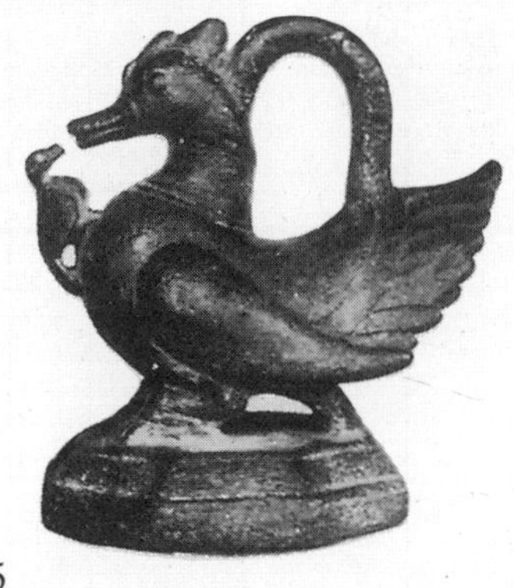
5

Numbers
indicate
style group

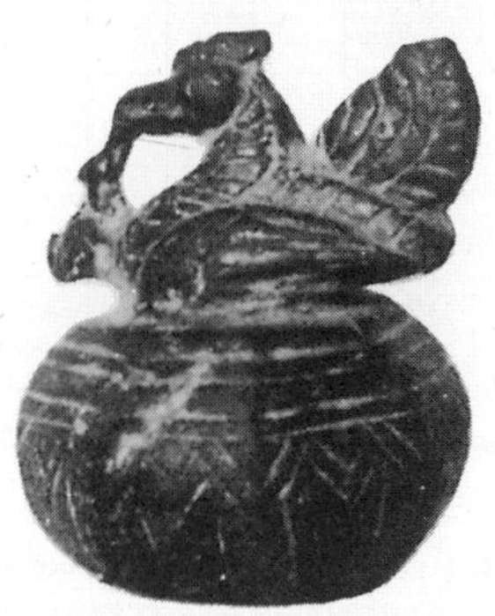
6

7

Plate 3:
Bird-shaped Mouth Appendages

TABLE 4: Comparison of Changes in the Masses (grams) of the Siamese *Baht* and Burmese *Kyat*

Date	South Siam		North Siam		Burma							Refs.
	Unit Mass Coin and Bullets	Locality	Locality	Unit Mass	Weight Groups							
					1	2	3	4	5	7	6	
1908	15,0	Bangkok	—	—								54
1908	15,1	—	—	—	16,3							55
1821												
	15,0 ±0,2	—	—	—		15,8	15,3	15,9				56
post-1782	15,1	—	—	—								57
	15,2	Ayudhya	—	—								58
pre-1782	14,5	—	—	—								59
1775 to 1739												
1752	14,6	Ayudhya	—	—								60
1693	14,2	Ayudhya	—	—				14,5	14,4			61
1615	14,1 – 14,9	Ayudhya	—	—								62
1615	—	—	Chiengmai	12,0 – 12, 7								63
	—	—		11,6 – 12, 0								64
1575 to 1559	—	—	—	—								
15th c to 14th c (?)	—	—	—	—							11.4 10.2	
14th c	14,0	Sukhothai	—	—							**14.0**	65
14th c	?	Sukhothai	Lanchang Lan Na	12,2								66
13th c			Lanchang	9.7								67

Bullets = pod duang currency ingots. Sometimes they also served as weights [55,61]
Horizontal lines = time limits of each unit mass, approximately.

Of the many hundreds of weights examined in Burma, very few had any form of adjustment. It appears then that the weights were issued in a standardised, sufficiently reliable form from a main manufacturing centre. It is evident that with the weight fabrication being under the supervision of the Chief Minister, approximate uniformity would have been simple to obtain and it would have been simple also to raise the *kyat* mass and change the style of the weight as desired by the ruler, an advantage which would have been obvious.

In view of the above, it seems likely that standardisation of the weights meant ensuring that the weights in use during a king's reign were based on the current *kyat* mass, at least so far as the weights used by the king's revenue collectors or his trade brokers were concerned. This standardisation would be likely to consist of verification that the weight conformed to the appropriate style and hence the appropriate *kyat* mass., ensuring that the weight was in unworn or 'unadjusted' condition and in cases of doubt or dispute, weighing them against standard weights kept at the local administrative centre. The presence of a cold struck sign on a proportion of the weights (official's weights?) may have been evidence that their reliability had been again confirmed. Some additional support for the foregoing description of the standardisation is that it is recorded[69] that the customs observed by the Nyaungyan or Toungoo kings (1531-1752) when first ascending the throne were:

(a) to survey and take over their domain;
(b) to order the re-writing and distribution of the Tripitaka (the three Buddhist scriptures);
(c) to record the weights. The weights and scales used were inspected and compared with those in the Treasury after which they were stamped and returned to their 'owners'.

In the case of the marble 'tortoise' weights (Appendix 1), the officials had to check them also to ensure that they were of the correct masses.

Notes

1 The Mesopotamian weights of mass one *mana* (about 495 grams) were made to an accuracy of about +/- 2% (Iwata, 1982, p. 1). The Indus valley weights deviate from their means by about +/- 12% at 1,7 grams to +/- 1% at 54 grams (Hendrickx-Boudot, p. 14). In both cases the masses of the mainly stone weights are believed to have remained constant over several centuries. An approximate estimate by the authors of the variations from their means of ancient Chinese weights yielded values between +/- 8%. Mitchiner (p. 10) mentions that in 19th century India, the mass of a *rati* seed varied from about 0,104 to about 0,117 grams. Thus the variation from the mean of the two extremes is about +/- 6%. This means that the silver *karshapana* of 32 *rati* seeds varied in mass between about 3,33 and 3,74 grams according to the location of use. (See also Mackay, p. 606). Another cause of difference between the exactness of weight masses is the accuracy of the weighing device. From 5000 BC onwards the Egyptians (and later the Greeks and Romans) used balances which could weigh to about 0,05 grams (Iwata, 1974, p. 7), so presumably the Mesopotamians and Indus valley peoples used balances of similar accuracy. There is no information known to the writers which concerns the quantitative sensitivity of the balances, steelyards and *bismars* used in south-east Asia. If Burmese weights were calibrated using *rati* seeds and perhaps also weighing devices of low accuracy, then the observed variations of masses from their means would be expected. Though not important to this work, in fact, two seeds were in use, the other *(adenanthera*

pavonina) being assumed to be double the mass of the *rati.* Temple, 1887, pp. 314 – 318 and 1899, pp 301, 302, 307, discusses the importance of these two seeds to the mass scales of South-east Asia.

[2] Moreland, pp. 87, 88.

[3] Le May, pp. 15, 16, 22, 59, 106.

[4] Cresswell, pp. 13 and 31.

[5] Brooks, p. 17.

[6] Warren, p. 114.

[7] Prinsep, pp. 21, 22.

[8] Mitchiner. Personal communication.

[9] Wu Ch'eng Lo, p. 60. The masses of the *liang* shown by the author on page 60 are based upon average masses of the coins of the relevant periods. The list is probably more reliable after 9-24 AD when a precise weight is available. Before then the estimates depend on relating the reliable information to the literary accounts of the changes in mass and their equivalences.

[10] Prinsep, p. 27.

[11] Yule, p. 256.

[12] Gear, Mean *kyat* mass of each style group.

[13] Ferrand, pp. 84-86, 264, 265, 273, 274, quoting Nunes ('Old' and 'new' ounces are used in the computations. Ferrand is a little confusing on the matter).

[14] Yule and Burnell, p. 918, s.v. Tical, quoting Gasparo Balbi's *Viaggio dell' Indie Orientale.* Venetia, 1585.

[15] Yule and Burnell, p. 918 s.v. *Tical* quoting Alexander Hamilton's *A New Account of the East Indies.* Edinburgh, 1727.

[16] A Group 4 bird weight bearing the date (Burmese era) of 1739 AD.

[17] Brooks, p. 11.

[18] Yule and Burnell, p. 919, s.v. *Tical* quoting Stevens, *A New and Complete Guide to the East Indies Trade.* 1775.

[19] Yule and Burnell, p. 919, s.v. *Tical,* quoting Forest, *Visit to Margui.* p. vii,1783.

[20] de Rochesnard, p. 33.

[21] Prinsep, p. 98.

[22] Prinsep, p. 120.

[23] Kelly, pp. 113, 115.

[24] Phayre, p. 38.

[25] Prinsep, p. 34.

[26] Harvey, p. 269.

[27] Wood, p. 239.

[28] Harvey, p. 194.

[29] Gardner, 1962, p. 2.

[30] Harvey, p. 171.

[31] U Kalar. From a partial translation made for the authors.

[32] Gardner, p. 2, quoting *Annals of Chiengmai.*

33 *Some Notes Upon the Development of Commerce in Siam*, p. 81.

34 Harvey, p. 49.

35 Luce, 'Economic Life ...', pp. 337, 338.

36 This is a *karsha* of about 700 AD. The *rati* may have weighed about 0,104 to 0,117 grams. There were 96 *ratis* to this particular *karsha* (one of four). Another system of silver weights included the ancient *satamana* which contained 100 *ratis*, that is, it weighed 10,4 to 11,7 grams. So it seems that the *klyap* could have been equivalent to either the *tola* (also 96 *ratis*), one of the *karshas* or the *satamana*. Codrington, p. 1, mentions that the name *karsha* has its equivalent in Old Persian and Susic. The lowest of the *karsha* masses, 32 *ratis* or 3,73 grams, is close to one of the common Indus valley masses of 3,45 grams (Hendrickx-Boudot, p. 12). The double of this, a common multiple, is approximately equal in mass to that of the light Babylonian *shekel*. The *satamana* is approximately equal in mass to the heavy Babylonian *shekel* and to the Achaemenid *shekel* (but compare Mitchiner, p. 15; Codrington, p. 15; Hori, p. 27 and Hendrickx-Boudot, p. 19). On the evidence of the foregoing there appears to have been a connection at first between Babylon, the Indus valley and so to the rest of India, then between Achaemenid Persia and India and lastly between India and Pagan.

37 Le May, p. 22.

38 Ridgeway, p. 162.

39 Le May, p. 61.

40 Yule and Burnell, p. 928.

41 Yule and Burnell, pp. 56, 807, 896.

42 Prinsep, pp. 115-121. Most of the many *maunds* in India fell into four main classes: Bengal about 36,5 kg (88 lbs); Central India about 18,2 kg (40 lbs); Gujerat, 12,8 kg (28 lbs); South India, 11,4 kg (25 lbs). See note in Fryer, v. 3, p. 126.

43 Harvey, p. 12.

44 Ten of these weights illustrated by Braun and Mollat yielded unit masses of between 10.9 and 11,6 grams, averaging 11,2 grams. One weight of 997 grams, Braun's no. 3, had a unit mass of 10,0 grams. However, the writers were informed regarding a weight of similar style in their possession that it was made at Sagaing at the end of the last century.

45 Wu Ch'eng Lo, p. 60.

46 Codrington, pp. 10, 11.

47 Decourdemanche, p. 122.

48 Warren, p. 144. See also Temple, 1899, pp. 303, 310, on the Graeco-Indian-Mahometan mass scale and mass units.

49 Robinson and Shaw, pp. 18-20 and 50-64. This implies a change from the light Indian weight standard, the 32 *rati karshapana*, to the heavier 96 *rati karshapana* or the 100 *rati satamana*, a change possibly due to the influence of Islam.

50 Braun's illustrations and accompanying masses, nos. 7, 13, 14 and 15, are referred to here. In comparison with the writers' own specimens, nos. 7, 13 and 14 are different in style and 13 and 14 look distinctly unusual. Possibly they may be genuine but of a type less subject to official demands for conformity. Their unit masses are all about 12,0 grams. Unless they are modern copies they may be ancient copies of 13th and 14th centuries A.D. from north Siam, (See Table 4)

51 E.g. Mitchiner, Petrie, Decourdemanche. The only information available concerning the

migration of the mass units through Central Asia ia that Sakan documents of about 2nd century and 3rd century AD found in Khotan (Tarim Basin) mentioned the use of the Greek Attic stater of about 8,8 grams and the *drachma* of about 4,4 grams (Jairazbhoy, p. 68). At about the same time, or a little earlier, it is said the Roman weights on a *stater* standard were being used by the Kushans whose empire abutted on that of the Han Chinese (Jairazbhoy, p. 148). It seems most likely that, subsequent to the occupation of Bactria by Alexander's men after 325 B.C., Greek weights would have been used there and the weight standards, if not the shapes, might have been adopted by neighbouring peoples.

52 **Wu Ch'eng Lo, pp. 73-74. The table below summarizes the variations in the mass of the Chinese liang in the past, as given by the author.**

Variations in the *Liang* Mass with Time

Date	*Liang* grams	Date	*Liang* grams
1121-225 BC	14,9	534-577	27,8
350 BC-9 AD	16,1	566-581	15,6
9-430	13,9	581-602	41,7
		603-618	13,9
479-502	20,9	618-1911	37,3
502-534	13,9		

The following notes have been summarised from the *Zhong Guo Gu Dai Du Lieng Heng Tu Ti*, pp. 1, 2, 46-49 and plates 153-237.

The Chinese first developed units of mass and mass scales between the 16th and 8th centuries BC. About 220 BC the Qin kingdom standardised its weights though elsewhere there was a considerable variety and variation. Between 25 and 220 AD bronze weights were increasingly replaced by iron ones. From 220 to 589 AD there was much strife and political dissension resulting in confusion in the weights and possibly scattered increases in the masses of the mass units. Standardisation of the weights and measures was ordered when the empire was re-united in 581 AD. Efforts to standardise were still being made during the period of the next dynasty up to 907 AD. At this time weights made of iron, and possibly wood, were being made. From about 220 AD until the accession of the Mongols in 1271 AD relatively few weights have been discovered and apparently none from the period 618-960 AD. From 1271 AD onwards there are many examples often bearing inscriptions in different languages.

Of the 87 weights illustrated none were found in Yunnan. The only adjacent state represented is Szechwan (Sichuan). All the other specimens were found to the north and east of Yunnan and its neighbouring provinces. That is, in the general region which was longest under the rule of one (or usually more) of China's many kingdoms and dynasties.

In the case of Szechwan, the mass of the *liang* was between 15 and 16 grams from 206 BC until between 25 and 220 AD. At some time about 200 AD the *liang* mass changed to between 12,5 and 13,7 grams. There is no more information for this region. Elsewhere the only certain information is that by 960-1127 AD the *liang* mass had risen to about 39 to 40 grams (2 weights available only) and after 1271 AD, it ranged between 35 and 40 grams.

53 Wyatt, pp. 118, 199.

54 Le May, p. 106. Standard silver coin.

55 Le May, p. 59, a weight.

[56] Gear, unpublished.

[57] Le May, pp. 15, 16.

[58] Cresswell, p. 15.

[59] Gear, unpublished.

[60] Brooks, p. 17, a weight.

[61] Le May, p. 59, quoting de la Loubere, *Du Royaume de Siam.* 1691. A weight?

[62] Le May, p. 22.

[63] Le May, p. 22, quoting a letter from L. Antheunis in 1615, published in vol. I of *Records of Relations with Foreign Countries, 1600-1700 AD*, Bangkok.

[64] Gear. Calculated from the metal composition.

[65] Gear. Unpublished investigation.

[66] Cresswell, pp. 13, 31.

[67] Cresswell, p. 31.

[68] Thiri Pyanchi, p. 3.

[69] Po Latt, p. 200.

[70] Robinson, p. 14, quoting Tomę Pires, and p. 21, quoting Pieter van Dam.

2
Mass Scales

Mass scale:

A MASS SCALE is one formed by placing in order the multiples or fractions of some particular mass which is chosen as the unit mass. In this case, the binary and decimal scales between 1 and 2000 grams are the most important, the unit masses being between about 8 and 16 grams.

The table 'Variations of East Asian Precious Metal Mass Scales with Time and Place' sets the Burmese mass scales and units in their regional context.[1-27] (See the plates 13 to 24 for weights arranged in accordance with the mass scales)

Pre-Pagan mass scales:

There is no information concerning masses greater than 10 grams in Burma prior to Pagan. However, in south Burma, since Indian monarchies and trade with India were dominant between the 2nd and 9th centuries AD,[28] it is likely that the Indian binary mass scales existed there among the Pyu and the Mons. Certainly the coins of the times, both Pyu and Mon, follow binary scales of the type 1, ½, ¼, ⅛, ?, and 1/128 with unit masses between 8 and 11 grams.

Pagan mass scales (1000-1300 AD):

Though no weights have been discovered at Pagan, sufficient data has been assembled from inscriptions there[29] to enable a provisional identification of a common metal

---

Table 5 notes

1. Date means date of European publication of information (bold type), date of indigenous record (brackets), or inferred date (unmarked).
2. The Indian -1200 scale incorporates two of the more common scales, one in brackets where the member is not recorded on both scales.
3. In 1959 A.D. the 16 and 8 multiples of the (+992 AD) Chinese mass scale were replaced by a single multiple of 10.
4. China, 992 AD. This year was that in which the mainly decimalised weight system was officiallyadopted. In fact it had existed for more than one thousand years previously. The older binary scale under the heading -992 AD continued to exist until the 16th or 17th Century AD.
5. Vietnamese (Annamese) weights do not appear on this table because, except for minor variations, they follow the Chinese scales[26, 27].

Table 5: Variations of East Asian Precious Metal Mass Scales with Time and Place

Country	Burma	Malaya	Cambodia	China	Burma		Pegu	Bengal	China	India	Siam	Laos	Cambodia
Groups	1 to 5, 7				6								
Approx date (AD)	–1554? to 1885	**-1892**	**-1893**	(+992) approx	-1554?		**1585**	pre-1821	(-992) approx	-1200	+1200	**-1885**	-1300
Approx mass of unit (g)	14 to 16	38 to 48	37	37	11	14	12,5	11,6	14 to 37	23 to 55	9 or 12 to 15	? to15	9 to 10?
Scale based on mass unit	100	100	?	100	100								
						80	80	80	80	(80)	80	?	
										(64)			64
	50		?										
									40	40	?	40	?
					25								
	20					20	20	20	20	20			
			16	16									
					12,5								
	10	10				10	10	?					?
				8					8	(8)			?
					6,2								
	5 or 4					5	?	5					
										(4)	4	4	4
					3,6		?	?					
						2.5							
	2								2	2			
					1,2								
	1	1	1	1		1	?	1	1	1	1	1	1
	½								½	½			
					0,4								
						0,3							
	¼								¼	¼	¼	¼	¼
											⅛		⅛
	⅒		⅒	⅒									
										1/16	1/16		
									1/24				
											1/32		1/32
												1/40	
											1/64		
			1/100	1/100									
References		1	2	3, 4, 5			6	7, 8	9, 10, 11	12, 13, 14, 15, 16	17, 18. 19. 20	21	22, 23, 24, 25

mass scale. This, at the time of assembly, was assumed to be, with reservations, a decimal scale. However, it is more likely to be part of, or mixed with, the old south Indian copper and silver scale which is known to have been already long in existence at this time. The scale is shown below with the equivalent Pagan names.

Table 6: Pagan Mass Scale

Pagan Name	S. Indian Name	Multiples and fractions of the unit
Pisa	Viss	40
—	Seer	8
Buih	Pollam	1
Klyap	Tola	⅓
Mat	Mat	1/12
Pay	Pa'i	1/96 (?)

Mat in Tamil means the "touch" of gold.

Pay is a Mon word meaning "bean" (and cowrie?)

About the middle of the 12th century Burma's first recorded weight standardisation occurred.[30] At this time at least ten different mass scales and/or units were in use at Pagan.

Moreover, in the last quarter of the 12th century, a Burmese chauvinistic reaction to foreign influence developed.[31] Also relevant may have been Pagan's adoption of the Sri Lankan (Ceylonese) version of the original (?) Theravada form of Buddhism. This may be seen in conjunction with the somewhat later weight standardisation, with the Buddhist moral injunction to use fair weights and the Jatakas (Buddha incarnation parables) illustrating this requirement Yet again Pagan's Buddhist ruler, the first uniter of Burma, would have been obliged to see himself as a Buddhist world teacher (shown on both weight forms), as a universal monarch (hence the *to* weight) and a champion of Buddhism (hence the *hamsa* weight). It seems possible therefore that at this time or somewhat later, the many Hindu binary scales presumably brought in by the Indians, Mons and Pyu may have been discarded in favour of a more uniform one. It has been suggested that the Siamese bullet (*pod duang*) and *kakim* ingots were introduced a little later for similar reasons.[32] These introductions also appear to have occurred about the same time as a Theravada revival in Siam.

Group 6 mass scale:

This is a scale which includes both decimal and binary elements between about 1120 and 35 grams and, apparently trinary and decimal elements between 35 and 4 grams.

The upper part of the scale has the ratios 1, (?), ¼, ⅛, 1/16, 1/32? while the lower part has ratios of 1, ⅓ and 1/10. If the unit mass is taken as 11,23 grams (the average of the largest weights) then the multiples on the scale become 100; 25; 12,5; 6,25; 3,6; 1,2; 0,4; If the unit mass is taken as 13,9 grams (average, assumed-*kyat,* mass) then the scale becomes 80; 20; 10; 5; 3; 1; ⅓. Referring to Appendix 1, it will be observed from the accompanying table that the 19th century marble weight scale and the 15th century Group 6 scale based on 100 are closely similar.

Because of the Indian influence on Burma, many Indian commercial precious metal and copper mass scales within a similar range of masses and of different ages were examined in the attempt to identify those similar to the Group 6 scale.[33] Some scales did show a closer similarity than others, notably the precious metal scale used in Magadha (north-east India) in the 5th century BC[34] and the grain and commercial scales of Bengal which are believed to have existed from about 1540 AD or before.[35, 36] Other similar scales, but which appear to lack members, are known from the 8th and 9th centuries AD[37] from northern and eastern India. It may be relevant to note that these scales appear to be most closely associated with the holy land of the Buddhists, i.e. Magadha and the last stronghold of Buddhism in India, i.e. Bengal.

The Chinese mass scales and mass units were also studied similarly.[38] The pre-992 AD mass scale which used as a unit the 4 AD to 530 AD *liang* mass of about 13,9 grams also showed a reasonably close similarity to the Group 6 mass scale and an identicality in one of the two possible unit masses. The table 'A Comparison of Chinese, Bengalese and Group 6 Mass Scales and Units' shows in detail the degrees of similarity in mass between the Chinese and Group 6 weights and, in scale, between the Bengal and Group 6 weights.

Turning closer attention to the time when this mass scale and unit originated, it is most likely that it existed before 1584 AD. A silver mass scale was recorded at Pegu[39] in which one *bize* (*viss*) equalled 100 *teccali* (*ticals*), one *agito* (*giro*) equalled 25 *teccali* and one *abocco* (*abuhi*) equalled 12½ *teccali.* The names *agito/giro* and *abocco/abuhi* and the scale 100: 25: 12½ are unknown to the writers. The ratios 100: 25: 12½ are equivalent to the ratios 80: 20: 10 referred to at the beginning of this section and so suggest the existence of the Group 6 mass scale at that time.

The kingdom of Lan-Na adjacent to Burma on its east existed from 1239-1558 AD. Its silver ingots, named *kakim,* followed the scale 10: 5: ½: ¼, with a unit mass of 12,2 grams.[40, 41] This scale has similarities to the lower part of the Group 6 scale below 130 grams. The ancient Indian Malabar mass scale may be of significance to the origins of the Group 6 mass scale especially when related to the discussion of the *man* or *maund* along the southern Indian coastline.[42]

The binary ratios and the unit mass of 11,2 grams suggest the Indian influence while the expression of these binary ratios in forms more akin to decimal scales, suggests the

Chinese influence. The change of scale between 35 and 40 grams could be due to the attempt to assimilate both the Indian *pala* and the Chinese *liang* while the trinary element might be due to the influence of Islam or even a still older one. It could also be due to the existence in the 13th – 15th centuries of two *liang*, one of about 35 to 40 grams and another, about ⅓ of these values, i.e. about 11 to 13 grams, so introducing a scale which may be expressible either as a trinary scale or a decimal scale or both together.

From the Pegu silver and the marble weights (Appendix 1) mass scales it seems that the scale 100; 25; 12½; was the one in use and therefore the unit mass would be about 11,2 grams, a value which agrees with those quoted in 1554 A.D. and 1515 A.D.

Group 7 mass scale:

Only five specimens of this bird weight group have been seen. From these it seems that the mass scales and units are similar to those of Group 5. However, other information[43] concerning possibly similar weights collected in Thailand indicates the possibility of a unit mass of about 12 grams yielding a scale as follows: ⅔, ??, 2½, 4, 5 and 10.

Groups 5 to 1 mass scale:

The most recent of the relevant Burmese mass scales, in *kyats*, is shown in the table below together with the names of its components. Weights have been seen, rarely, which were estimated to be 500, 1000 and 2500 *kyats* in mass. Weights of similar and still greater masses were obtained from the marble 'tortoise' weights which (in 1972) were still to be seen occasionally lying around village market places. These weights continue to lesser masses. The beast weights of Groups 2 (r) and 5 sometimes have the five *kyat* weight replaced by a four *kyat* weight. The four *kyat* member may have been used when these weights were used by peoples accustomed to using low mass weights on a binary scale.

The alternative name for *kyat* is *tical*, a name which is absent from the pre-1300 AD Pagan inscriptions but was in use at Pegu in 1515 AD. Similar names are attached to similar masses in Bengal (*tuckah*) and Nepal (*takka*). The name *tical* could have come from the old Mon weight system in use in south-east Asia from about the 6th century AD. Still earlier it may have been derived from the Sanskrit-Persian-Turki name *tanka* through the Mon *taka* or *t'ke*. It could also have originated from the Paganese *ta klyap* (one *kyat*). Still other origins have been suggested for the word.[54]

The *tanka* was the chief silver coin of India in the 13th and 14th centuries AD, had a mass of about 11,4 grams approximately, equivalent to the mass of the *tola* (11,3 to 12,3 grams) which was used in north-east India to weigh gold and silver. At this time the largest Indian mass unit used to weigh gold and silver, amounted to 100 *tolas* thus

Table 7: A comparison of Chinese, Bengalese and Group 6 Mass Scales and Units

European Name	Chinese Mass Scales and Units					Group 6	Bengal		Note
	Mass Scale Post 992 AD Mass Unit Post 618 AD		Mass scale	Pre 992 AD From AD 4 to AD 530	Mass grams	As observed As observed	pre 1752 pre 1821		
	Mass grams	Chinese names and ratios		Chinese names and ratios		Mass grams	Mass grams	Bengal name	
			2400	1 shih = 40,0 chün					
Picul	59680	1 tan/shih = 100 jin/chin	1600						
			640	1 chün = 8,0 huan	8909				
			80	1 huan = 2,0 lieh/kin	1114	1123	931,4	Seer	
			40	1 lieh/kin = 2,0 chin (A)	556,8				
			20	1 chin (A) = 2,5 yuan	278,4	282,3	232,8	Pouah	
Catty	596,8	1 jin/chin = 2 liu	16						
			10			137,0			
	298,4	1 liu = 8 liang	8	1 yuan = 4,0 chin (B)	111,4				
			5			69,5	58,2	Chittack	
			3 to 2,5			42,0 to 38,5			b
			2	1 chin(B) = 2,0 liang	27,8				
Tael	37,3	1 liang = 10 ch'ien	1	1 liang = 2,0 che/tse	13,9	13,9	11,6	Sicca	a
			½	1 che/tse = 2,0 hua	7,0				
			⅓						
			¼	1 hua = 6,0 chu	3,5	4.1			c
	3,7	1 ch'ien = 10 fen	1⁄10				1,2	Massa	
			1⁄24	1 chu = 0,6 ?	0,6				
Candarin	0,4	1 fen = 10 li	1⁄100						

(a) The pre-1752 mass of the sicca reported by Brooks, p.9, was about 10,6 grams. That shown is from Prinsep, pp. 96, 97.

(b) Three Group 6 masses only available, 38,5, 40,0 and 42,0 grams, average 40,2 .

(c) One Group 6 mass only available, 4, 1.

(d) The Group 6 masses are shown here on an 80:20, scale similar to that in use in pre-992 AD China and in Bengal. See Table 5 for Group 6 displayed on a 100:25:12½, scale.

having a mass of between about 1130 and 1230 grams. Similarly, between 1515 AD and 1585 AD the reported mass of the Pegu *viss* of 100 ticals ranged between 1140 and 1260 grams approximately (Table 2). On the other hand, the south-east Indian *viss* at this time had a mass of about 1360 grams. The distinctly different mass systems of north-east and south-east India were related through the *rati*, the *tola* and the *seer*. In particular, in Bengal, the ancient, and later, mainly South Indian name, *viss (visa)* was used for the 5 *panseri* weight of similar mass to the *viss*.

From the foregoing it seems that mass system of Pegu (Ramannadesa) may have been derived from north-east India, the terms *kyat* and *peikta* not coming into common use until after the Burmese conquest of Ramannadesa in 1539 AD.

Above one *kyat* the Groups 1 to 5 mass scale is a decimal one while below it may be equally well either decimal or binary.

Concerning the mass scale below one *kyat*, in north India from about 200 BC and until an unknown date after 1030 AD, the *karsha* (an old Persian name) of about 14,9 grams was employed as a unit in the weighing of gold and copper, *karsha* masses of about 11,2 and 9,3 grams being used elsewhere. This *karsha* unit was sub-divided into weights based on the *gunja* seed, the scale being as follows: 1 *karsha* = 4 *tankahs* = 16 *masha* = 128 *gunja*. The recorded Burmese scale of 1 *kyat* and less for Groups 1 to 5 is very similar, i.e. 1 *kyat* = 4 *mat* = 8 *mu* = 16 *pay* = 64 *ywe-gyi* = 128 *khin-ywe*. The same ratios with a unit mass of 11,3 grams were in use in Nepal from some unknown date before 1830 AD.

Table 8: The Groups 5 to 1 Mass Scale and the Names of its Parts

Mass	**Names**		
(kyats)	**Specific**	**Multiples of mass units**	**Notes**
1/20 (or 1/16)	Pay		
1/10 (or 1/8)	Mu		
¼	Mat		
½	Khwe	Ngamu (5 mu)	44
1	Kyat, Tical, Tual		45, 46
2		Hnit (2) kyat	
5	Bo, Bol		47, 48, 49
10		T'say (10) kyats	
20		Hn'say (20) kyats	
50		Nga'say (50) kyats	
100	Beittho, Peikta }		47, 50
	Viss, Peitha }		45, 51
250			
1000	Kwet	T'toung (1000) kyats	47, 52
2500	Kwet ahseik		44, 53

Origins of the mass scales:

The Burmese scales are wholly decimal with two, possibly three, exceptions. The mass scale below one *kyat* can be expressed either as a binary scale or a decimal scale. The Group 6 mass scale above about 5 *kyats* is partially binary and partially decimal. Below about 5 *kyats* it could be binary, trinary or decimal.

The Indus valley cubical stone weights of about 2000 BC followed a binary or decimal scale above 54 grams and a binary scale below while the fractional weights occurred in thirds.[55] The old weights of Malabar and some Indian scales after approximately the 11th century AD were modified later by Mahometans by a factor of three as in Persian weights of the time.[56, 57] The Persian connection appears also in the use of weight names which were originally Persian. There are also Greek connections. The lower part of the Pagan mass scale is a binary/trinary scale. These items complete the evidence from all of the Indian, South-east Asian and Chinese mass scales examined which suggest an influence of the Babylonian/Assyrian sexagesimal mass scale on the Burmese weights.

The great majority of Indian weights are of a binary nature, even though by using a different mass unit some binary scales can be converted partially into decimal scales and some scales are binary/decimal. The decimal system was not widely used in India until after the 5th century AD.[58] This came about at the time when Buddhism was at its peak in China and there was much Indian-Chinese contact. The inferred pre-Pagan and the upper part of the Pagan mass scales follow the Indian binary system. Though this may have migrated north from the Indian dominated Pyu people or south-east from Bengal, it could have been brought in from the Shan-Siam-Cambodia region on the east. The Group 6 mass scale appears to be based on an ancient Indian mass scale utilising a unit mass equivalent to that of one of the *karshas* (from about 11 to 14 grams).[59] Since the scale may have been in use in Pegu in 1554 AD it may have entered southern Burma from north and east India. It could have entered northern Burma from Bengal, though after 1200 AD the Mahometan conquest of north-east India and the accompanying destruction of Buddhist relics in the holy land of Magadha may have diminished contact between Burma and India.[60] The fact that the Shan of Assam lost contact with their eastern neighbours after about 1300 AD indicates a degree of likelihood for this possibility. The scale could also have entered Burma from Ta-li (the successor state to Nanchao in the Yunnan region) bearing in mind that China itself had been hampered from entering Burma from about the 7th century AD until 1287 AD when the Mongol conquerors of China, entered Burma. This implies the possibility of the use in Nanchao and Ta-li of a mass less than the Chinese *liang* mass of about 13,9 grams. Since the Chinese binary scale officially was abolished in 992 AD but nevertheless continued until the 16th or 17th century AD, it seems likely that in the independent or semi-independent region of

Yunnan, this mass also could have continued. It seems possible that a common weight system or similar weight systems had been brought into use throughout the north-eastern India, northern Burma and Yunnan region during the period of the rise of Buddhism in China. This was a time when many things Indian were adopted by the Chinese including perhaps the mass unit and binary mass scale of the Buddhist home-land, Magadha. Relevant to this matter is that the fact that the peoples of all three of these regions were mainly Mongoloid, that they used cowries for a currency[61], that they did not use a coinage or, at least, not their own, and they were on a common trading route. See also the origins of the Burmese mass units.

The Yunnan region and northern Burma, occupied mainly by Shans (western Thais) from south and west China were under the nominal or direct rule of China from 1287 AD until about 1530 AD or later. In 1368 AD the Chinese lost control of one of their main trade routes, the silk route across Central Asia. They turned their eyes southwards for new outlets.[62] At this time the powerful Maw (Mao, Miao) Shans on the then Burma-China border were creating disturbances and hampering trade from China down the Irawaddy River. The Chinese-controlled Yunnanese warned them to desist in 1382 AD. Shortly afterwards, in 1391 AD and again 1403 AD, China (Yunnan?) supplied both the Cambodians and the Siamese (Lan-Na?) with copies of her mass scales and units. It would seem unlikely that she did not also supply the Shans with similar standards, particularly when she wished to pass part of her trade through Shan territory. In fact, a mission was sent from China (Yunnan?) to Burma at some time between 1403 and 1425 AD.

China's trade was passing down the Irawaddy River and the Salween valley to the sea and the Chinese must have been made arrangements through Yunnan with the Mons or Ramannadesa (Lower Burma). Did they agree to use a mass unit and scale which was in common use over a wide area as would be expected? If so, then this was probably the Group 6 mass unit and scale inferred to be in wide use. Moreover, there had been a long-standing connection between the Mons of Lower Burma and those of central and north Siam, particularly in the field of religion. From the end of the 14th century AD a Buddhist revival had begun in north Siam. Was the former association of Chinese Buddhist monasteries with fair practices in trade and contracts applied in this case to guarantee agreements and weights? Was this why Yuan Thai (or Old Mon) script was placed on the Group 6 weights together with a bird shape?

Turning to the decimal mass scale, although in much of what is now China metrology has been based largely on the decimal scale for the past 3000 years, the pre-992 AD mass scale is evidence that there were important exceptions. About this time there was renewed interest in the decimal system in China proper. Decimal paper currency had been produced (and failed) sometime about 900 AD.[63] The abacus, though known in China by the 6th century AD,[64] came into common use in the 13th century AD.[65] The invading Mongols of the 13th century AD counted their soldiers decimally as did the people of Nanchao (in the armies of which many Burmese served) and the people of Burma in 1630 AD, and probably before, also counted their armies decimally. The binary mass scale apparently disappeared from China about the 16th or 17th century AD. It may not be coincidence that the Burmese adopted the decimal scale not later than about the 15th or 16th century AD.

Notes

1 Ridgeway, p. 171. See also Temple, 1899, p. 307.

2 Ridgeway, p. 160, quoting from J. Moura, *Le Royaume de Cambodge, vol. 1.* Paris, 1883, p. 323.

3 Needham, vol. 3, p. 85, quoting Wu Ch'eng Lo, *History of the Weights and Measures of China.* Shanghai, 1957.

4 Kann, p. 236. Sigler, pp. 7-9, states that the Chinese *tael* (a *liang* weight of silver) was neither a constant weight nor a constant standard of fineness. Some were both, others were either a weight unit or a standard of fineness. The Shanghai *tael* was a mass of 35,3 grams of silver, 98% fine, the Tsaoping *tael* was a mass of 36,6 grams of silver, 94% fine and the Haikwan tael was a mass of 37,8 grams of silver, 100% fine. According to Yule and Kendall, p. 175, *catty* or *kati,* the *tael* (1/16 of a *catty*)as agreed by treaty between China and the European powers, was 39,1 grams (37,8 in 1858 according to Schroeder, p. 181). In Chinese trade it varied from 7,1 grams (tea vendors, Peking) to 49,7 grams (coal merchants, Honan).

5 Yule and Burnell, p. 175, *catty* and p. 690, *pecul.*

6 Yule and Burnell, p. 918, quoting Gasparo Balbi's *Viaggio dell' Indie Orientale.* Venetia, 1585.

7 Decourdemanche, pp. 135-136. A scale which became dominant in north India and Bengal in the early 16th century, though perhaps originating between the 12th and 14th centuries (p. 122). In fact, the scale itself is much older. (Gear).

8 Prinsep, 'Useful Tables', p. 116.

9 Coole, p. 11.

10 Temple, 1898, pp. 29, 30, 31, 61, 63, 90 and1899, pp. 310, 311. Quoting Terrien de Lacouperie, Temple refers to an old Chinese popular scale that existed up to the 7th century AD and partially for several centuries later. The part of the scale quoted is identical with that assembled here by the writers from different data. Moreover, he comments that this scale is similar to the old Indian popular scale. In fact, he implies that his Chinese popular scale originated from the Indian scale. Whatever the origin, it seems probable that a common mass scale, and probably a mass unit also originally was in use by traders in India, China and north Burma. A common mass scale was used by all nationalities trading in the Far East from about the 16th century AD or before, though the mass unit varied See Temple 1898, pp. 1 – 8 and 91 and 1899 pp. 312 – 313.

11 Zhong Guo Dai Du Liang, p. 60.

12 Ridgeway, p. 177, quoting from H. Colebrooke, *Algebra with Arithmetic and Mensuration translated from the Sanskrit of Brahmagupta and Bhascara.* London, 1817.

13 Prinsep, 'Useful Tables', p. 31. Quotes the 80 unit.

14 Prinsep, 'Useful Tables', pp. 17, 18. Quotes the 16 unit.

15 Codrington, pp. 1-11.

16 Elliott, pp. 45-52.

17 Le May, pp. 57-60.

18 Decourdemanche, pp. 165, 166.

19 L'Hermitte, p. IV. The Siamese also had a commercial mass scale which was wholly Chinese.

20 Temple, *As. Qu. Rev.*, p. 305.

21 Ridgeway, p. 160, quoting E. Aymonier, *Notes sur le Laos.* Saigon, 1885. See also Sale and L'Hermitte.

22 Panish, pp. 167-168, quoting from A. Cabaton, *Brève et véridique relative des évènements du Cambodge.* Paris, 1914, p. 9.

23 Sahai, p. 98.

24 Groslier, p. 28. Groslier believes that the Cambodians never had an indigenous weight system. They employed only the Chinese commercial scale, certainly in use by the 13th century AD, and a Hindu precious metal scale, in use before 1000 AD. They received Chinese standards in 1391 AD and again in 1403 AD.

25 Ridgeway, p. 160.

26 Des Michels, p. 201.

27 Schroeder, p. 181.

28 Harvey, 1969, p. 440.

29 Luce, *Economic Life ...*, pp. 337-338.

30 Harvey, 1967, p. 49.

31 Than Htun, p. 12.

32 Le May, pp. 10, 13.

33 Codrington, Decourdemanche, Mitchiner, Warren, Jervis, Yule and Kendall, Prinsep, Brooks, Kitsch, Ridgeway, Maity, et al. See also Temple 1897, p. 318 and 1898, pp. 57 – 67, 91, 169 – 171. On p. 210 (1897) and on pp. 116 – 118 (1898) he records that during King Dammazedi's reign the priests used the names of the Indian weights, *pala* and *tula.* The approximate weight of the former was probably between 60 and 70 grams and of the latter, between 4000 and 5000 grams.

34 Codrington, p. 2.

35 Decourdemanche, pp. 122, 135, 136.

36 Prinsep, p. 116.

37 Codrington, pp. 10, 11.

38 Mills, Coole, Temple, Ridgeway, Kann, Decourdemanche, Yule and Burnell, Needham, Kitsch, et al.

39 See reference 13, 'Mass Units and Standardisation'. The names quoted are not Italian nor Portuguese nor do they appear to be from any Romanic language. They may be Italianised versions of Mon names or versions of the Persian coin names *aguo* and *abussi* (see Fryer, v. 3, p. 138 and Mundy, v. 3, p. 309).

40 Cresswell, p. 13.

41 Le May, pp. 22-23.

42 See 'Unit Mass', Group 6.

43 Braun's illustration no. 17 shows a mass of 48,3 grams which could be a 4 *kyat* weight if the unit mass were 12,1 grams. This is not impossible but his weights 7, 13, 14 and 16, which also yield unit masses of about 12 grams, are of unusual shapes and so raise questions. Could they be Lan-Na weights or post Lan-Na copies?

44 Kyaw Tun.

45 Markets, etc.

46 Brooks, p. 111.

47 Scott, 1910, p. 557. *Bo* may mean weight like *ale* (pronounced arlay). It can also mean either 1/20 or 1/25 of a *viss*, these amounts corresponding to 5 and 4 *kyats* respectively. *Kwet* means *beittho* or *beittha.* It is used mostly to express multiples of 10 beittho. It may be spelt as *ak'wet. Awettha* mean one half and *aseit*, one quarter, e.g. of a *beittho.*

48 Temple, *Asian Quarterly Review*, p. 304.

49 Wilson, Appendix, p. lxi. A 4 *kyat* weight sometimes is found in the Group 2r (rod-bearded)beast weights, a fact first recorded in 1827.

50 Yule and Kendall, p. 967, s.v. *Viss.*

51 Judson, p. 63.

52 Ferrars, p. 214.

53 Scott, 1906, p. 324. Temple, 1897, pp. 318 – 329 discusses in detail both the names of the weights forming the Groups 5 to 1 mass scale, and the scale itself.

54 De Campos, pp. 119 ff. Temple, 1897, pp. 245, 253 – 256.

55 Piggott, 1979, p. 259. Hendrickx-Boudot, pp. 14-16. Iwata, 1975, p.14.

56 Codrington, p. 10.

57 Decourdemanche, p. 122.

58 Neugebauer, p. 1102.

59 Codrington, p. 1.

60 Coomaraswamy, pp. 169-172, stresses the Burma-Bengal/Bihar connection. There are *stupas* of Pala (Bengal) design at Pagan and many of its stone reliefs of the 11th century AD may be importations from Bihar, especially Nalanda, or Bengal. Frescoes of the same age have stylistic affinities with Bengal and Nepal. One of the Pagan temples, Mahabodhi (1210-1234 AD), is fashioned after the much older one at Bodhgaya in Bihar. Tagaung, the earliest seat of organised rule in the north Burma region received its Indian culture through Assam and Manipur rather than from the south. Harvey, p. 42, states that King Kyanzittha (1084-1112 AD) was the first Burmese monarch to restore the shrine at Bodhgaya. Griswold, *Siamese Art*, p. 17, and Harvey, p. 119, record that the king of Pegu (Dammazedi), like the king of Lan-Na in 1472 AD, sent his architects to Bodhgaya to obtain plans of the Mahabodhi temple. Dammazedi also exchanged envoys with Yunnan. Presumably this was a trade delegation to the Chinese governor of Yunnan concerning the passage of Chinese goods through his territory. China needed to use the southern land routes, having lost the northern ones, and Yunnan, at that time, had been under the firm control of the Mings for about one hundred years.

61 In 1855 cowries were not used in Burma (Yule, *Mission to Ava*, p. 259). They are not

mentioned by other late 18th and 19th century AD writers. However, they were a Burmese currency before the mid 18th century AD (But see Temple 1897, pp. 290, lines 5, 6, 7 and 1898, p. 178, pp. 290, lines 5, 6, 7 and 1898, p. 178, lines 10, 11). Their use demonstrates another connection between north-east India, Burma and Yunnan.

While cowries were used as currency in many parts of southern Asia, the Pacific and Africa, the places where their use has been recorded as being most common and extensive were China, Siam and Burma, Bengal, Bihar and Orissa, and the Maldive Islands. They were also used throughout most of the rest of India along the main trade routes though but little off them. Nor were they used along the Malabar coast. Bengal was the main distribution centre. (Quiggin, pp. 29, 193 ff, 202). Pyrard (v. 1, pp. 237-240) recorded 30 to 40 ships in one year laden only with cowries, travelling from the Maldives to Bengal.

Cowries were used in China from the 14th century BC until recent times alongside or in place of the *cash* coins. Probably they originated from Borneo or other islands south-east of China. However, in the 13th century those being used in Yunnan came from Bengal which obtained its supplies from the Maldive Islands (Polo, pp. 263, 265, 267. Maldive cowries differ in colouring from China Sea cowries). Cowries were the chief currency for normal domestic purchases in Bengal from long before the 13th century AD and remained so until the 19th century AD (Barbosa, v. 2, p. 105. Manrique, v. 1, p. 29. Pyrard, v.1, p. 78. Yule and Burnell, pp. 269-271).

Cowries are recorded as being in use in Lower Burma in 851 AD (Harvey, p. 10, quoting Suleyman), in Upper Burma in the 13th century AD (Polo, p. 283), in Arakan, Martaban, Pegu and throughout Siam in 1514 (AD (Pires, v. 1, pp. 94, 100, 104). They may have fallen out of favour in Burma when the capital was moved to Ava in central Burma in 1635 AD. However, they may have continued in use until the mid-18th century AD. Between 1740 and 1757 AD Lower Burma, the old state of Rammanadesa, regained and held its independence. In so doing it commanded the seaports and prevented the movement to Upper Burma by sea, river and land of cowries from the Maldives and the China Sea, of *ganza* from Canton and copper from Japan. About this time there was a shortage of cowries in Siam (Le May, p. 125), suggesting that there may have been a similar shortage in Burma also. The shortage in Siam was so severe that baked clay coins (*prakab*) were used in the place of cowries. The 1765-1769 war with China must have inhibited the flow of copper into north-east Burma. See 'Copper' in the chapter 'Weight Materials and Manufacture'. Because Burma had gained control of the most important source of lead and silver in south-east Asia between 1750 and 1763 AD, i.e. the Bawdwin mines, it seems possible that the Upper Burmese turned to the Chinese practice of using their now abundant lead as a currency, so dispensing with the need for the unavailable cowries. With the capture of Pegu in 1757 AD and the preceding interruption of the cowrie imports, all of Burma's peoples would have been required to turn to the use of chopped lead for domestic purchases and abandon the use of cowries. The former long period of use of cowries would have been associated particularly with the Mons, their seaports and their southern Siam connections. This alone may have been a sufficient reason for the Burmese conquerors to cause the abandonment of their use. Cowries continued to be employed in Yunnan until approximately the 18th century AD and in Siam and Malaya until the 19th century AD.

In 16th century Pegu about 500 cowries would buy a chicken. About 15 000 cowries were worth one *viss* of *ganza* or ¼ to 2½ *kyats* of silver, the cowries not being counted but measured in coconut shells supposedly holding either 1000 or 500 shells, each man having his own coconut shells for this purpose. (*Le May*, Arcady, p. 250. *Crawfurd*, Siam p. 331. *de la Loubere*, pp. 72, 73.) In Siam, and probably in Burma, cowries were used throughout the country for daily domestic purchases, metals being used only for larger purchases. Their advantages were that they were clean, cheap and could not be forged. It was for the last reason that in China about 340 BC the copper *cash* were withdrawn in favour of cowries because even *cash* were forged.

Ingots and lumps of gold, silver and copper alloys, rarely iron, were formerly used as currency throughout Eurasia. The silver cowrie-like *larins* used in Persia and the hook-like *larins* once commonly used in the Persian Gulf region, India and Ceylon ceased to be made in medieval times. China, Laos, Siam and Burma are the countries most noted for their ingot currencies, the use of these persisting until modern times. There was a long-established importation of Chinese silver *sycee* to south-east Asia and Indonesia. In the 13th century AD flowered silver ingots were in use both in Yunnan and Burma.

The foregoing facts are further additional indications of the long existence of an important trading route from north-east India to Yunnan. Along this route similar currencies and manners of effecting payment existed. Hence it would not be surprising if one mass scale predominated along the route.

62 Harvey, pp. 101, 102. However, there was another reason. In the 12th and 13th centuries south China was buying far more luxury goods than it could afford, just as Rome had done about a thousand years before. In addition, its production of gold, silver and copper cash had declined to between 10% and 50% of that required. This situation brought the prohibition of the export of these metals, among other necessary measures. It also led to the issue of banknotes. In the 13th and 14th centuries the printing of excessive numbers of banknotes resulted in inflation and rapidly-rising prices. In turn, these stimulated the search for, and production of, the metals, e.g. in the Shan states. See Rockhill, pp. 419 – 426.

63 Goodrich, p. 145.

64 Needham, vol. 3, p. 77.

65 Flegg, p. 159.

3
Mass Frequencies, Goods and Masses Weighed, Weighing Procedures

Mass frequencies:

The weight masses occur in the following frequencies, the ranges being related to frequency differences exhibited by each style class. The proportions shown represent their occurrence on the stalls in 1972.

Table 9: Frequency of Occurrence of Weights By Mass

Kyats	Less than 1	1	2	5	10	20	50	100	250
% of 1078 weighings	very small	1	5 – 10	10 – 20	20 – 30	40 – 50	4 – 8	3	2

Goods and Masses weighed:

In much of post-14th century India the *seer* of about 903 grams (about 80 *kyats*) was the upper limit of the precious metal scale, while the *chattack* of about 58 grams (4 *kyats*) was the lower limit of the gross weights. The *seer* was in use in Pegu before the 1750s. In Burma the precious metal scale may have been used within the range 2 to 1500 to 4000 grams while the marble "tortoise" weights may have been used solely for ordinary commercial transactions. For gems it is likely that seeds were used rather than the animal or marble weights.[1]

From the earliest days of European traders it was reported that gold, foreign gold and silver coins, rubies, pearls, coral, personal ornaments, musk, costly medicines, spices and camphor were weighed by the smaller weights. Gums, resins and waxes were also weighed in small quantities. In the case of gold, usually about 10 *kyats* of silver would purchase one *kyat* weight of gold leaf, though there were the usual fluctuations in exchange value. Weights of less than one *kyat* would appear to have been used to exchange small quantities of silver for lead or *ganza* since one could obtain about 1500 *kyats* of lead for one of silver.

The silver 'flower' money, commonly used before and after 1287 AD[2] in Burma, the Shan states, northern Siam and Yunnan, was most frequently made in ingots, supposedly about 75% silver, weighing about 300 to 450 grams (20 to 30 *kyats*). There were several grades of silver alloy in use and those of high silver content were made in ingots of 50 to 100 grams (5 to 7 *kyats*).[3](See plate 4). Today the most frequent masses of the Siamese and Laotian ingot currencies appear to be the following: bullets (*pod duang*), 15 grams or less; *lat hoi*, 50-70 and 100-120 grams, sometimes greater than 120 but rarely less than 50 grams; *kakim* or *chieng*, 60 grams. After about 1539 AD these silver and copper ingots probably occurred in Burma because they were indigenous to territories occupied by the Burmese.[4] Also weighed were the Chinese silver ingots brought in by caravans from Yunnan both to Siam and Burma.[5] These, called *sycee* or *hsi ssu*, latterly only weighed between 160 and 370 grams but ranged up to 3700 grams.[6] *Sycee* ingots *(See Plate 5)* of about 375 grams (10 *liang*) appear to have been first made about the 10th century AD[7] while the 50 *liang sycee* (1875 grams) came into use in the late 13th century.[8] For weighing individual Chinese and Burmese ingots, weights below 50 kyats most often would be required. For Siamese and Laotian silver ingots, weights below 10 kyats would be needed most frequently. The silver in the ingots used as currency was alloyed and the alloy could vary between a silver content of 25% and 95%. Each kind of alloy received a particular name.[9] The differences in proportions of silver were identified by differences in colour, surface texture and sound (ring) when dropped, though the touchstone was also in use. When a silver article was used as a currency it had to be weighed and its purity had also to be determined. This determination was usually done by inspection of the ingot by the market assayers (*pwezas*) who though claiming an accuracy of 0,5%, were often in error by 10%.[10] They were held responsible for their estimates. They also made silver ingots as required and for this purpose had furnaces and crucibles with them. In the 19th century the market assayer's earnings were 1% of the bullion weighed.[11]

Very voluminous or heavy materials would be weighed by the steelyard (*datchin*)[12] or the marble weights. (See Plates 56 and 57) These materials would include elephant tusks, rhinoceros horns, large quantities of lead, copper, tin, *ganza* and rice.

Ganza was an alloy of lead, copper and tin[13] used as a low value currency. About 1669 AD 2800 *kyats* of *ganza* were worth one *kyat* of silver. Though it seems that the value of *ganza* was extremely low yet in 1569 AD the market assayers of silver bullion were paid at the rate of 200 *kyats* of *ganza* a month, a sum which was presumably a living wage. One *kyat* weight of silver in the Siam of 1688 AD would buy enough rice for a man to live for two and a half months.[14]

Ganza was in use in and before the early 16th century. Lead had replaced *ganza* mainly by 1786 AD but some *ganza* may still have been in use in the early 19th century.[15] The capture by the Burmese of the Bawdwin lead mine between 1750 and 1763, the cessation of the importation of cowries, the Burmese need for copper for its several wars from 1750 onwards and the likely cessation of Chinese copper imports during the1765-1769 AD war with Burma, (Siam was short of copper at this time) may have brought about the change from the use of *ganza* to the use of lead.

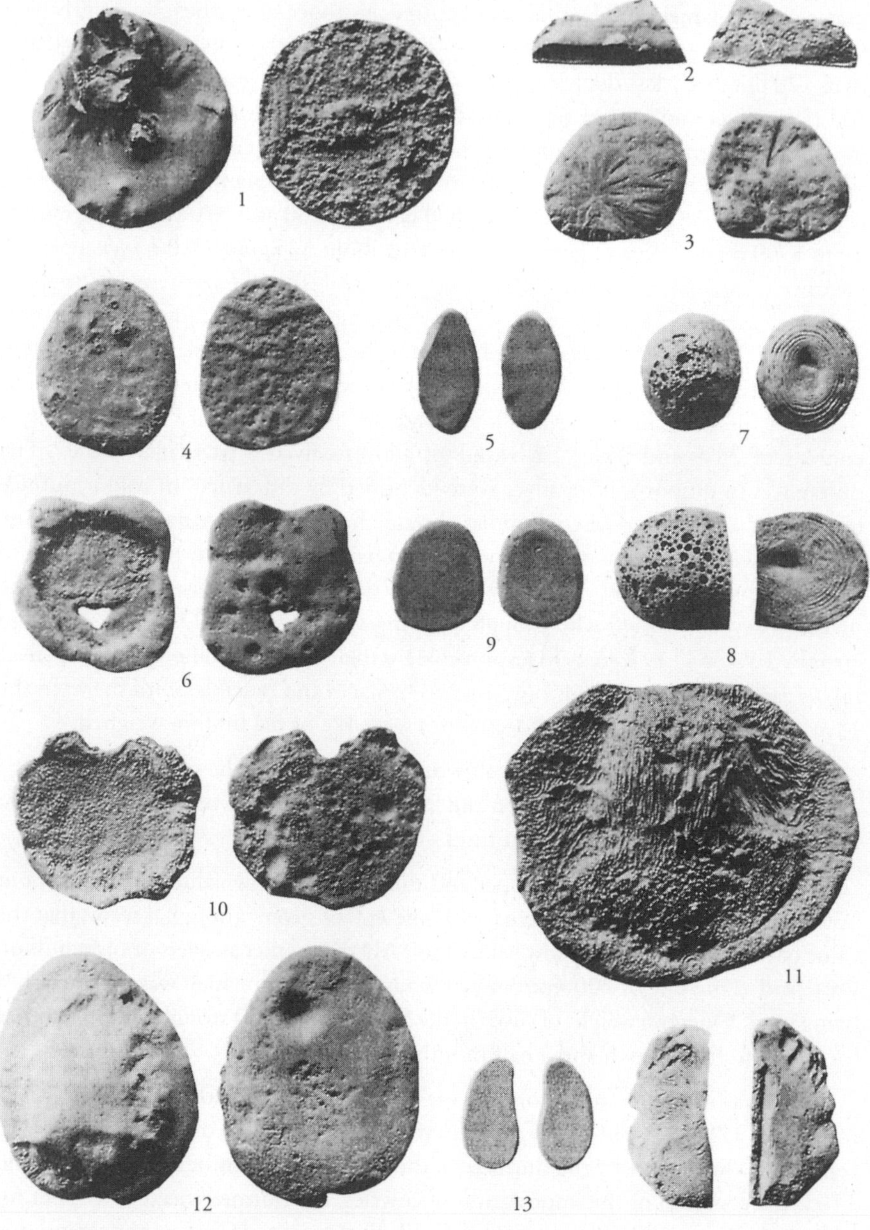

Plate 4: Silver Alloy Ingot Currency (actual size) Burma, 19c AD

Description of Plate 4 (Figures 1 – 14)

Fig. 1 ***Baw.*** Approximately 98% silver. Under surface: Often with reddish-yellow spots of litharge. Usually shows a local "bulge" of silver efflorescence. The *kayubat* variety has concentric rings on its upper surface. *(See fig. 7).* Used mostly for European trade.

Fig. 2 As for Fig. 1 but chipped for use.

Fig. 3 ***Dain.*** Approximately 92% silver. Upper surface: Flowered (crystallised) in long crystals or striations. Commonly about 20 – 30 kyats or more. Used mainly for Chinese trade. Most common form in use in early 19th century.

Fig. 4 ***Ngwelon.*** Approximately 92% silver.

Fig. 5 ***Maingyon-ngwe.*** Approximately 92% silver. A Shan silver ingot.

Fig. 6 ***Ywetni*** (Flower money). Approximately 85% silver. Upper surface: Flowered like dain but crystals shorter. Under surface: Often with reddish-yellow spots of litharge. The standard ingot type from 1700 AD. Burmese and Chinese names both meant "flowered silver".

Fig. 7 ***Thakwa.*** Approximately 85% silver. Upper surface: Marked by concentric rings. Under surface: Spongy appearance. Used mainly in Bhamo for the China trade. Probably made either in Yunnan or Shan states.

Fig. 8 As for Fig. 7 but chipped for use.

Fig. 9 ***Lezege.*** Approximately 60% silver.

Fig. 10 ***Sengajatke.*** Approximately 85% silver.

Fig. 11 ***Asekke.*** (Oyster-shell money, Chiengmai tok). Approximately 75% silver. Upper surface: Hairy or feathery. Under surface. Often broken. Often stamped on edge. Widely used in Burma.

Fig. 12 ***Ngwema.*** "Mother of silver". Valued as a charm because of the sound it made when shaken.

Fig. 13 ***Lezege.*** Approximately 60% silver.

Fig. 14 ***Kege.*** A piece of lead chipped off a larger ingot

---

(a) The plate is that from R.C. Temple, Indian Antiquary, June 1897, courtesy of the Pitt Rivers Museum, Oxford.

(b) The descriptions have been adapted from Temple, Indian Antiquary, April and May 1919, pp. 41 – 54. Temple, 1931, p. 75. Yule, Ava, p. 260.

(c) The names shown were well-known names referring mainly to the alloy composition of the ingot. In addition, there were names referring to the ingot itself (Temple, 1897. p. 10, n. 17). These were different in Burmese and Mon. In the late 1880's when the ingots were no longer in use, people were "hazy as to differentiating the standards".

(d) The silver percentages are reasonable estimates but the proportion of silver varied, e.g. by up to ±10% approximately. Percentages are given here but it was customary to value the alloy in terms of *baw* or *ywetni.*

(e) The appearance of the ingot served to identify its composition, like a coin. Weight was often assessed by handling only.

(f) Ingots with less than 50% silver were confiscated by the king but nevertheless were used outside the capital and its neighbourhood.

When lead was used as a currency, a piece of about the required weight, usually between 15 and 500 grams, was cut off with a hammer and chisel (See Plate 6) and tendered in payment. Chopped lead could also be purchased, as in China, and baskets of it were prominent objects in the markets of Burma in 1855.[16] The exchange rate at this time was 100 *viss* of lead to 6 1/2 *ticals* of silver, i.e. 1500 to 1. Obviously the larger weights would be required for the larger amounts.

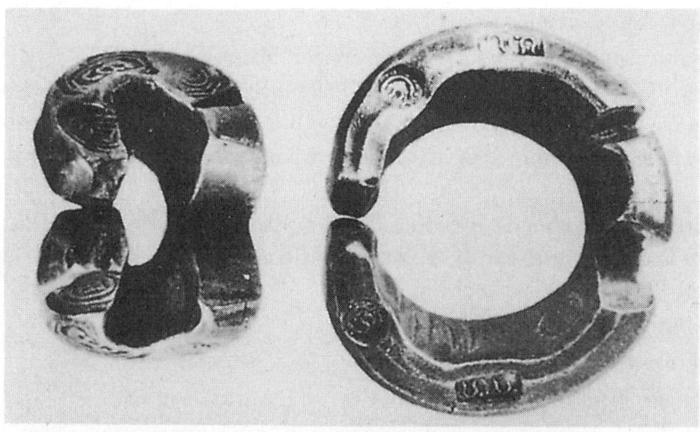

Kakim, NW Siam: 13 c – 16 c AD

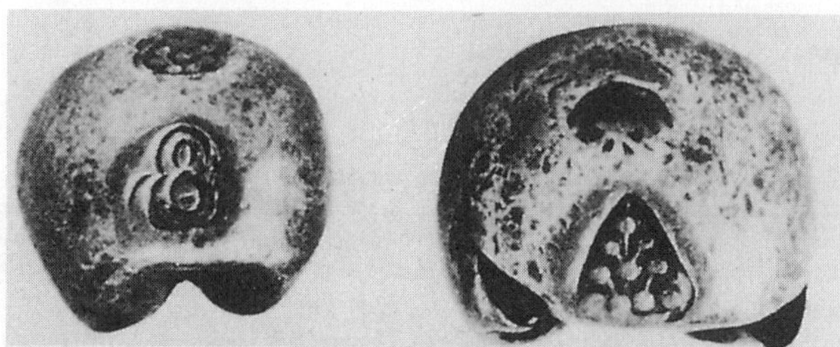

Pod Duang, Ayudhya: 13 c – 18 c AD

Sycee, Yunnan: Post 13 c AD

Plate 5: Silver Ingot Currency (North Siam and Yunnan)

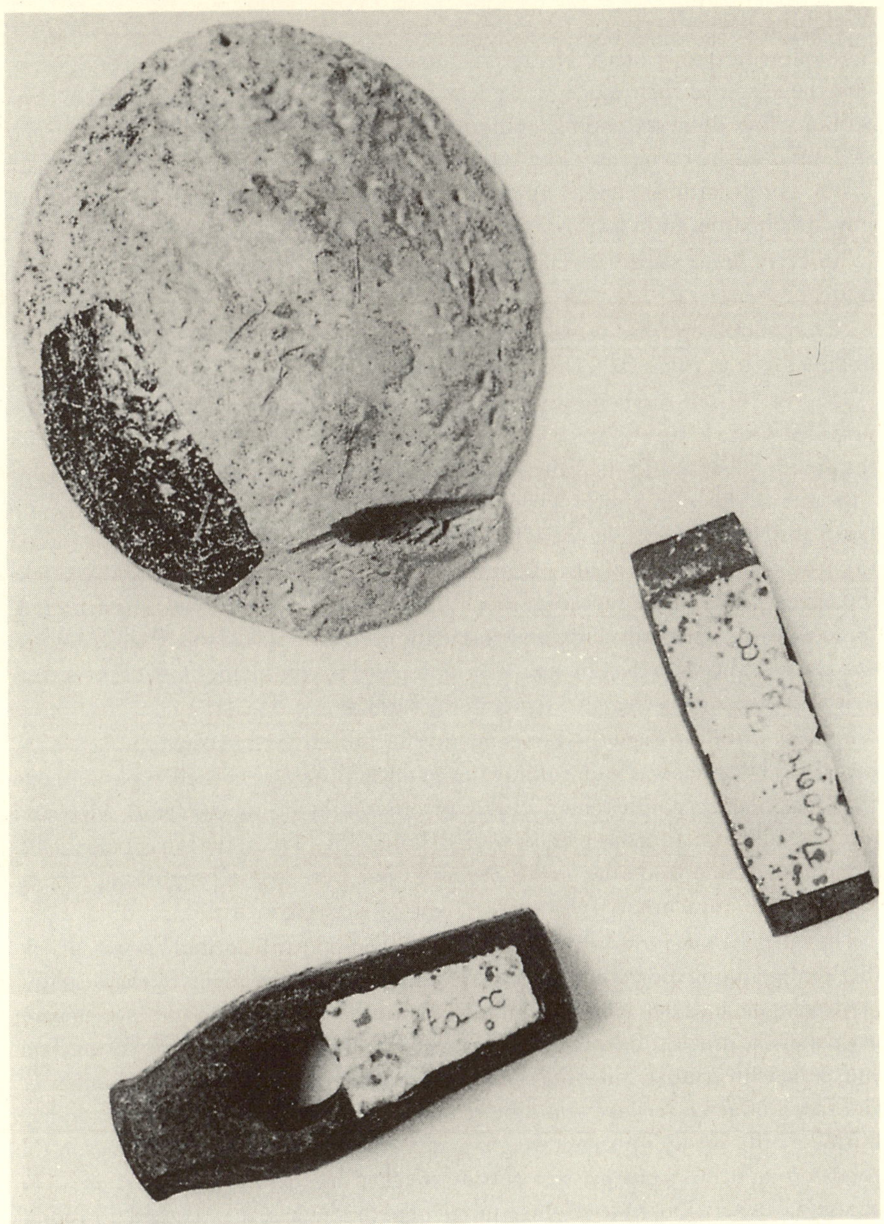

Plate 6: Lead Ingot Currency, Hammer-head and Chisel, Burma, 19c. AD

Weighing procedures:

The following quotations add to the foregoing description of the weighing procedures and the extent of their usage in the late 18th and 19th centuries. However, most within-village domestic trading for the daily necessities of life both in Burma and Siam still was done by barter, as it had been for centuries past, and little money changed hands. For the ordinary family money would have been needed only to pay taxes and buy little luxuries from local or travelling merchants.

"In every home there are scales and weights and the house is engaged in regular trade."[17]

"Every shopkeeper has a small box containing scales (See Plates 1 and 7) to weigh bullion given in payment for commodities."[18]

". . . in 1861-2, anybody wanting to transact business on the market must be provided with a lump of silver, a hammer, a chisel, weighing scales and weights. When the potential buyer asks the seller to state his price, the latter responds by asking his customer to show his money. Having formed an opinion about the fineness of the lump of silver produced, he then quotes a price in weight. Thereupon they proceed to cut off from the lump of silver a corresponding piece which is then weighed. Often the operation has to be repeated several times until the correct weight is achieved. Needless to say, much silver becomes lost in the process. To avoid this, the buyer often asks the seller how much of his goods he is prepared to give for his lump of silver such as it is. Alternatively, change is given in the form of rice."[19]

Flouest, writing of Pegu and Rangoon in 1786, stated that the current money in the bazaars and markets was lead, cut into lumps of various sizes. The sellers put into one scale the goods they were selling and in the other scale the lumps of lead. Meat and fish were sold sometimes weight for weight. For 25 lbs. of meat the buyer paid 25 lbs. of lead. Vegetables and other goods of lower value were sold in proportion. Flouest remarks that "this currency was seldom used for large payments."

The Burmese weights were commonly used in Laos, north Siam and Yunnan. Partly this was because all these countries had a history of making payments by the weighing of silver ingots and later, lead. Partly it was because the north of Siam and Laos, or parts of them, were usually under the domination of the Burmese from 1557 AD until the end of the 18th century. Since the Burmese had destroyed the local weight standards[20] and the conquered regions would be required to pay tribute through the resident Burmese officials, the Burmese weight system would have been used to weigh the goods brought in, goods perhaps already weighed by the indigenous system. The occasional use of a four *kyat* weight, equivalent to the Siamese *tamlung* and the Indian *chattack,* in beast weights Groups 2 and 5 supports this view. Another reason for the use of the weights outside Burma may have been that the Burmese weights probably were the only ones which had had an uninterrupted history and upon which reliance

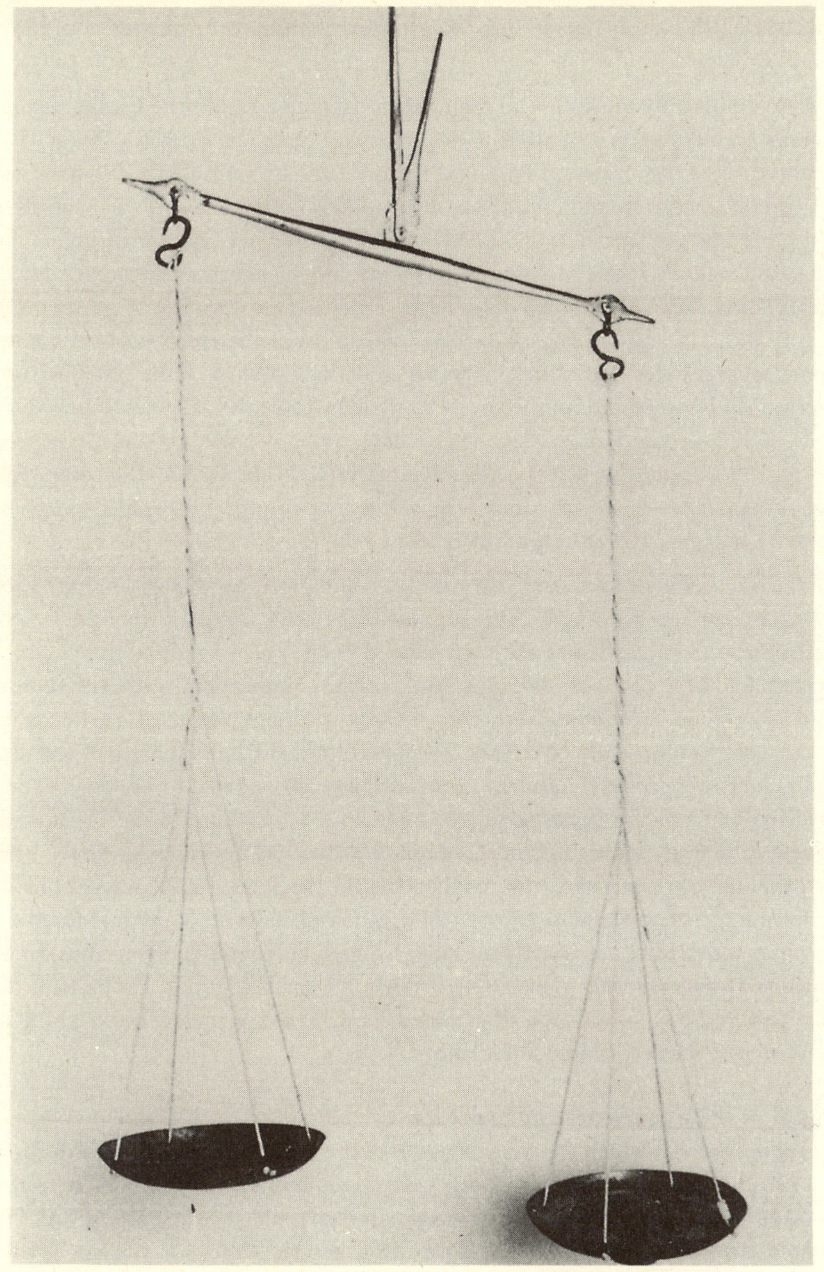

Plate 7: Burmese Scales, 19c. AD

could be placed. The obvious animist/Buddhist symbolism of the weights would have been a connected reason.

It was almost only the king who required to accumulate and store wealth, e.g. for religious and irrigation structures and certain war materials such as cannon. The villagers formed the armies, providing their own handweapons. They also provided the war boats and were responsible for their upkeep. Criminals and slaves supplied labour not due from the villagers. The king accumulated personal luxury goods and issued laws to restrict their purchase by his subjects but his actual needs for money and other treasure were small. Much of his palace expenditures were met by payments to him in kind. He enjoyed monopolies in trade and was the chief, if not the only, import/export broker. To ensure that his payments were received at the entrepots, tax-collecting and storage depots, there were royal officials to assess quantities and weigh goods, collect dues and make sales and purchases on his behalf.[21] It would have been the king's agents and the Burmese merchants who used the Burmese weights while foreign merchants would use those from their own countries. Disputes would be settled by reference to the king's standardised weights.

In the moist tropics the normal person's life could go on happily without the need to acquire artificial wealth. Nature supplied his needs, except during periods of catastrophe, so long as the population remained small. The few natural things which were required from other people which could not be counted or measured by volume could be weighed. A problem arose when man-made things were used. In the case of coins, no trust could be placed either in the mass of the coin or its content of precious metal. Coins were often fraudulent or debased so each one had to be weighed and tested like ingots. The Burmese people must have been aware of the Indian and Chinese difficulties with coin and paper money issues. They must also have been aware that the Chinese had discovered that only the low-valued 'cash' was reasonably safe from fraud of one kind or another though best of all was the cowrie. Moreover, the coin manufacturer, i.e. usually the king, required his commission on coins which immediately lowered their value relative to a home-made ingot. It was perhaps the desire to avoid the commission that caused the Chinese to prefer the weighing of ingots even at the cost of fraudulent alloys.

Origins of the frequencies and procedures:

The proportions in which the weight masses occurred in 1972 may represent those in which the various weight magnitudes were required and made. They may also represent the remainder after various vicissitudes have affected the proportions. For example, the smaller weights are more easily lost and the larger weights would have been more easily collected for other purposes at times of high copper prices or copper shortages.Moreover, above 100 kyats the marble weights are said to have been used. [22]

Mass Frequencies, Goods and Masses Weighed and Weighing Procedures

The use of weights, described previously, must have been familiar to the Mons and Pyu, dominantly influenced by the Indians for approximately the first eight centuries of the Christian era and to the Shans, Burmese and other peoples who entered south-east Asia from the north over a long period at about the same time.

Silver coins[23] were in use (though probably only by the rich) between about the 3rd and 8th or 9th century AD in Arakan (in north-west Burma) and by the Pyu and Mon peoples, mainly of south Burma. They came to an end along with Indian political influence and the succeeding dominance of Nanchao (Yunnan) influence. Thereafter they appear in Arakan about the 16th century AD following a Bengal pattern, i.e. under Indian political influence. In Burma lead coins might have appeared briefly about the 16th century AD, silver and copper coins appeared briefly at the end of the 18th century AD and a modern coin system came into being in the mid-19th century. So before the end of the 18th century payments presumably were made by the weighing and testing of currency ingots and by cowries. In Pagan between the 11th and 13th century AD, this appears to have been the case.[24] Silver was not popular in China until about the 15th century AD. Before this, silver coins, though experimented with, were soon withdrawn. In Yunnan no coins have been reported but flowered silver ingots were used there at the same time as in Pagan. These ingots certainly would have been weighed, just as the Chinese *sycee* of the same period were.

It seems then that the use of silver coins is associated with early Indian influence of before 8th century AD. The use of silver and other ingots together with the related paying procedure is connected primarily with Chinese influence from the 7th century AD or before.

Notes

1 Elliott, pp. 49, 50. They would be used in conjunction with the Chinese goldsmith's scales which were quite accurate. (Annandale, p. 197).

2 Temple, April 1919, p. 49.

3 Kneedler, p. 5. Yule, p. 260. Prinsep, pp. 35-36, 61. Symes, Ava, pp. 325 – 327.

4 Accounts of the ingot currencies are given by Cresswell, Le May, Guehler, Kneedler, Coins in Thailand, Phayre, Temple (1919 and 1928, especially good on Burmese silver ingot currencies), Kann, Sigler, Laughlin, Morrison, Robinson and Shaw, Quiggin, Ridgeway and many others. The statement concerning the occurrence of exogenous ingots in Burma is borne out but for different reasons by Pires, v. 1, p. 100. There he mentions that in Pegu there was in use in 1514 AD "a round silver coinage marked with the mark of Siam because it all came from there". This might have been one of the bullet ingots. It weighed a *tael* and a half (about 56 grams) and was called a *caturna.* 56 grams is about the mass of a 4 *baht pod duang* which apparently was rarely, if ever, made in central and south Siam. The objects referred to may have been called *"kub"* or *"kud",* considered to have been weights during the Sukhothai period. they were made of tin or a copper-nickel alloy. Those members quoted by Le May appear to have a mass unit of about 11,9 grams and be on a unitary scale. See Le May, 1932

pp. 17,18 and 57-62: Coins in Thailand, pp 30, 31: Ferrand, pp. 113 – 114 quoting Sparr de Homberg. This observation by Pires illustrates the close connection between Lower Burma and Siam and suggests the origin of the script on the Group 6 weights.

5 Le May, pp. 10, 12.

6 Sigler, p. 12. About the 14th century AD the name for *sycee* was *balish* which translates as 'shoe', the silver ingot being in the form of a Tartar shoe. A common mass of the balish at that time was 500 mithcals, i.e. about 2500 grams (Yule and Cordier, v. 2, pp. 197, 198, **quoting Odoric, 13th century AD). Another term applied to ingots of both gold and silver** which were used from the Black Sea to China in the 14th century AD was *sommo* or *saumah.* These ingots weighed between 5 and 8,5 Genoese ounces, i.e. between 110 and 190 grams. (Yule and Cordier, v. 3, pp. 148-150, quoting Pegalotti, mid-14th century AD.) Temple, 1928, pp. 2-9, is especially good on *sycee* in Burma. He also comments on *balish* and *sommo.*). In the 16th century AD at Kanchow and Suchow in China the currency in use was that of gold and silver in small rods which were cut into fragments for spending. (Yule and Cordier, v. 1, p. 293). The Mongol term *toman* (10 000) was applied by the Persians and Arabs to a weight of silver equal to ten thousand *mithkals*, i.e. about 45 kg. Staunton (v. 1, p. 35, quoting J.G. Mendoza of 1585) states that the Cantonese silver and gold money of that time had no sign or print and went by weight. On the other hand the weights did bear a seal or sign.

7 Kann, p. 238.

8 Laughlin, pp. 287, 288. Sigler, p. 10. Morrison, pp. 68-94. Temple, *Ind. Antiq.*, 1928, pp. 2-8.

9 Prinsep, pp. 35-36, 61. See also W. Desai, *History of the British Residency in Burma, 1826-1840.* Rangoon, 1939, H. Malcolm, *Travels in South East Asia, vol I, Burman Empire.* Boston, 1839, and Temple, *Ind. Antiq.*, 1919, pp. 49-55. Hunter, p. 85. Yule, pp. 256 – 261. Symes, Ava, pp. 325 – 327. Crawfurd, Ava, p. 188. Flouest, p. 41.

10 Yule, pp. 258-259.

11 Temple, 1898, p. 145, quoting J. Alexander, *Travels from India to England.* London, 1827. In Siam, rather than weigh low-valued pieces of metal, buyer and seller often would agree on its value merely by inspection. (Quiggins, p. 210 quoting L. de Carne, *Travels in Indo-China.* 1872). Linschoten, v. 1, p. 132, states that, in China, soldiers were paid every month with pieces of cut silver with which they made payments using a small balance and the instrument to cut the metal which they always carried with them. See also Malcom, p. 270. Symes, pp. 325 – 327, has a good account of market practices when using silver. Crawfurd, Ava, p. 188, has interesting comments on the associated silver losses.
The *pwezas* would also receive a merchant's money and pay it out as necessary like a banker, charging a fee of 1% for this service. The *pwezas* were often called *"pymon"* (foreigners). Possibly these were Indians and Chinese who long had been settled in Burma. In both India and China the merchants had been accustomed to trading on both small retail and large wholesale scales since, perhaps, 500 BC. (Basham, pp. 221 – 223. Ball (Tavernier), pp 28, 37). For example, the Yunnan-Ava trade was mostly in the hands of Chinese merchants who were settled in Ava, Bhamo and Yunnan and who arranged both the goods and the payments. Probably those Indians settled in Yung-chang in Yunnan also participated. (Davies, Yunnan, p. 168. Crawfurd, Ava, p. 191.)

12 Yule and Kendall, p. 298. Bowrey, p. 241, remarks that in Janselone (Junkceylon), off the west coast of Malaya, masses of up to 420 lbs. or more were weighed either with scales or steelyard. In Ayudhya, mid-18th century, masses up to 129 lbs. only could by weighed by the *datchin*, Brooks, p. 17. *Datchin* (*dachein, dodgeon*, etc.) is a corruption of the Cantonese *toh-ch'ing* from *toh*, 'to measure', and *ch'ing*, 'to weigh'. (Mundy, v. 3, p. 311). It was not only a steelyard but a weight. (Pires, v. 1, pp. 101 and 207-208). The weight may have been the counterpoise. Scales (*dala*) apparently were used in the Bengal ports (Pires, v. 1, p. 100). In Cambodia there were always two kinds of balances in use. One was for daily commodities and the other for precious metals (Groslier, p. 27).

13 de Campos, p. 112. See also 'The Voyages of Caesar Fredericke' in *Purchas his Pilgrimage or Relations of the World*, v. 2, p. 96 London, 1626. Pires, v. 1, p. 97, identifies *ganza* (*caca, camsa, gamca*) as *fruseleira*.. Nunes, p. 38, refers to the *gamca* currency of Pegu being of dishes, pans and other domestic utensils made of metal like *frosyleyra* (*fruseleira* = an ingot made of brass scrapings broken in pieces). According to Temple, 1898, p. 177, an almost identical situation still existed in Manipur in the 19th century. On p. 125 Pires states that among the merchandise imported to Pegu from Canton were vases of copper and *fruseleira*. Elsewhere he writes that *fruseleira* of copper and tin is better than that of copper, tin and lead, while the worst is that of copper and lead. That of Martaban was believed to be of best quality. The names *fruseleira* and *gamca* also appear to have been applied to an alloy of copper, silver and tin (de Campos, p. 122, quoting Nunes). See also Temple 1919, pp. 149, 158, 159.

14 de la Loubere, p. 72.

15 Kelly, p. 113. However *ganza* might have been a term that continued in use even after the metal itself had been replaced by lead (Hall, 1928, p. 129).

16 Temple, 1919, p. 109-111, gives and excellent account of the Burmese lead ingot currency. See also Phayre, p. 38, Yule, pp. 258-260 and Malcom, p. 277. See Temple 1897, pp. 309 – 311 for other exchange values of lead to silver, silver to gold and silver to tin.

17 Ferrars, p.77. This may be an exaggeration since scales and weights would have been expensive for the ordinary household. It seems that the widespread use of scales and weights was a phenomenon which came about during the second half of the 18th century. Before this the use of weights and scales may have been confined to the king's agents, full-time merchants and *pwezas*. Support for this view comes from the authors' 1972 survey when it was found that the bird weights of Group 2, periods C and D amounted to about 71% (456) of all the accepted weights counted. This proportion was obtained after the removal of those weights judged to be non-royal or unofficial, most of which, though different, had styles generally similar to the bird weights of period C. Taken together, the proportion would have been much higher. That is, there was a great increase in the numbers of Group 2, periods C and D style weights compared with those of the periods which preceded it. Considering the bird weights only, since they are continuously characteristic of Burma, unlike the beast weights, the percentages of all bird weights within the Groups 1, 2, 3, 4, (5, 6 and 7) were as follows: 0,3; 86,2; 2,0; 8,8; 2,8. Approximately the durations of each group in years were estimated to be 20, 100, 10(?), 200, 150. Of the possible explanations for these figures, the only reasonable one is that there was an 'explosion' in weight-making and hence in weight usage with the introduction of the Group 2 weights. This probably came about as a result firstly, of the cessation of the supply of the cowrie currency; secondly a shortage of *ganza*;

thirdly, the replacement of both cowries and ganza by lead and fourthly, by the passing of the ownership of the Bawdwin lead/silver mines into the hands of the king of Burma. He was now able to obtain an income from the rental of the mines to Chinese operators, to buy the lead at his own price and sell it at a much higher one. His profit on the latter, in fact, was 400% (Yule, P. 256).

These more or less contemporaneous changes would result in an increase in the numbers of scales and weights required by ordinary folk compared with the numbers required formerly. Moreover, there would have been less demand for the more expensive *ganza.*

For further information on this topic see the comments on *ganza* in this chapter, the sections entitled 'Copper' and 'Lead' in the chapter 'Weight Materials and Manufacture', the comments in the chapter 'Signs and Style Systems' dealing with the significance of the signs and note 61 of the chapter 'Mass Scales'.

[18] Temple, 1898, p. 21, quoting J. Alexander, 1827.

[19] Einzig, pp. 95, 96, quoting A. Bastian, *Reisen in Birma in den Jahren 1861-3.* Leipzig, 1866, pp. 57-58

[20] Gardiner, 1968, p. 3.

[21] This was the practice in the 13th century AD (Polo, p. 218) and in the 17th century AD (Floris, XXV). In the Europeanised system operating in Malacca after its capture by the Portuguese, several merchants would jointly assess the value of imported merchandise. The importers would be expected to abide by their judgement (Pires, v. 1, p. 98). Foreign trade in Siam of 1688 AD was reserved almost entirely to the king. He had the first choice of imported goods for which he would pay at a rate fixed by himself when he saw fit (de la Loubere, pp. 71, 112). In Burma of the late 18th and early 19th centuries there was a uniform inport duty of 12½% on all goods of which the king took 10% and the local **governer 2½% (Sangermano, pp. 176, 177). Imposition of even more adverse conditions of** trade would cause foreign merchants not to enter the ports.

[22] Forbes, p. 132.

[23] Phayre, Temple, Robinson and Shaw, Mitchiner, Aung Thaw and others give comprehensive summaries in English of the Burmese silver coins.

[24] Luce, 'Economic Life . . .',.pp. 337, 338.

4
Weight Materials and Manufacture

Colours:
The normal colours of the matt surfaces of the weights are almost black, dark brown, bronze and silvery bronze. Brassy colours are found only on modern weights. The patina, if present, may be an overall or patched greyish green, occasionally giving an attractive colouration to some of the older weights. A dark greyish-white or glossy black patina often occurs in the interstices. An overall ashy white patina is found only on modern weights.

Materials:
The animal weights mostly are made of a solid metal alloy. The Group 6 weights are frequently of an alloy with a stone-cored (pegmatite, etc.) base. The non-animal weights are of white marble.

The weight alloy:
The chemical composition of the alloy forming the animal weights is variable. Probably this is due to the type of material available to the founders. From the north this would be in the form of Chinese cash (and ingots?) brought in by caravans from Yunnan. From the south, ingots from a variety of sources were brought by sea to the ports and then transported by boat up the main river. Usually there are three main elements in the alloy: copper, lead and tin. Zinc became important in the19th century. Iron, nickel, silver and arsenic occur as minor constituents.

The ranges of percentages of the main elements comprising the alloy are shown below. These are from eight analyses made in 1977. By their side are shown the ranges obtained from fifteen analyses in 1983.

Table 10: Chemical Composition of the Weights

Tin	3 — 14%		0 — 18%
Copper	54 — 77%		80 — 78%
Lead	14 — 31%		3 — 31%
Zinc	0 — 1%		0 — 25%
Iron	0 — 8%		0%
TOTAL	100%[1]		100%[2]

Comparison of limited published information concerning the copper alloys used in China,[3] India[4] and Yunnan[5] shows that the weight alloys appear to be closer in composition to those used in India, except Bengal. Indians preferred to use copper. Only in south India does lead seem to have been used as an additive to any extent. However, some of the Bengal artifacts do have compositions similar to those of the Burmese weights perhaps because alloys originating in Yunnan were reaching Bengal. Normally it would be expected that the copper mines of adjacent Bihar would have supplied most of Bengal's needs.

Copper:

Economic deposits of copper ores existed neither in Burma nor in Siam.[6] Both Burma and Siam had to import copper and had done so since before the 16th century.[7]

Copper was plentiful in China and had been mined since about 3000 BC. It was one of its important exports to south-east Asia, perhaps mostly in the form of Chinese cash. The export of unwrought metal was forbidden. Mines were worked throughout Yunnan which was China's main source of copper. Tung Chhuan in north Yunnan, for example, was a source of copper-nickel ore. Shansi and Nanking also mined copper.[8] So did Laos.

By the 16th century China had developed its copper trade with Europe.[9] Copper, probably from Bihar, also reached Burma in Pagan times, having been shipped from Tamluk (Tamralipti) in the Ganges delta. Possibly the trade continued later but on this matter there is but little information.

Although Chinese and other copper imports were so important to Burma, the total trade of all kinds with China was not large relatively , at least not after about the16th century AD. This apparently was one of the reasons why the Chinese abandoned the 1765-9 AD war with Burma,[10] a war which must have affected Burma's copper imports. The imports of copper from China fluctuated. All mining was stopped during several dynastic periods because, according to Confucian teaching, it was considered it disturbed the earth spirits, ancestral graves and the repose of the public.[11] In the 17th century Burma was critically short of copper for its cannon[12] and in the 18th century Siam unsuccessfully requested China to relax its restrictions on the export of copper.[13] On at least one occasion a severe earthquake disrupted mining and when it was resumed, supplies were diminished.[14] Like silver, copper was sometimes popular in China and at other times it was in but little demand. Demand for copper increased during wars and so normal supplies became inadequate. One instance was mentioned above. Another would seem probable during the period 1750 to 1800 AD, approximately when the Burmese were either fighting Siam, and so losing the copper-rich region of Lan Chang (Laos) or invading the Arakan, Manipur, Bengal regions. They also had to fight off the Chinese invasion in 1765-9 AD. Internally there was the siege of Pegu and decimation of the Mons in 1757, several revolts in central and upper Burma between 1768 and 1769 and the Arakanese revolt of 1794-7 AD.

Shortages of copper may have been partially responsible for the change from the use of *ganza* as a currency to that of lead in the mid-18th century and for the lower frequencies of the larger weights.

Tin:

An ancient source of tin was Chhang-sha in Hunan. In Central Asia the first tin mine was in operation before the 14th century BC while the first copper mine in the Altai mountains of Central Asia was being worked at about the same time.[15] Burma had small deposits of tin in various places, perhaps the most important today being at Mawchi near the river Salween east of Toungoo. All of them are on the tin-bearing zone that stretches down into Siam and Malaya. Tin was also readily available from Siam,Laos and Malaya.

Lead (and silver):

The main concern with lead originally was the silver in the ore, yields up to 18 ounces per ton of lead being common. However, the lead itself was used by the Burmese as an alloy with copper for the currency metal *ganza* and later used alone as a currency. European traders bought it.

Lead and silver were available from Yunnan, notably from near Chhang-sha and from the Kungshan mines but of these only the silver is reported to have been exported. Lead was obtainable from the Shan states and further south into the tin-bearing region of Tavoy. The Thein-ni and Bawdwin localities of the eastern Shan states were anciently worked, the latter, the most important lead mine in south-east Asia, being mined by Yunnanese convicts in 1412 AD and probably earlier.[16] Bawdwin was under the control of the Shans until between 1750 and 1763 AD when it passed to the Burmese.[17] Siam and Cambodia had but small supplies of lead and silver, Patani being one local source.

Silver occurs in small amounts up to about 0,3% in some weights. The main use of silver was for currencies. Cambodia and Siam had but little silver and purchased large quantities from Yunnan, especially after its conquest by the Mings in the late 14th century. However, from this time on, China's demands for silver were large and shortages occurred in Siam sometimes making silver more expensive than gold. Before 908 AD silver was used mainly in commerce with barbarians, i.e. aboriginal tribes.[18] This and its relative abundance in Yunnan may have been an influence governing the importance of silver to the barbarians of Yunnan and south-east Asia.

Zinc:

Zinc is often associated with lead in its ores and may occur in the older weights as an impurity in the intentionally added lead. Brass, an alloy of copper and zinc, was known to the Chinese in the 5th century AD. The most recent weights, i.e. of Group 1, are distinctly brassy in colour and composition due to an increased zinc content. In this they are similar to the more recent Chinese cash.[19] Zinc was introduced into the cash made by the mints in the neighbourhood of Peking (Beijing) in 1505 AD. Its amounts ranged from 12% to 37%.[20] Similar information for the Yunnan region is not available, perhaps because cowries and silver alloy ingots appear to have been preferred for domestic purchases until the Chinese effective control of the region after 1368 AD. Zinc was imported into Burma from Laos.

Nickel:
Nickel is occasionally a minor constituent of the alloy. Nickel is present in the Bawdwin ores, in copper ore on the Siam-Laos border[21] and in Yunnan, e.g. at the Tung Chhuan mines of north Yunnan. By the 16th century the Chinese were exporting a copper-nickel alloy.

Stone:
The Group 6 weights frequently have a stone-cored (pegmatite) base covered with a thin skin of bronze about one millimetre thick.

Metal working:
In India the civilisations of the Indus valley of about 2500 BC were making use of copper and bronze but did not spread their knowledge. When the waves of bronze-working Aryans entered India, perhaps from south Russia and Uzbekistan between about 3000 and 1500 BC, especially the latter, they brought with them their bronze tools and weapons and spread their knowledge of them across much of India. At about the same time or earlier, bronze-working had spread from Mesopotamia into Persia, with its centres in Luristan, Urardhu and Zawiyeh.[22] Also at about the same time (the 18th to 10 century BC) the Tazabagyab bronze culture south of Lake Aral developed.[23] On the Huang-ho River and on the eastern limits of the steppes between about 1500 BC and 600 AD was the bronze-working culture of Ordos.[24] Going north again to the Lake Baikal region there was the Andronovo culture (18th-13th century BC) and that of Glazkovo (17th-12th century BC) which may have transmitted the secret of bronze metallurgy to China.[25]

In south-east Asia as a whole the beginning of knowledge of bronze and its working has been variously dated to between 3000 and 500 BC.[26] The first bronze-working centre in south-east Asia was Dong-S'on in north Vietnam,[27] active between the 6th century BC and the 1st century AD and was probably connected with Yunnan. Its works spread to all countries of south-east Asia and to Indonesia. In particular it was associated with the bronze drums that are spread throughout the same regions and were still being made or used in the 19th century AD in eastern Burma.[28]

On Lake Dian (Tien) near Kunming in Yunnan was another bronze -working culture which produced the earliest known bronze drums about 300-200 BC.[29] The craftsmen of Lake Dian between 100 BC and 100 AD are also notable for the high standard of their metal-working particularly in using the lost wax technique,[30] just as the Indians were in Bihar and East Bengal during the first millenium AD.[31] One of the several major achievements of the Yunnan craftsmen was the technique that they developed for producing a silver-grey exterior while casting, an exterior which tended to resist corrosion and green patination.[32] Relatively few Burmese weights are patinated. Another of their noted skills was their ability to cast very thin-walled bronzes.

Bronze work is known from about the 3rd century AD in south Burma so that the capability of obtaining materials for the weights and making them certainly existed then. Little information on this matter exists for north Burma but a trade and Buddhist pilgrim route from Bengal through north Burma existed from the 2nd century BC and probably before. Moreover, the Mons, Burmese, Shans and other peoples living in north Burma had arrived there after passing through Yunnan which

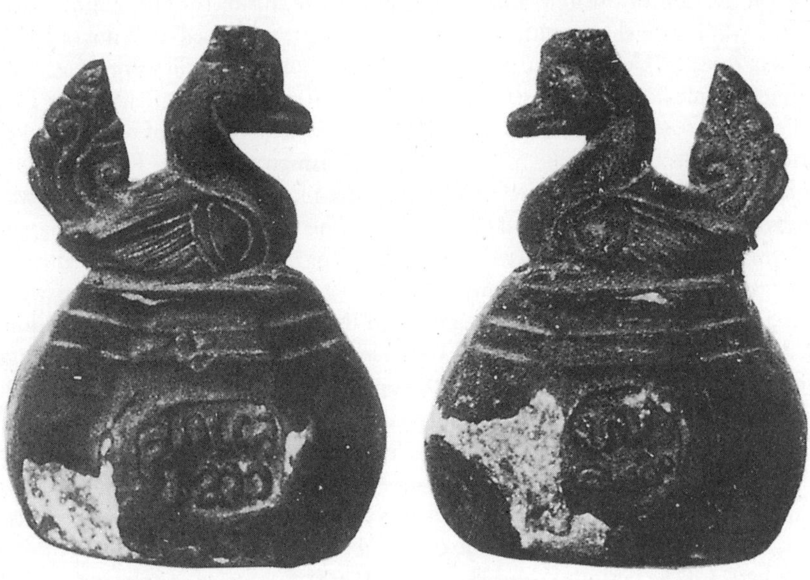

Plate 8: The Thin Metal Skin over the Stone (pegmatite) Core of a Group 6 Weight

(This technique is known to bave been used in 2c. BC Yunnan and 14 c. AD south Siam)

had been a bronze working centre for many centuries. Again, in the Pagan museum there is, or was, a bronze drum typical of the region, apparently a war drum of the Chinese general who marched to the Chinese-Burmese border at Yung-ch'ang in 220-230 AD.[33] Hence the peoples of north Burma and Siam would have known about the materials and products of the Yunnanese. See also the chapter entitled 'Trade'. In the case of the Yunnanese thin-walled bronze manufacturing technique, the Siamese, especially of Ayudhya, between 1400 and 1475 AD had a similar skill, either inherited from the Yunnanese or developed because of copper shortages.[34] The thin-walled bronze coating over the stone base of the Group 6 weights has been mentioned. This construction was a weak one and frequently the base broke away from the overlying animal effigies. One sometimes sees the upper parts only for sale. This is a unique kind of happening and it suggests a link between Siam (and Burma?) and Yunnan. (See Plates 8 and 34)

The fact that the Ming Chinese became enthusiastic developers of the Yunnan lead-silver mines after 1368 AD and were mining the Bawdwin deposits in 1412 AD, supports the suggestion made elsewhere that the Chinese weight standards reached the Shans about 1391 to 1403 AD just as they did the Siamese and Cambodians.

Weight manufacture in Burma:

The lost wax technique could have reached Burma either through Bengal or Yunnan, perhaps at the time of the first entries of the peoples of Burma into the country. The wax used was a special one with a high melting point, produced in Yunnan by the white wax insect.[35] Near the end of the last century the weights still were being made by this technique in a "weight-making" village near the capital, Mandalay. (See Plate 9). The village was still known as such in 1972. Perhaps this village of weight-makers was not unconnected with the metal-working craftsmen of Ava (or Kyaukse?) who were famous in 1290 AD. In this year five hundred families of metal-workers were given by Ava to the king of Lan-Na to promote their skills throughout the latter's kingdom.[36] The Shans, in particular, were noted for their metal-working skills, the contrary being said about the Burmese.[37] In the 19th century the main "brassware" working centres were Amarapura, Sagaing and Rangoon.[38] During King Mindon's reign (1853-1878 AD) the weights were made to the order of the Kinwunmingyi (Chief Minister) or the Shwetaikewun (Treasurer), stored in the royal treasury at the capital and distributed thence upon request of the king's officers in the large towns.[39] Weights were similarly distributed in 1795 AD. Probably these procedures were similar to those which had been long in use. However, it is possible that with the introduction of the Group 2 bird weights this distibution procedure may have been followed only for those weights in the standard style used by the king's agents and some others. Of the periods C and D, more varieties of unusual, non-standard style are found than of any other periods between A and J inclusive. This suggests that these non-standard varieties may have been (allowed to be?) produced by makers who were not officially supervised.

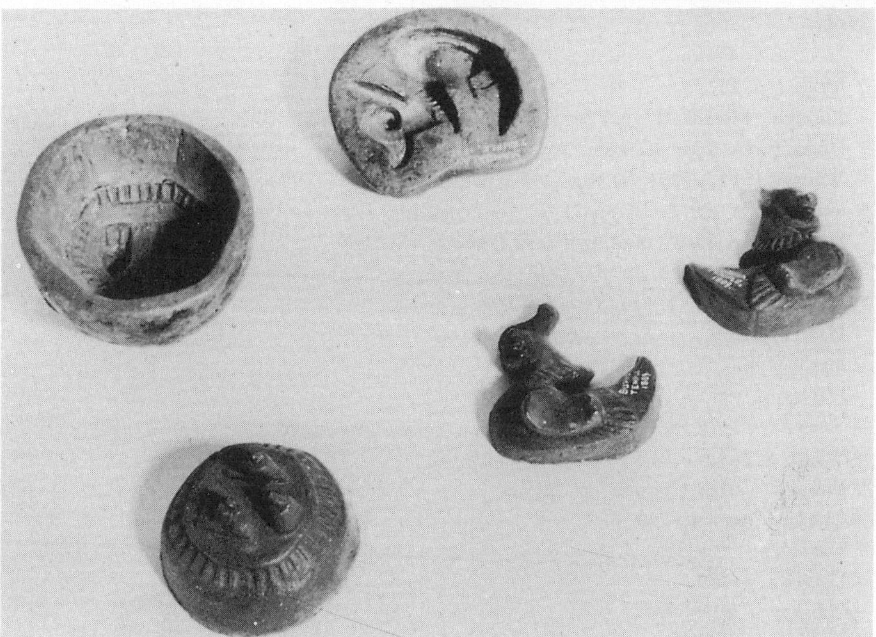

Wax castings of the parts of the whole weight were made, the castings then being united. Smaller weights were made in two halves. Larger ones were made in two halves and the base. The moulds were of lead. Lead moulds of the base and the right half appear at the top of the illustration. Wax castings of the base and the two halves appear below.

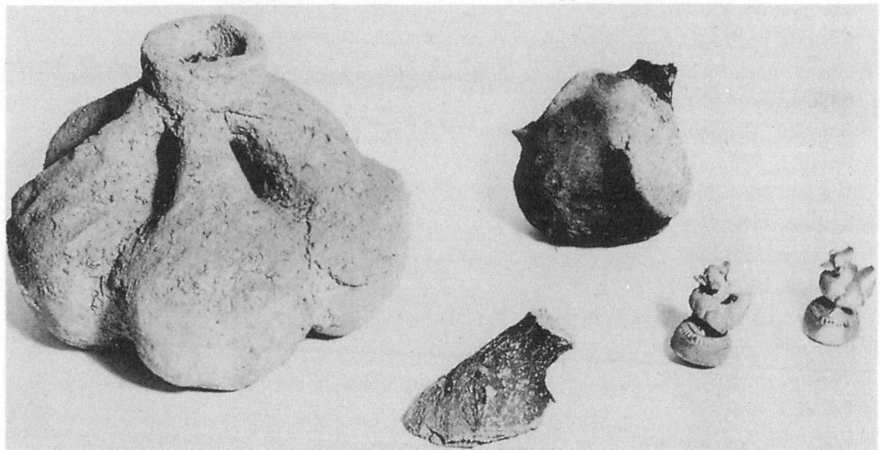

The completed wax models were enclosed in moulds of clay and chaff (left and centre), up to four wax models being enclosed (left). When heated the wax melted and escaped through ducts at the bottom, so leaving hot negative moulds to take the molten copper alloy poured in at the top. The baked clay was broken away from the cooled, solidified metal to obtain the rough castings (bottom right). These were then smoothed to produce the finished weight (extreme bottom right).

Plate 9: The Making of Burmese Metal Weights, 19c. AD.

From R.C. Temple's donations to the Pitt Rivers Museum, Oxford, in 1889 and 1892.
By courtesy of the museum authorities.

Notes

1 Houben G. Personal communication.

2 Mollatt, p. 407.

3 Bussagli, "Metalwork", p. 820. Smith B. p. 32. Gyllensvard, p. 222.

4 Goetz, p. 33. Coomaraswamy and Rowland, p. 96. Willetts, p. 625. Kar, p. 1.

5 Watson, 1981, p. 601. Rawson, 1983. p. 24.

6 Chhibber, p. 229. Le May, p. 17.

7 Furnivall, pp. 63-65, quoting Duarte Barbosa. 1518 AD.

8 Needham, v. 5, pt. 3, pp. 199, 227, 232, 263.

9 Hall, 1939, pp. 112, 114. Temple, 1898, p. 141.

10 Harvey, p. 258.

11 Sigler, p. 3.

12 Harvey, p. 159.

13 *Some Notes upon the Development of Commerce with Siam*, p. 81.

14 Wagel, p. 268.

15 Bussagli, "Steppe Cultures", p. 380.

16 Chhibber, pp. 140-150.

17 Harvey, pp. 350, 358.

18 Laughlin, p. 287.

19 Quiggin, p. 229.

20 Bowman, Cowell and Cribb, pp. 25-30.

21 Le May, p. 17.

22 Grancsay, p. 246.

23 Bussagli, "Metalwork", pp. 814-815. Elisseeff, pp. 23-25.

24 Loehr, 1965, pp. 774-782.

25 Elisseeff, pp. 21-23.

26 Glover, *Early South-east Asia*, p. 176. In the same volume see also Bayard, p. 16; Watson, p. 62; Davidson, p. 100; Lyons, p. 353.

27 Bezacier, "Vietnam", pp. 777-779. Heine-Gelder, pp. 45-48.

28 Rosati, p. 738.

29 Watson, 1981, p. 601.

30 Rawson, 1983, p. 8. Bunker, 1967, p. 319.

31 Loehr, p. 262.

32 Rawson, 1983, pp. 23, 24.

33 Taw Sein Ko, p. 56. Se also Parker, p. 7.

34 Griswold, "Siamese Art", p. 18.

35 Davies, pp. 169, 179, referring to Hosie's account in his "Three Years in Western China".

36 Wyatt, p. 48.

37 Crawfurd, Ava, p.104

38 Bird, p. 47

39 U Mya, a retired officer of the Archaelogical Survey in Rangoon, who was 92 when the writers met him in 1971. See also Symes, Ava, pp 325 – 327.

5
Dimensions, Shapes and Decorations

Introduction:

The animal-shapes on the weights of Burma are examples of the animal shapes used as weights in regions of the Eurasian land mass during the past 4000 years.[1] Animal art as an expression of religion was important until the anthropomorphic religions became dominant between the 5th century BC and 1st century AD. It continued until about 1000 AD in India and possibly Scandinavia. South-east Asia, perhaps Tibet and Nepal are the last regions where this form of art continued until modern times, either in the form of weights or other artifacts. Not all the weights of South-east Asia have been of animal shape, though few examples of weights of geometric shape remains.[2]

The weight shapes have been separated into numbered style groups based on common characteristics, e.g. proportions, body attitude, etc. Within each group, style classes have been identified which are based on characteristics which are not common to the group, e.g. mouth append!age and base plan shapes, etc. Since the styles fall into a sequence, they can be assigned to periods and/or places. The style classes of each of most periods fall into systems, e.g. in period A, there are bird weights which differ in the number of their crest knobs, 1, 2 or 3. Each of these is associated with none, one or two neck mane knobs, so making a style system of nine classes. The group and style system characteristics are tabulated.

The following notes summarize in a simplified way the more obvious shape characteristics by which a weight can be recognised and classified and/or which have a symbolic meaning. They are accompanied by tables and figures of the animal shapes.

Shape overall:

Each weight consists of one beast or one (usually) bird representation mounted on a base, the whole being of sturdy construction. The shapes are illustrated by period on Plates 53, 54 and 55 and by partial sets on Plates 13 to 24.

With the exception of the marble 'tortoise' weights (see Appendix 1), no weights of other shapes were discovered in Burma during the writers' two three-year sojourns there. Nor were those many people questioned (villagers, dealers, *phongyis* (monks), academics) aware of any other shapes. Elephant-shaped weights of North Siam are very rarely found in Burma itself (See Appendix 3).

Dimensions:

The table below sets out the linear dimensions of the weights, the dimensions varying with the style.

Table 11: Linear Dimensions of the Weights

Weight Magnitude	Kyats	¼	½	1	2	5	10	20	50	100	250	Ratio
.Height	mms	14 to 16	18 to 20	24 to 26	30 to 34	38 to 45	50 to 80	70 to 80	90 to 100	115 to 120	165 to 180	1,00
Base width/ diameter	mms	8 to 13	12 to 14	13 to 16	19 to 21	25 to 29	30 to 40	45 to 50	55 to 70	100 to 120	100 to 120	0,64

Handle:

Handles are found on the 50, 100 and 250 *kyat* weights of most but not all classes, running from head to tail. Handles are found on some square-based 10 *kyat* beast weights of Groups 2. All the Group 2 weight handles are knobbed or have deep indentations.

Horns:

On the beast weights two horns are located on the sides of the head top in front of the ears. They may be bulbous at the top or columnar (Group 2) or columnar to conical (4 and 7). Rarely in Group 7 they may be but low mounds. In some classes of Groups 2 and 5 there is a deep groove on the front of each horn which occasionally becomes a distinct sub-knob or tine. One or two near-vertical nose bridge horns are present on the Group 2 beast weights. On Groups 4 and 5 where such knobs occur they slope less steeply and in Group 5 are often low mounds.

Bird crests:

All the bird weights have either head crests (Groups 6 and 7) or head knobs (Groups 1 to 5). The head knobs are either clearly separated on a low to very low crest (Groups 1 to 4) or overlap each other along most of their lengths (Group 5). In Groups 1, 2 and 3 the knobs are located on the head top. In Group 4 and some classes of Group 2 they are positioned on the rear of the head top and/or on the back of the neck. In Group 5 they occur on the front and top of the head. Groups 1 and 4 weights bear 1, 2 and 3 knobs. Group 2 U-tailed weights mostly have one knob and Group 2 V-tailed weights mostly have three knobs. Groups 3 and 5 have three knobs though rarely 250 *kyat* Group 5 weights may have up to seven.

Mouths:

The beast weights almost always have an open mouth set in a characteristic almost flat-fronted nuzzle. The two mandibles of the beaks of the birds are shaped as follows:

straight, closed or slightly parted, with rounded tips, i.e. duck-like (Groups 1, 2 U-tailed, 4, 5, 7); straight, closed with a pointed tip, i.e. fowl-like (Groups 3 and 6); slightly arced downwards, closed, pointed, i.e. predator or fowl-like (Group 2 V-tailed).

Mouth appendages:

On different styles the mouth appendage may be a beard, an emanation from the mouth, a tongue or a chest attachment touching the mouth. It may be partially separated from the chest or adhere to it along its whole length. Both bird and beast weights may or may not bear these appendages. The Groups 4 and 5 beast weights may have two. Some variation in the mouth appendage shape may take place in some classes as the magnitude of the weight changes. For example, in class 2v of the beast weights, the triangular beard present on weights of 20 *ticals* and below changes into a rod-like shape on weights of 50 *ticals* and over. The main mouth appendage shapes are listed here.

Table 12: Mouth Appendage Shape

Bird Weights		Beast Weights		
—	Rod	± Adhering	Rod (r)	Also twinned
Adhering	Spatula	Adhering	Vee (v)	Sometimes vertically striated
Adhering	Trefoil	± Adhering	Trefoil (t)	Also twinned
—	Bird	Adhering	Strips	Twinned

Notes: Adhering = Appendage adheres to chest along its whole length.
— = Appendage adheres to chest only along part of its length.

The rod on the beast may be vertical (all groups) or directed to the left or to the right (Group 4 only). Twin appendages are directed both left and right (Groups 4 and 5). The mouth appendages are illustrated. (See Plate 9).

Manes:

These occur in different numbers, usually no more than three except sometimes in Group 5. They appear as lines, ridges, knobs and/or indentations on the neck back.

Wings:

These occur only on the bird weights, folded against the body, sloping downwards and with a semicircular leading edge. Their shapes, useful for classification, are:

(a) oval and concave (upper wing only or both upper and lower wing);
(b) comma and convex;
(c) trapezoidal and convex. Group 2 wings may be prominently knobbed. Group 5 may bear a few lines sketchily indicating feathers while Groups 6 and 7 weights usually bear moulded feathers. The wings are illustrated on Plate 11.

Feet:

On the beasts the foot is marked by a horizontal line near the lowest part of the leg, suggesting a hoof. On the 'hooves' of the beasts there may be none to six vertical lines indicating toes. Five or six vertical grooves may appear on the feet of the beast weights of Group 2 (r) and on some weights of Group 5 (horse and lotus bud tail). On the Groups 2(v) and 2(t) and the Group 4 beast weights there are no, or rarely, vertical grooves.

On the bird weights the hoof is common. However, three elongated talons sometimes occur and less often a rearward directed claw. The occurrence is as follows.

Table 13: Bird Feet

Group	Period	Foot Shape	Number of talons
1	A	Elongated shape	1
2	B	Elongated claw	1 mostly, 2, 3, rare
2	C, D	Elongated claw	3 mostly, 1, 2, rare
3, 4, 5	—	Hoof	1 one-toed
6, 7	—	Elongated claw	3

Tails:

The bird tails may be spread laterally or vertically. In the first case the tail-shape as viewed from the rear and also from above, may be either an inverted V or an inverted U with spreading arms. The V-tail is associated with a fowl-like bird, the U-tail with a duck-like bird. In the second case the vertically (and longitudinally) spread tail is deeply scrolled and appears in the form of a triangle or a strip. It is associated with the hen-like bird of Group 6 and the duck-like bird of Group 7. See Plate 11.

The beast tails mostly appear as long straight pendent rods, each almost always with an elevated base, which adhere to the rumps along the whole of their lengths, with rare exceptions. The weights of 100 and 250 *kyats* sometimes show a tail with flattened sides in a strip-like form. Another type of flat-sided tail appears as a short erect looped rod or strip but only in Group 5. A somewhat similar Group 5 type is curled over the rump in an inverted S-shape. The last Group 5 type is a short erect cone segmented by grooves. Both the last two do not adhere to the rump. (See Plate 11.)

Back covers:

The back cover is a flat-lying structure which appears on the back of the bird weights of Groups 1, 2, 3 and 5. Its termination near or on the tail may be straight or gently arced to acutely ogival. See Plate 11.

Table 14: Decorative Motifs

Group	Birds							Beasts		
	1	2	3	4	5	6	7	2	4	5
Mouth appendage decoration (Weights over 2 kyats)										
Separated from chest	•	•	X	•	X	•	•	X	•	R
Adherence to chest	•	•	•	X	•	•	•	X	X	X
Vertical striae	•	•	•	•	•	•	•	X	X	X
Double chevron	•	•	•	X	X	•	•	•	•	•
None	•	•	X	X	X	•	X	X	X	X
Mane decoration										
Lines bordering and parallel to edge	•	•	•	X	•	•	•	X	•	•
As above with lines at 90° to edge	•	X	•	X	X	•	•	X	X	X
Dots	•	•	•	•	X	•	•	•	X	•
Chest decoration (sets)										
Raised capes	•	•	•	•	•	•	•	•	X	X
Surrounding lines	•	•	•	•	•	•	•	X	X	X
Bosses	•	•	•	•	•	•	•	X	•	X
Dots	•	•	•	•	•	•	•	•	X	X
Radials between sets or on capes	•	•	•	•	•	•	•	•	X	X
Set base rounded	•	•	•	•	•	•	•	X	X	X
Set base pointed	•	•	•	•	•	•	•	X	R	•
None	X	X	X	X	X	X	X	•	X	•
No chest decoration was seen on the bird weights in Burma. Two Group 2 bird weights in Europe have decoration. (modern additon?)										
Wing decoration										
Knob(s) on wing front	•	X	•	•	•	•	•	•	•	•
Groove(s) parallel to front edge	•	X	X	X	X	•	•	•	•	•
Frontal strip with loop	•	•	•	•	X	•	•	•	•	•
Groove parallel to lower edge	•	•	•	R	X	•	•	•	•	•
Incised feathering radiating from centre line	•	•	•	•	X	•	•	•	•	•
Front nr. vertical grooves. Rear oblique grooves	•	•	•	•	•	X	X	•	•	•
Oblique grooves	•	•	•	•	•	•	X	•	•	•
None	X	X	X	X	X	•	X	•	•	•
Tail side decoration										
Lines/grooves parallel to edge	•	•	•	X	X	•	•	X	X	X
Oblique grooves, flutings	•	X	X	•	X	•	•	•	•	•
Dropped tail sides	•	X	X	•	X	•	•	•	•	•
None	X	X	X	X	X	•	•	X	X	X
Beast tail base always elevated except for one class in Group 4										
Base decoration										
Approximate maximum number of horizontal lines	6	4,3	3,2	1	2,1	?	?	4,3	2	3,1
4 – 5 horizontal grooves	•	•	•	X	•	•	•	•	X	•
Horizontal grooves/flutings	•	•	•	•	•	X	•	•	•	•
Horizontal rows of bosses	•	•	•	X	X	•	•	•	•	•
Short verticals	•	X	•	•	•	•	•	•	•	X
Long verticals/obliques	•	•	•	R	X	X	X	•	R	X (a)
8 pairs vertical grooves, a pair to each corner	•	•	•	•	•	X	•	•	•	X
Incised single/double normal/inverted, Vs, Us, ogives,semi-circles	•	•	•	R	•	X	•	•	•	X
None	•	•	•	X	•	X	X	•	X	X

(a) Only on circular base plan weights except for a rare octagonal Group 4 bird weight.

R = Rare x = Seen • = Not seen

Base shapes:

Bases unique to the beast weights are, in plan, square, rectangular or rectangular without the corners (lozenge). Unique bird weight base plan shapes are hexagonal and polygonal. Both bird and beast weights may have octagonal or circular base plan shapes. In cross-section most sides taper upwards. In older weights circular plan bases may have vertical or convex sides. The base shapes are illustrated on Plate 12.

Decoration:

The knobbiness of the weights of Groups 1, 2 and 3 is an obvious feature. The more important base decorations are the vertical or obliquely grooved bases, usually on circular plan bases of Groups 4, 5 and 7 bird weights respectively; the eight pairs of vertical grooves on the corners of the high octagonal bases of the Group 6 bird weights and the usually increasing numbers of horizontal lines with decreasing age of the weights of Groups 1, 2, 3, 4 and 5. Weights of 100 *kyats* or heavier are usually more elaborate than the lighter weights. Weights of two *kyats* and below are deficient in detail.

The beasts on the weights always show chest decoration while the birds never do so if genuine. The decoration takes the form of elevated (rare and old), ridged or incised capes or manes, either plain or decorated with one or more of one, two or three parallel incised lines; rows of punched dots; rows of circular bosses; short vertical lines between the long incised lines; long vertical lines on the capes between the other sets of decoration of which there are one, two or three. The decoration passes around the chest in a semicircular fashion, the lower part being rounded or taken to a point. Their terminations may be on the upper chest but are usually near the ears.

The manes may show lines either bordering their edges or at right angles to them. The decorative motifs are tabulated.

Notes:

1 The following list displays the names of animal representations possibly used as weights on the Eurasian land mass other than those considered in this work. It is unlikely to be complete.

DUCKS (or other anserines): Ancient Babylonia, Assyria, Persia and Egypt. 25th century to 6th century BC.

SNAKE: India (?). Ancient Egypt.

TURTLE, TORTOISE: Israel, 6th-5th centuries BC. Ancient Egypt.

FROG: Ancient Egypt.

COWRIE SHELL: Ancient Egypt.

SCARAB: Ancient Egypt.

HIPPOPOTAMUS: Ancient Egypt.

LION: Ancient Babylonia, Assyria, Persia and Egypt. 15th13th to 6th century BC.

HORSE: Scandinavia (medieval, doubtful weight). Luristan, 1000 BC.

BOAR: Pompeii, Mesopotamia (head only), 2500-550 BC.

GOAT: Pompeii.

CATTLE: Ancient Egypt. India (?).

BULL: Bactria, 2500 BC. Also engraved on handbag-shaped Irano-Afghan weights of 2000? BC.

GAZELLE: Egypt. Mesopotamia, 2500-550 BC.

VARIOUS: Greece.

OSTRICH: Mesopotamia, 2500-550 BC.

WATERBIRD: Mesopotamia, 2500-550 BC.

The above list has been assembled mainly from Kitsch, pp. 121-123, Petrie, 1926, p. 6, Quiggin, p. 255, Iwata, 1982, pp. 8-16 and Hori, pp. 16-23. Inquiry of the Indian Metrological Society and the Heras Institute of Indian History and Culture yielded the information that they knew of no weights in the shapes of animals being made in India, Nepal or Tibet. Inquiry of the curator of the Weight Museum, Yale University, resulted in the information that their one specimen of a supposed Indian weight in animal form, a snake, was marked as doubtfully identified as a weight. Some shapes may not be weights but figurines.

2 Truncated cones of lead may have been used in Pyu times since such shapes were found at Beikthano of the 1st -4th centuries AD (Aung Thaw, 1968 , p. 55). They are similar to one of the shapes of Szechwan weights used between 206 BC and 8 AD when the *liang* mass was approximately similar to the earliest (14th/15th century AD) Burmese animal weight masses of 11 to 14 grams (*Zhong Guo Gu Dai Du Liang Heng Tu Ji*, pp. 47, 48).

A cube of lead was found in north Thailand which may have been a local weight of ancient times (Kneedler, p. 13). In Cambodia some of the known shapes of the weights of 800-1400 AD were octagons and eight-pointed stars (Groslier, p. 27). In Siam of 1688 AD the silver 'bullets'(*pod duang*or *kub*?) were in use as weights to measure the masses of 'coarse' silver, i.e. the tok currencies(Temple, 1919, p. 51). The Chinese, outside of south-west China, always used weights of various geometric forma (*Zhong Guo Dai Du Liang Heng Tu Ti*, pls. 153-240) mostly spheres, hemispheres, cylinders and truncated pyramids of different cross sections, rings and slabs of circular and figure eight plan-shape. There is no information known to the writers concerning the weights of south-west China. In India, Elliott (pp. 48, 49) refers to jewellers' weights of geometric shape. The cities of the Indus valley of about 2000 BC used cube-shaped weights mainly, though other geometric shapes were used also (Hendrickx-Boudot, pp. 7, 8). Weights have been recorded by Stein (pp. 121, 316) from Khotan in the Tarim basin, probably pre-1000 AD. These are rectangular, of lead and bronze, and octagonal of bronze.

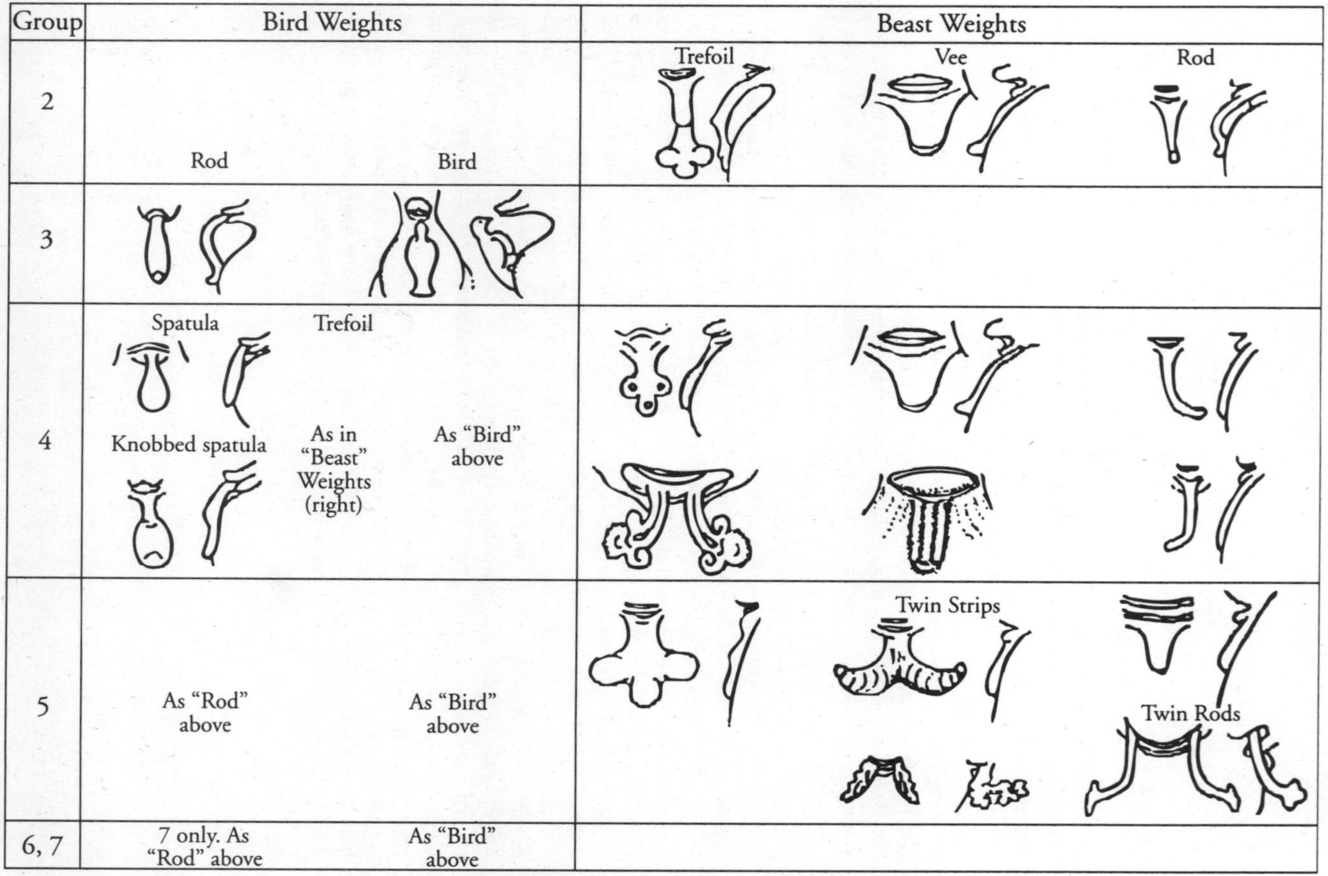

Group	Bird Weights		Beast Weights		
2	Rod	Bird	Trefoil	Vee	Rod
3					
4	Spatula Knobbed spatula	Trefoil As in "Beast" Weights (right)	As "Bird" above		
5	As "Rod" above	As "Bird" above		Twin Strips	Twin Rods
6, 7	7 only. As "Rod" above	As "Bird" above			

Plate 10: Mouth Appendage Shapes

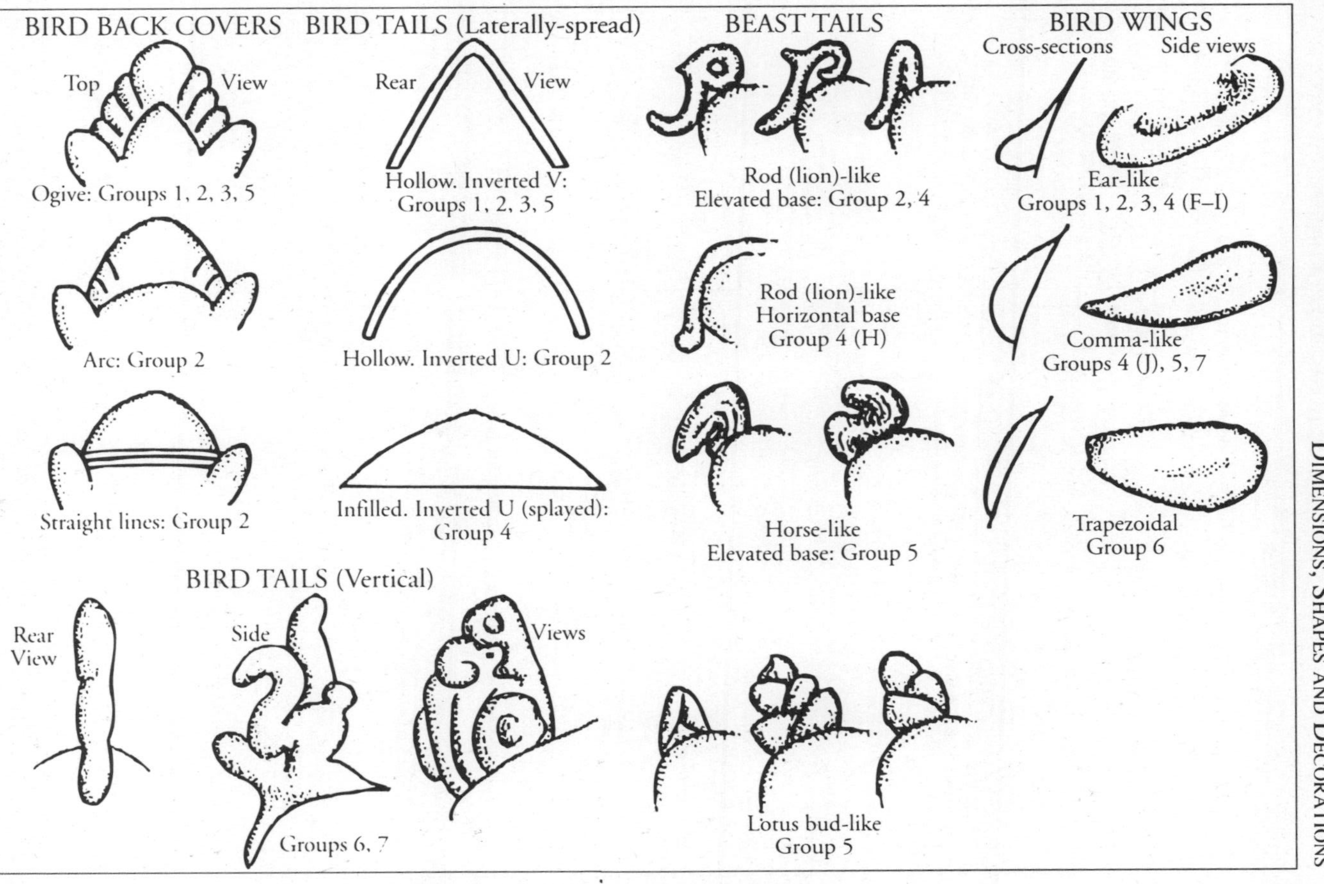

Plate 11: Tail, Back Cover and Wing Shapes

	Plans						Cross Sections				
	Bird	Birds	Beasts	Birds	Beasts	Beasts	Birds	Beasts	Birds	Beasts	Birds
Groups	Hexagon and Polygon	Octagon		Circle		Other Shapes	TRUNCATED PYRAMID/ CONE		VERTICAL (V) and CONVEX (X) SIDES		Other Shapes
1	HEXAGON - ELONG.	—	—	—	—	—	H	—	—	—	—
2	HEXAGON	OCTAGON	Octagon	—	—	SQUARE	H	H	—	—	—
3	—	Octagon	—	—	—	—	H?L	—	—	—	—
4	—	Octagon	Octagon	CIRCLE	Circle	—	L	L	—	—	—
5	POLYGON (9, 10)	Octagon and	Octagon	Circle	Circle	RECTANGLE	L	H?L	V	—	—
	—	OCTAGON – SQUARE	—	—	—	OCTAGON – LOZENGE	—	—	—	—	—
6	—	Octagon as for Group 5	—	Circle	—	—	—	—	VX	—	DOUBLE PYRAMID/ CONE (A)
7	Polygon	Octagon as for Group 2	—	Circle	—	—	—	—	X	—	TRUNCATED HEMISPHERE

H = High. L = Low. V = Vertical sides. X = Convex sides. A = Altar.

PLATE 12: Base Shapes

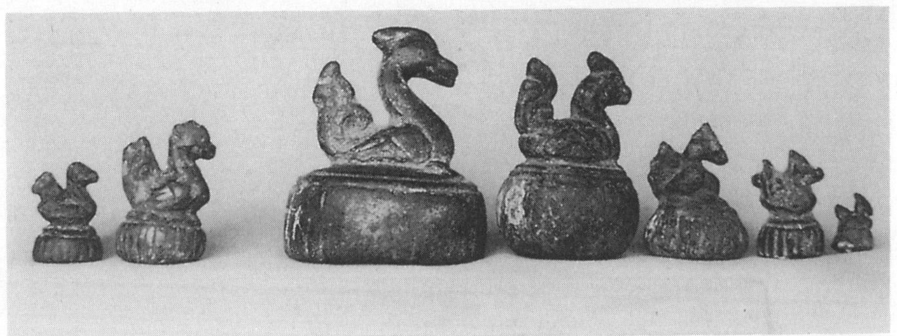

Period M 20 to ½ kyat

Two styles with circular plan bases and vertically-striated sides

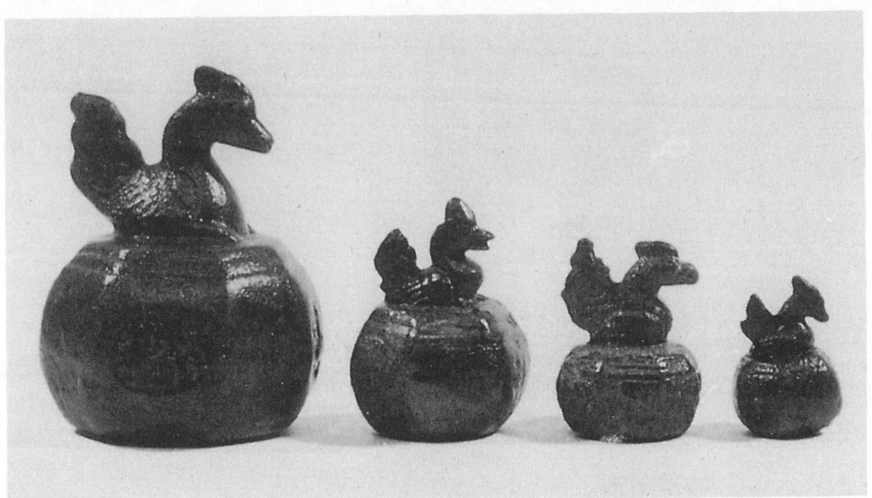

Period M Octagonal plan base 80 to 5 kyats

Plate 13: Bird weights, Group 6. Partial sets

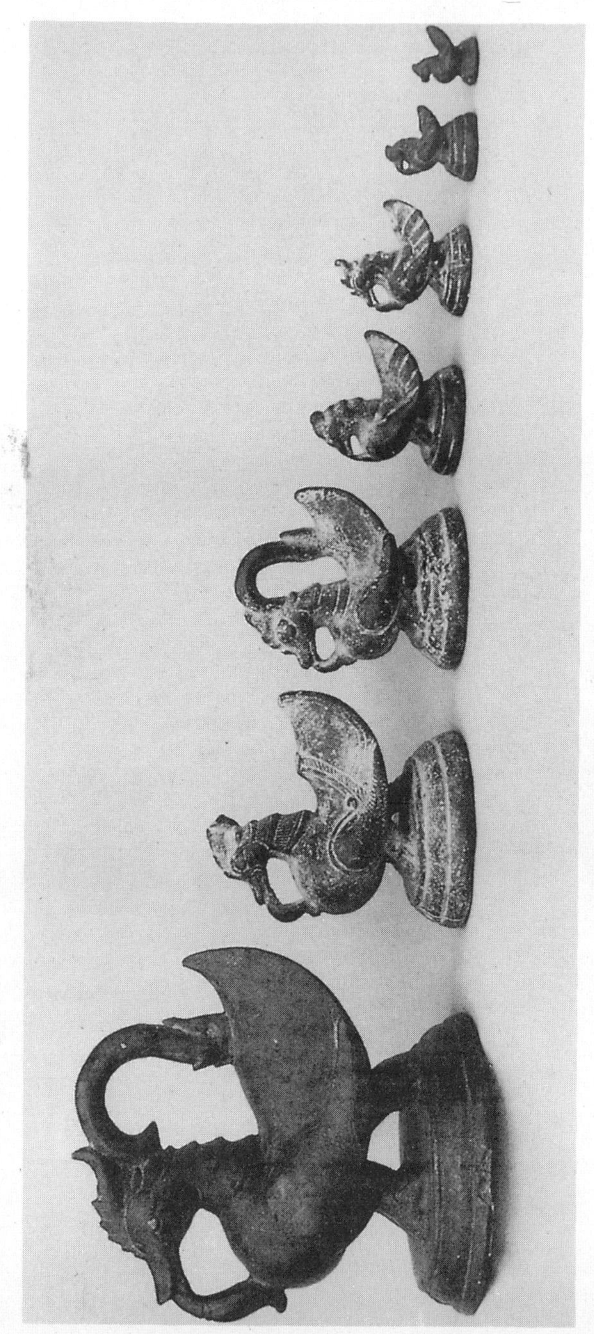

Period K Mouth appendage, rod-shaped. Head knobs, overlapping. 250 to 2 kyats

Plate 14: Bird weights, Group 5. Partial set

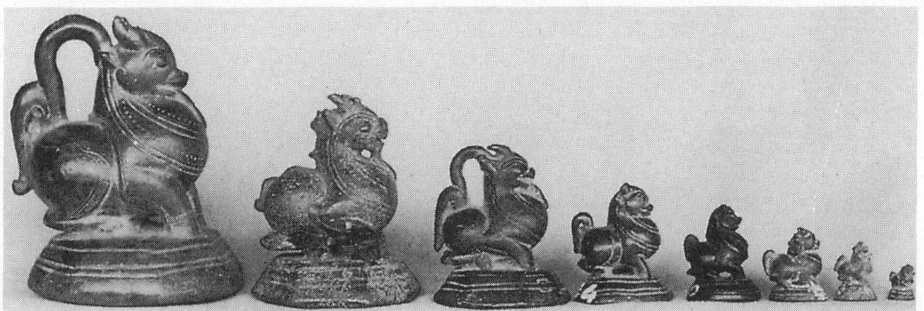

Periods J and I 250 to 1 kyat

Mouth appendage present. Tail base elevated.

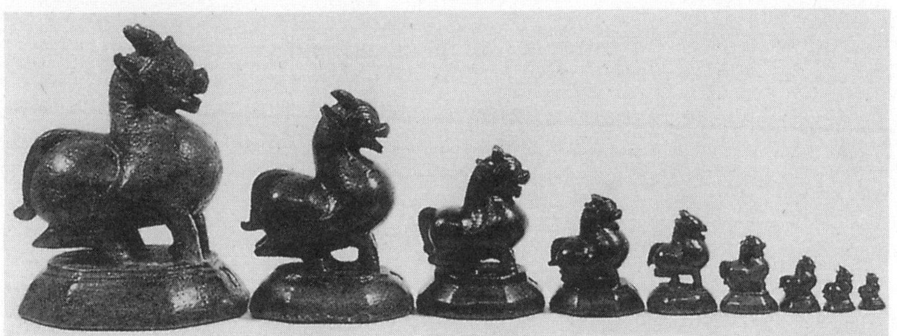

Period H 250 to ¼ kyat

Mouth appendage absent. Tail base horizontal.

Plate 15: Beast weights, Group 4. Partial sets

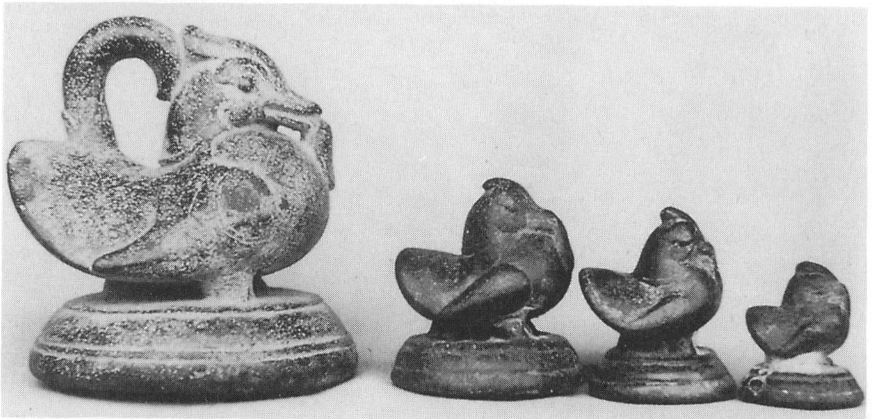

Period J — 100 to 5 kyats

Spatula-shaped mouth appendage

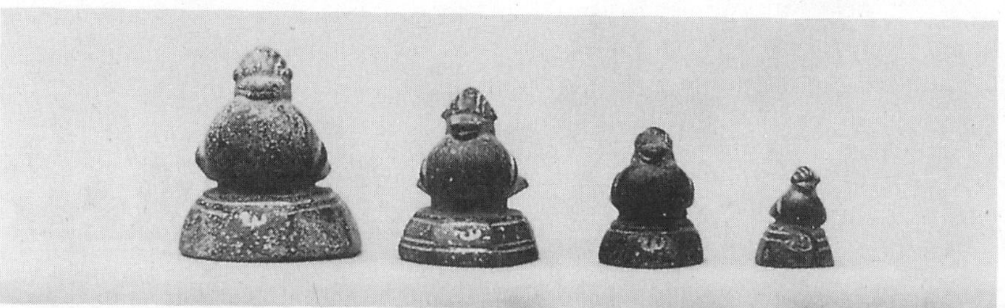

Periods I and H — 20 to 2 kyats

Mouth appendage absent.
Gha sign on base front

PLATE 16: Bird weights, Group 4 . Partial sets

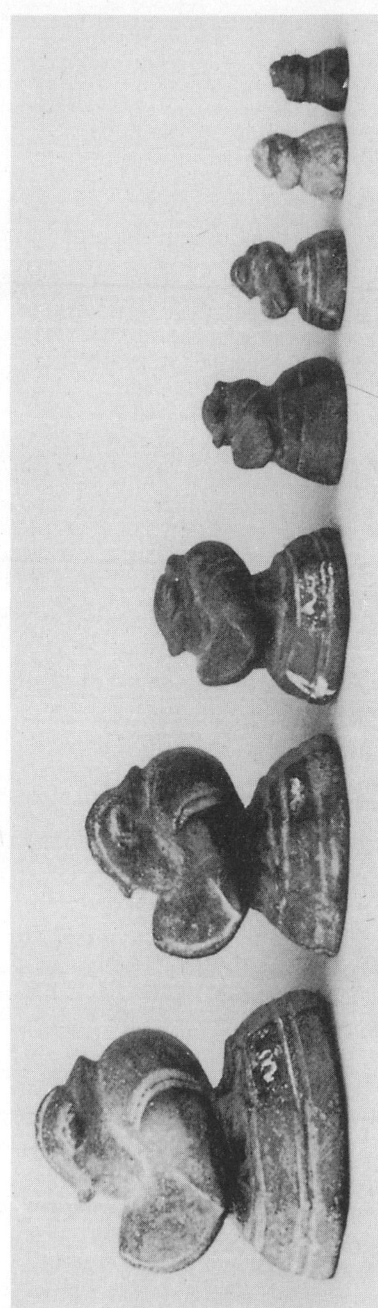

Period F

20 to ¼ kyat

Mouth appendage absent.
Gha sign on base right front

Plate 17: Bird weights, Group 4. Partial set

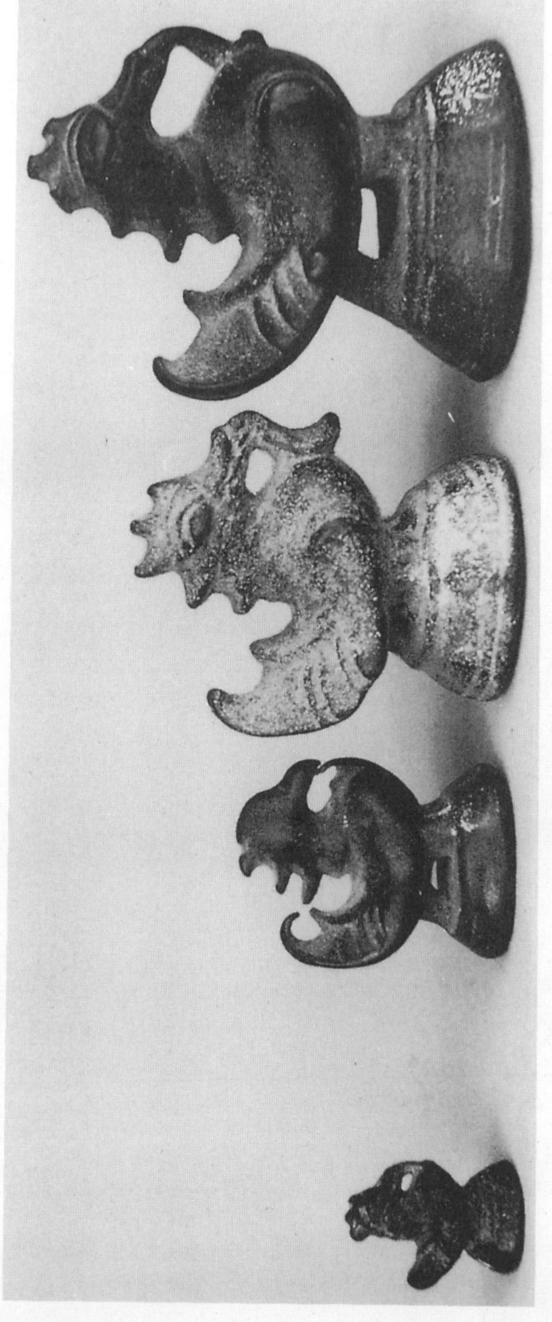

Period E 100 to 1 kyat

Mouth appendage rod shaped.
Head knobs separated.
9-rayed star on base right front

PLATE 18: Bird weights,Group 3. Partial set

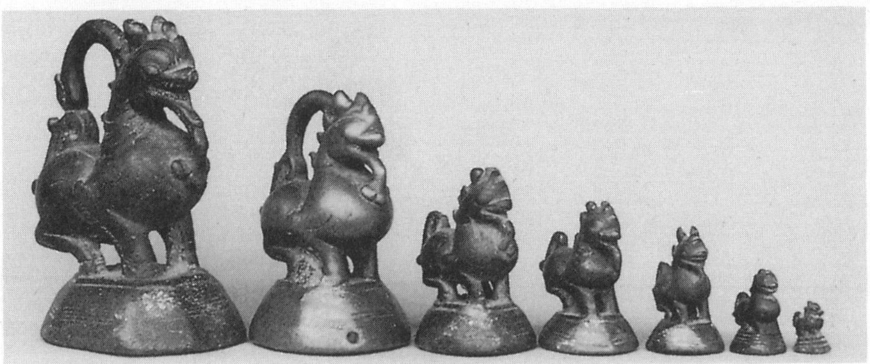

Period E

Mouth appendage with trefoil.
4-rayed star on base right front

100 to 1 kyat

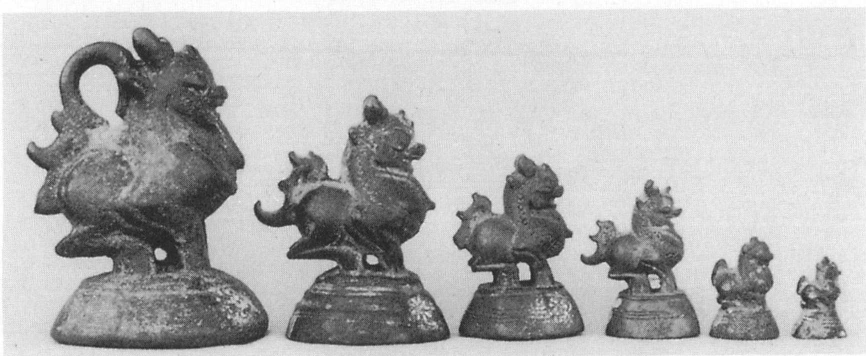

Period D

Mouth appendage vee-shaped
9-rayed star on base right front

50 to 1 kyat

PLATE 19: Beast weights, Group 2. Partial sets

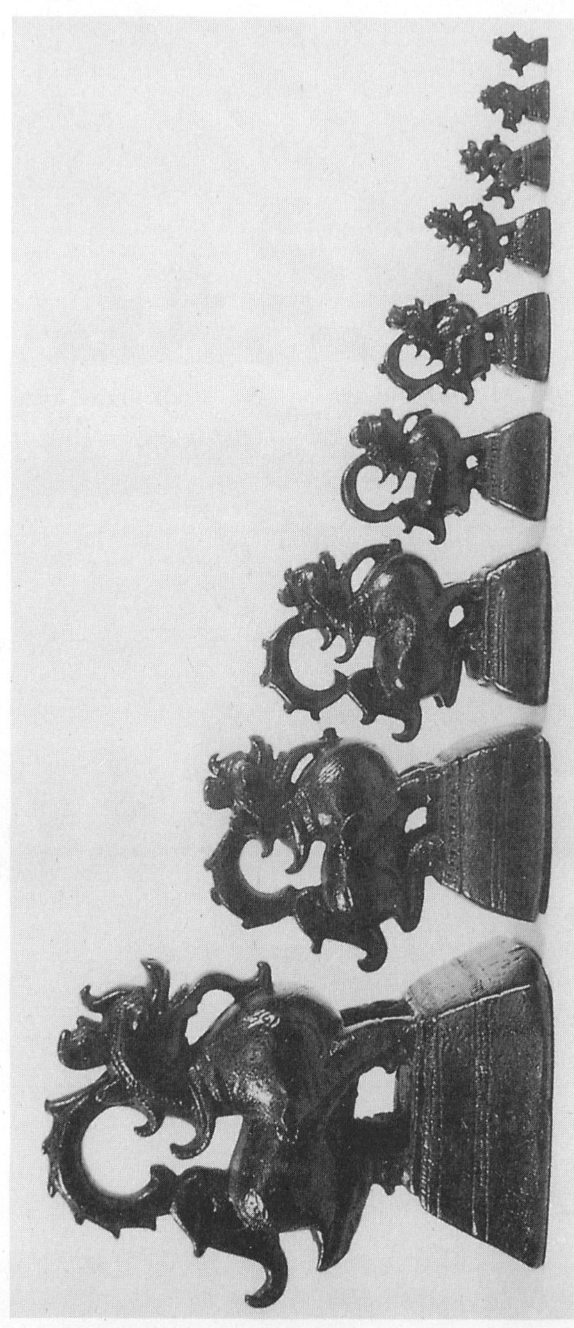

Period B and C

250 to ¼ kyat

Mouth appendage rod-shaped
9-rayed star on base right front

PLATE 20 : Beast weights, Group 2. Partial set

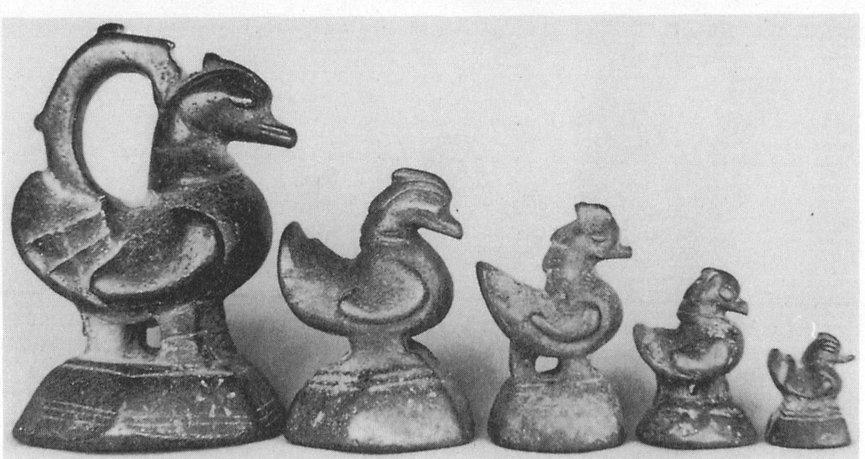

Period D 50 to 2 kyats

Tail shape, inverted U.
Head knob, one.
6-rayed star on base right front

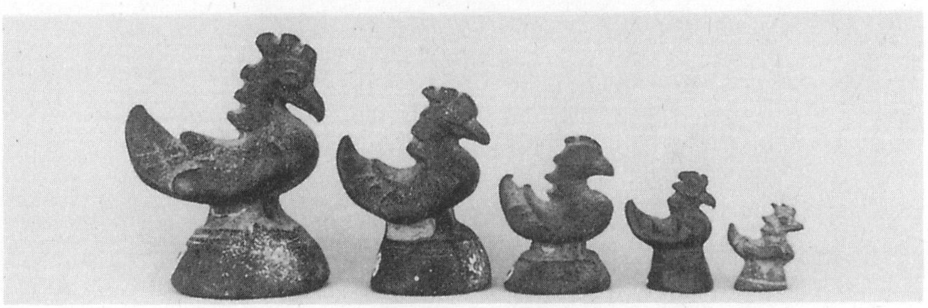

Period D 20 to 1 kyat

Tail shape, inverted V.
Head knobs, three.
Bird sign on base right front

PLATE 21: Bird weights, Group 2 . Partial sets

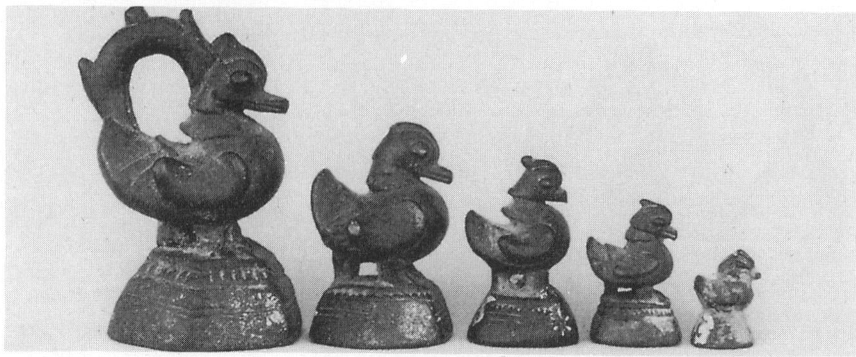

Period C 50 to 2 kyats

Tail shape, inverted U.
Head knob, one.
Base decoration, short verticals
6-rayed star on base right front

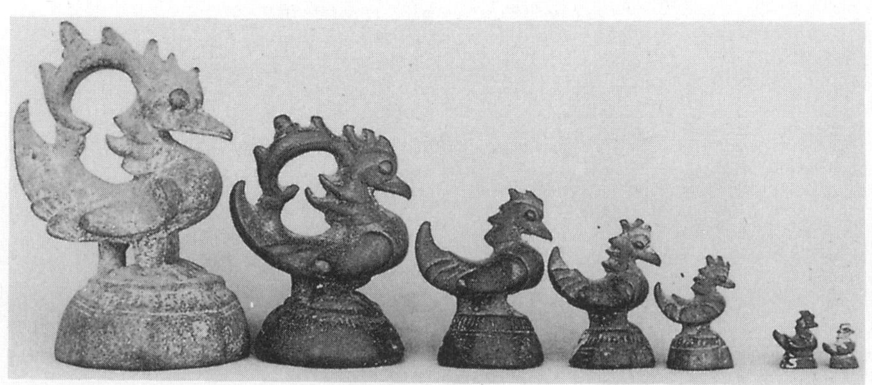

Period C 100 to 1 kyat

Tail shape, inverted V.
Head knobs, three.
Base decoration, short verticals
Bird sign on base right front

Plate 22: Bird weights, Group 2. Partial sets

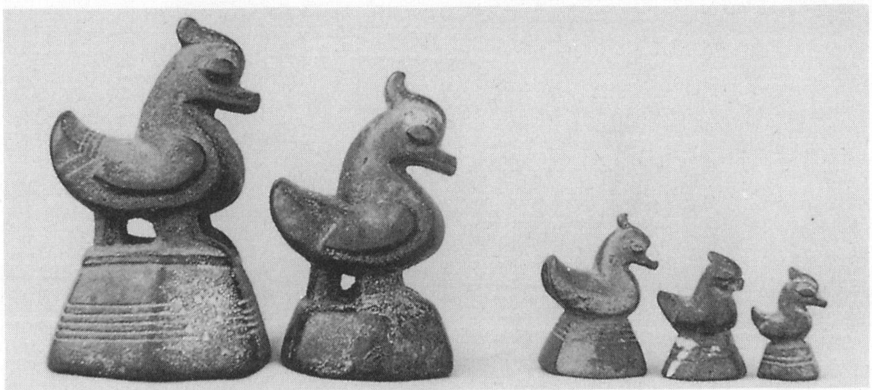

Period B 20 to ½ kyat

Tail shape, inverted V.
Head knob, one.
6-rayed star on base right front

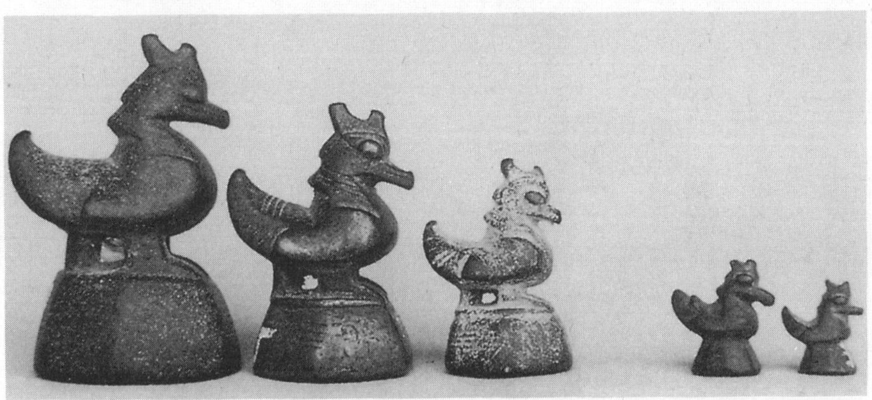

Period B 20 to ½ kyat

Tail shape, inverted V.
Head knobs, two.
Bird sign on base right front

PLATE 23: Bird weights, Group 2. Partial sets

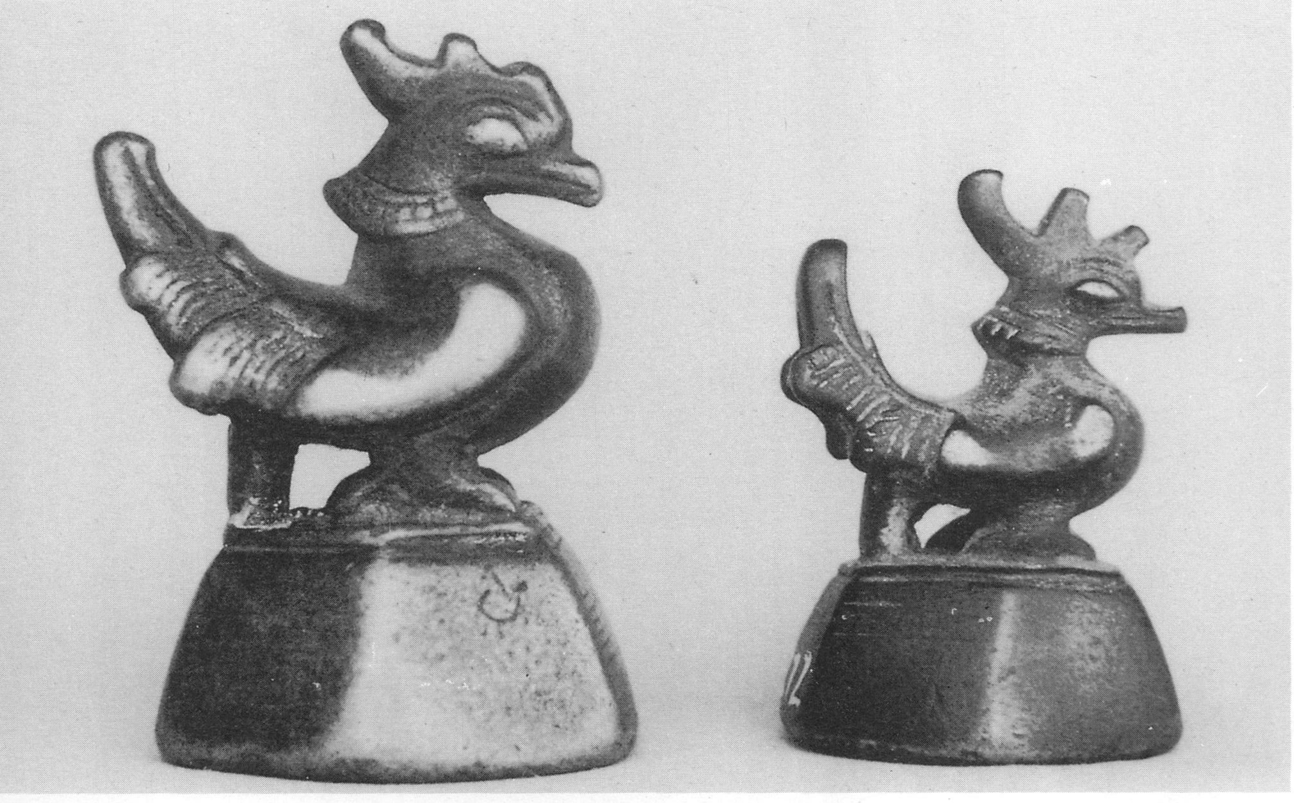

Period A — Tail shape, inverted V. — 20 to 10 kyats

Head knobs, three.

Bird sign on base right front

PLATE 24: Bird weights, Group 1. Partial sets

6
Signs and Style Systems

The Signs:

Signs, i.e. marks with an intended meaning, are classified as main, i.e. the only or first emplaced sign, and auxiliary, i.e. signs emplaced after and which accompany the main sign.

The weights often bear one or more recognisable signs almost always on the sides of the basal block. Weights of less than five *kyats* usually have the main sign on the base underside, though often the sign is omitted from them. The five-rayed star always occurs on the base underside. The table 'The Signs on the Weights' shows the relationship between the signs and the style groups.

The Group 7 bird weights and the round-based Group 6 bird weights never bear signs. The bird weights of Groups 5 and 4 which normally have frontal main signs almost always have them. Those of Groups 4, 3, 2 and 1 which normally have main signs on the right front may lack them in about 70% of specimens. Two bird weight classes of periods C and D are notable for having a large proportion of weights without signs. They are also the most frequently copied. The octagonally-based bird weights of Group 6 always show four main signs but on the 80 *kyat* size only. The beast weights of Group 5 and 4 almost always have frontal main signs and those of Group 2 bear them on the right front. Partial obliterations of the signs were estimated to amount on average to about 50% of the bird-shaped signs, 35% of the star and *gha* signs and 5-10% of the frontal punch signs. In some classes up to 80% of the signs may not be clearly recognisable. Weights of the same style class without signs are not detectably different in mass from those with signs.

The main signs are a bird facing to the front or rear, a star pattern with 4, 6 or 9 rays and ဃ (the Burmese character *gha*). These all occur on the right front of the base and are about 9-11 mm in size. The following main sign shapes are to be found on the front of the base: circular and square, often rather deep depressions 4-10 mm in diameter; rectangular panels (5 x 15 mm) with an internal beast or polygonal panels with an internal bird or, on Group 6 bird weights, old Mon/Yuan Thai script. Bird and *gha* signs never occur on beast weights.

The auxiliary signs, mostly about 6 mm in diameter, are four, five and six-rayed stars, a small ဃ, the circle (4 and 6 mm) and the broad arrow. They occur on different parts of the base according to a complicated but recognisable system. For example,

Table 15: Signs on the Weights

Style group	Birds					Beasts					Relative frequency per 1000
	Main Signs		Auxilliary Signs			Main Signs		Auxilliary Signs			
	Shape	Loc'n	Shape	Loc'n	No.	Shape	No.	Shape	Loc'n	No.	
	Below here the signs were probably emplaced cold										
1	Bird	RF			0						
2	Bird	RF			0	9-star	RF	Circle and dot	Back	0-1	
2	6-star	RF			0	4-star	RF			0	
3	9-star	RF			0						
4	6-star	RF			0						
4	ω	RF			0						
4	ω	RF	Circle	F	1						
	Below here it is uncertain whether the main sign was cast or emplaced cold										
4	ω	F			0						
4	ω	F	5-star	U	1						
	Below here the main signs are small, frontal, cast (?) depressions of circular or square shape. The auxiliary signs were probably emplaced cold.										
4	Circle	F	6-star	RF	1	Circle	F	6-star	RS	0	
4	Circle	F	6-star	RS	1	Circle	F	6-star	RF		
4						Circle	F	6-star	LF		
4						Circle	F	4-star	LF		6
4						Circle	F	Circle	RS		
4						Circle	F	Circle	RF		4
4						Circle	F	Circle	LF		
4	Circle	F				Circle	F				13
4	Circle	F	Hti	F	1	Circle	F	Hti	F		
						Square	F	6-star	RS		
4						Square	F	6-star	LF		
4						Square	F	6-star	RF		
4	Square	F	4-star	RS	1	Square	F	4-star	LF		7
4	Square	F	4-star	F	1	Square	F	Circle	RF		6
4	Square	F	Hti	F	1	Square	F	Hti	F U		
4	Square	F			1	Square	F			6	50
	Below here the cast main frontal signs contain an internal bird or beast.										
5	Circle	F	4-star	LK	1	Rectangle with internal beast	F				
5	Circle with internal beast										
6	Signs are found only on the 80 kyat octagonal-based weights. Large signs on four sides, two of birds, two of script. Round-based weights exhibit no sign.										
7	No sign				0						

F = front. S = side. K = back. L = left. R = right. Hti = arrowhead those signs occurring on the left sides of the beast weights are absent from these locations on the bird weights.

The stars are patterns of rays from a central point, the rays sometimes becoming dots on the circumference of a circle. Some rare weights may bear the date in Burmese era years and some Group 1 weights have the owner's name on them. Some signs have been mis-struck, e.g. the 6 rayed star may sometimes appear as a 5 or 7 rayed star, the mis-strike being recognisable because of the rarity and asymmetry.

One auxiliary sign may occur on the bird weights of Group 4 (10%) and on those of Group 5 (rare). No other bird weights have them. About one quarter of the beast weights of Groups 2(r) and 4, period H bear an auxiliary sign and about 6% have more than two auxiliary signs. Five such signs have been seen on one weight. Rarely may one auxiliary sign be seen on other Group 4 beast weights.

Most of the signs are illustrated on Plates 25 to 30.

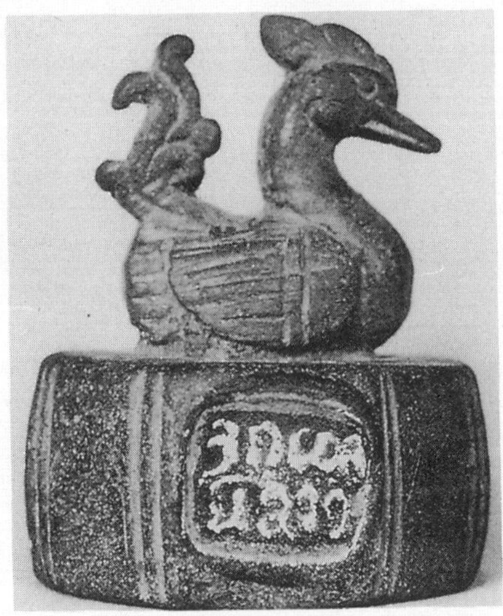

One of the four signs.
Possibly Old Mon or north Thai Script shown here

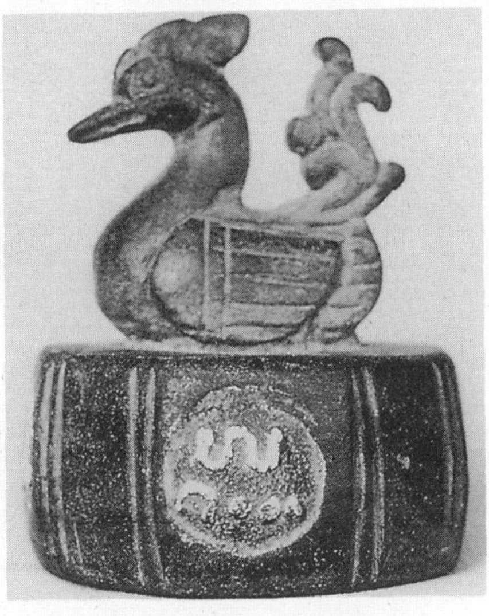

One of the four signs.
Possibly Old Mon or north Thai Script shown here

Plate 25: Bird weights, Group 6. Signs on the bases

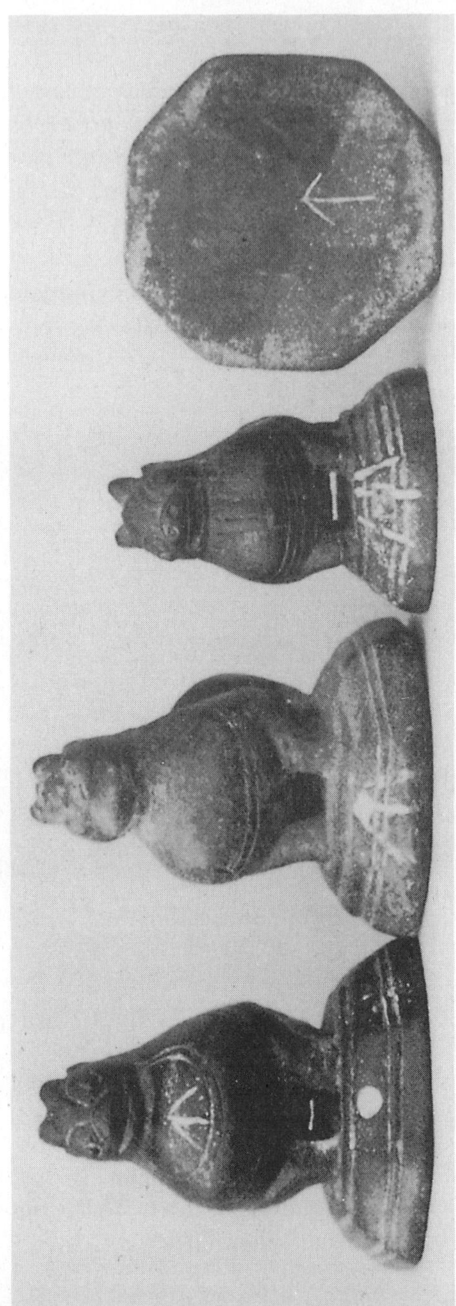

Periods Ib, Ic, J — Auxiliary sign. Arrow head shapes in different locations

Plate 26: Beast and Bird weights, Group 4. Signs on the weights

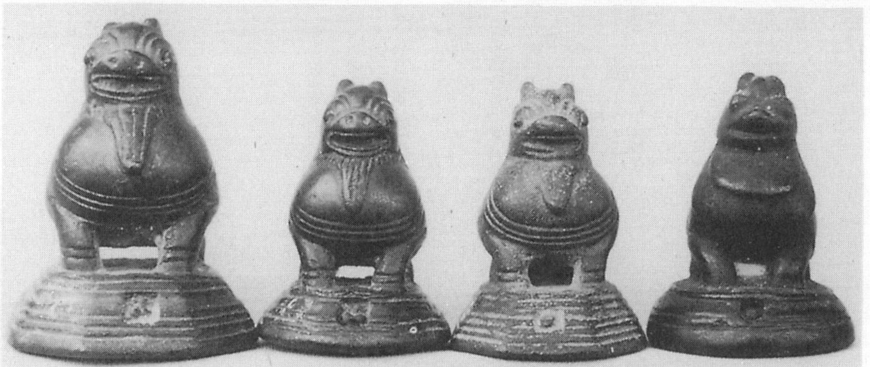

Periods Ib, Ic, Beasts and bosses within square and circular depressions

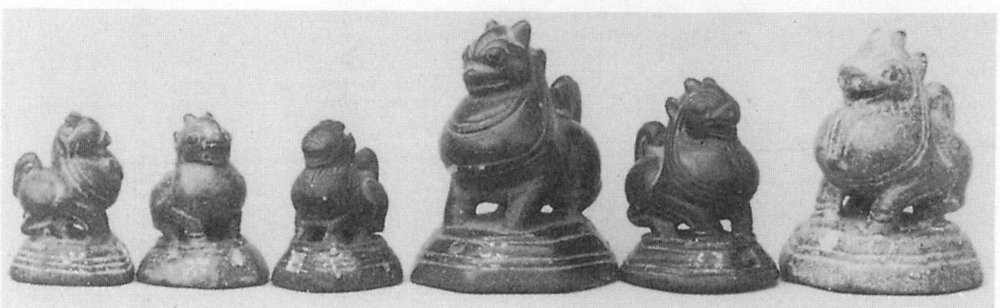

Periods Ia and Ib Auxiliary signs 6-and 4-rayed stars
Circle. Left- or right-front of base

Plate 27: Beast weights, Group 4. Signs on the bases

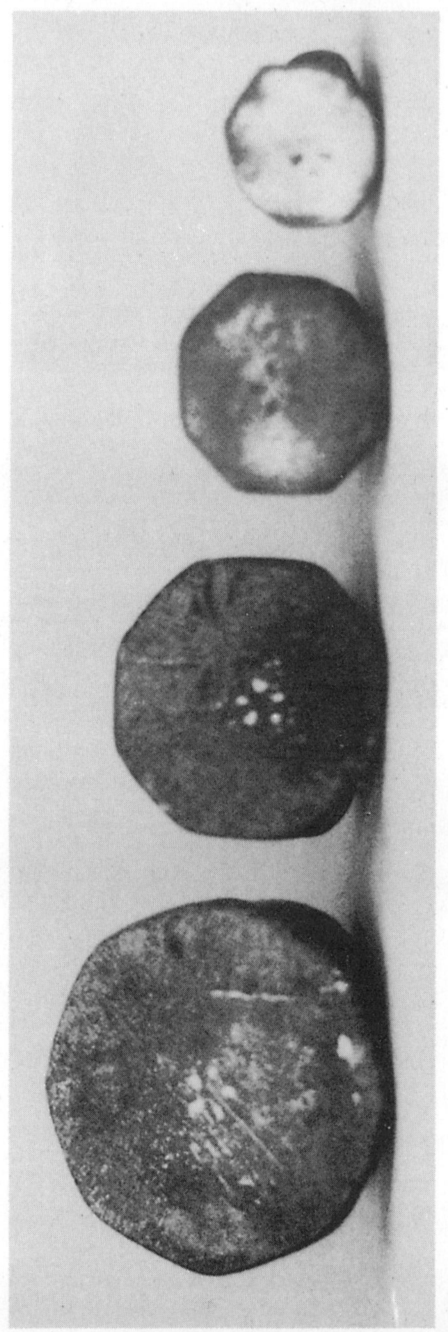

Period H or Ia

5-rayed star on base underside

Plate 28: Bird weights, Group 4. Signs on the bases

Other sign illustrations appear as below:

6-rayed star on right front, see plate 40

Gha sign on right front, see plate 17

Gha sign on front, see plate 16

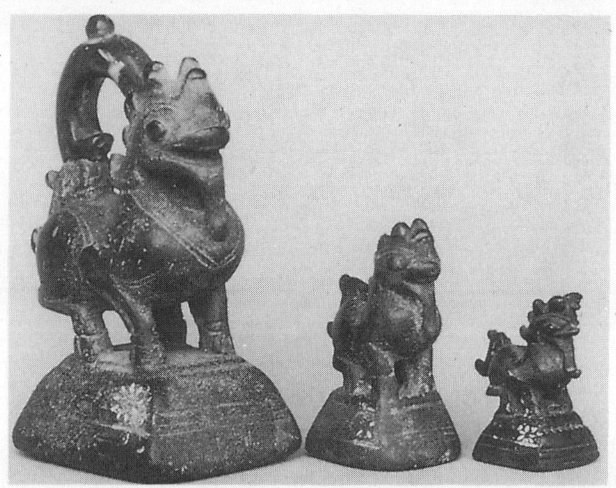

9- and 4-rayed stars on base right front.
9-rayed star also occurs on Group 3.
4-rayed star also occurs on Group 4.

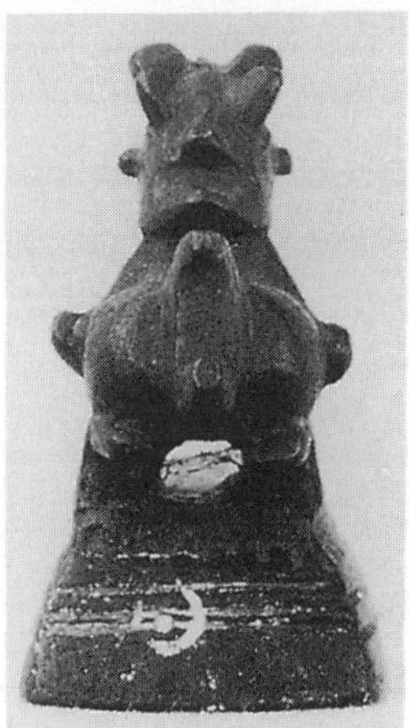

Auxiliary circle and
dot on rear of base

Plate 29: Beast weights, Group 2. Signs on the bases

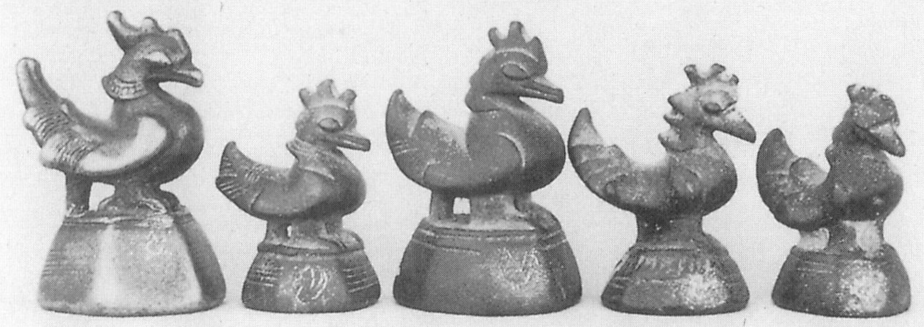

V-tails Bird on base right front Groups 1,2

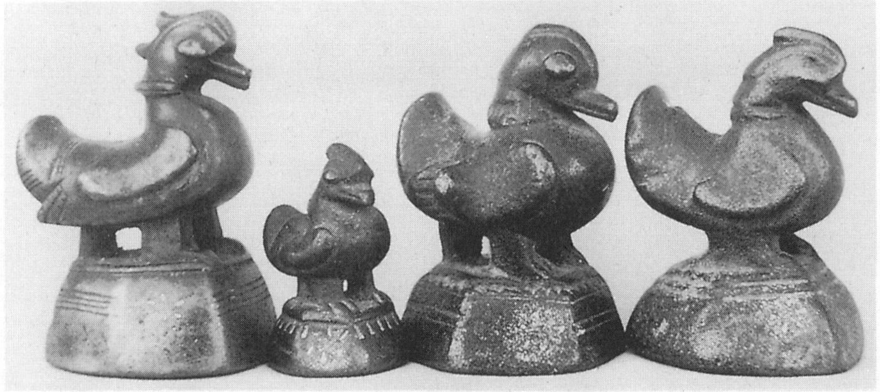

U-tails 6-rayed star on base right front Group 2

Plate 30: Bird weights, Groups 1, 2. Signs on the bases

The Style Systems:

The shapes of the parts of the weights were changed in systematic ways, the changes characterising the group styles being the most obvious. Within the groups there are changes in the shapes of the minor parts of the weights by which they can be classified. Commonly the bird weights of a period fall into assemblages of classes which have one or more parts differing in systematic ways. For example, the Groups 1 and 2, and later Group 4, bird weights differ systematically in their number of head knobs and manes. The bird and beast weights of Group 4 show systematic variations in the mouth appendages and base shapes. The weights of Groups 5, 6 and 7 also show such systematic style variations but because fewer of these are to be found and because of occasional doubts regarding the authenticity of some specimens available, less reliance can be placed on the existence or type of system determined. The table 'Style Systems' shows the characteristics which form the basis for the style system of each period, where such exists.

Many of the style systems are illustrated on Plates 31 to 43.

Table 16: Style Systems

Group and Period	Animal	Variables							No. of System Classes
		Tail shape	Head knobs number	Manes number	Mouth Appendage		Base plan shape	Other	
					number	shape			
1A	Bird		1 2 3	0 1 2					9
2B	Bird	U V	1 2 3	0 1 2					18
2BC	Beast							(a)	2
2C	Bird	U	1	1 2 3			H	(a)	3
		V	3	? 2 3			O	(a)	3?
2D	Bird	U	1	1 2 3?			H		3
		V	3	3			O		1
2D	Beast								1
2E	Beast								1
3E	Bird					R B			2
4E	Bird				0 1	L			2
4F G	Bird				0 1	L B			6
4H Ia	Bird				0 1	L B			2
4H	Beast								1
4Ia Ib	Beast				0 1	R L V	O C		8
4Ib Ic	Bird				0 1	Sk	O C		4
4Ic	Beast				0 1	? L V	O C		6? 8?
4J	Bird				0 1	S	O C		4
4J	Beast				0 1	R	O C		4
5KL	Bird			Variable		R B	O C P	(b)	3?
5KL	Beast	R B S D		Variable	0 1 2	R B	O C L R		?
6M	Bird				0 1	R B	O C	(c)	6?
7M?	Bird				0 1	R B	O C P		9?

Notes

Tail Shape: U = U-tailed. V = V-tailed. R = Rod. B = Bud. S = Single loop. D = Double loop.

Mouth Appendage Shape: R = Rod. L = Trefoil. B = Bird. V = Vee.
Sk = Knobbed spatula. S = Spatula.

Base Plan Shape: O = Octagonal. C = Circular. P = Polygonal.
L = Elongated octagon. R = Rectangle.

(a) With or without short verticals as base decoration.

(b) Rare weights with more than one creature, excluded.

(c) One, two or three birds on the base.

No. of System Classes: Weights with bird-shaped mouth appendages are not included.

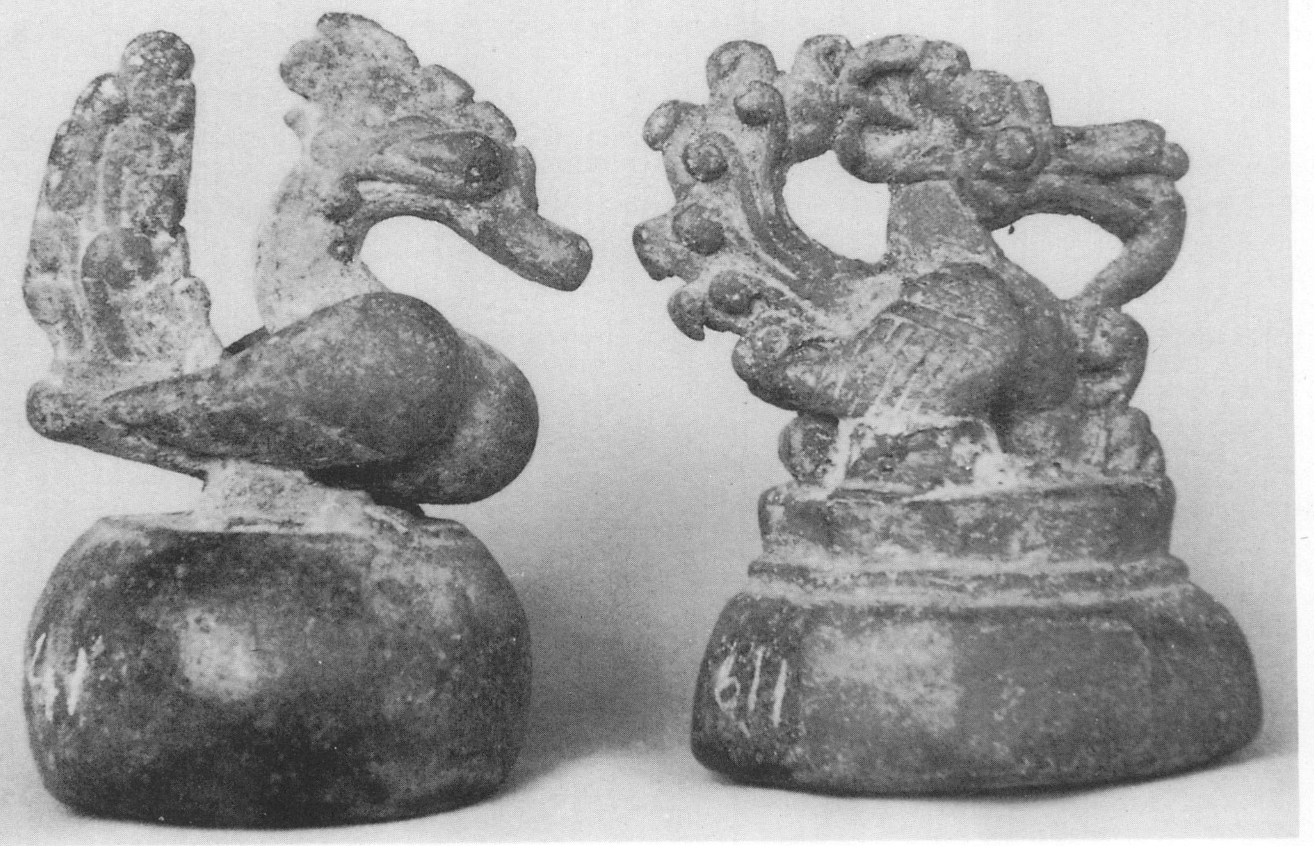

Period M — Groups 1,2

Circular base with no rod mouth appendage • Polygonal base with rod mouth appendage

Plate 31: Bird weights, Group 7. Style systems

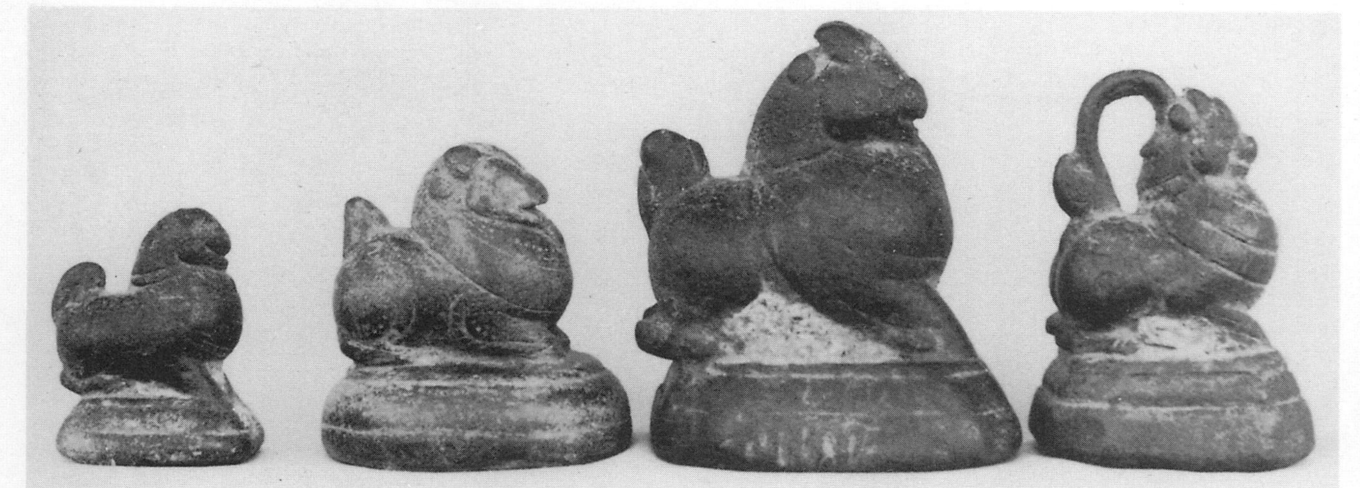

Period KL

Rarely there is a long pendent horse tail which is associated with both a circular and an octagonal base and one or two mouth appendages.

Period M and KL (?)

There are three main tail types, lotus bud, short erect horse with one loop, short erect horse with two loops. Each of these tail types is found on both circular and rectangular bases, the latter with or without cut-off corners. Each of these nine classes appears to be associated with 0, 1, 2, mouth appendages.

Plate 32: Beast weights, Group 5. Style systems

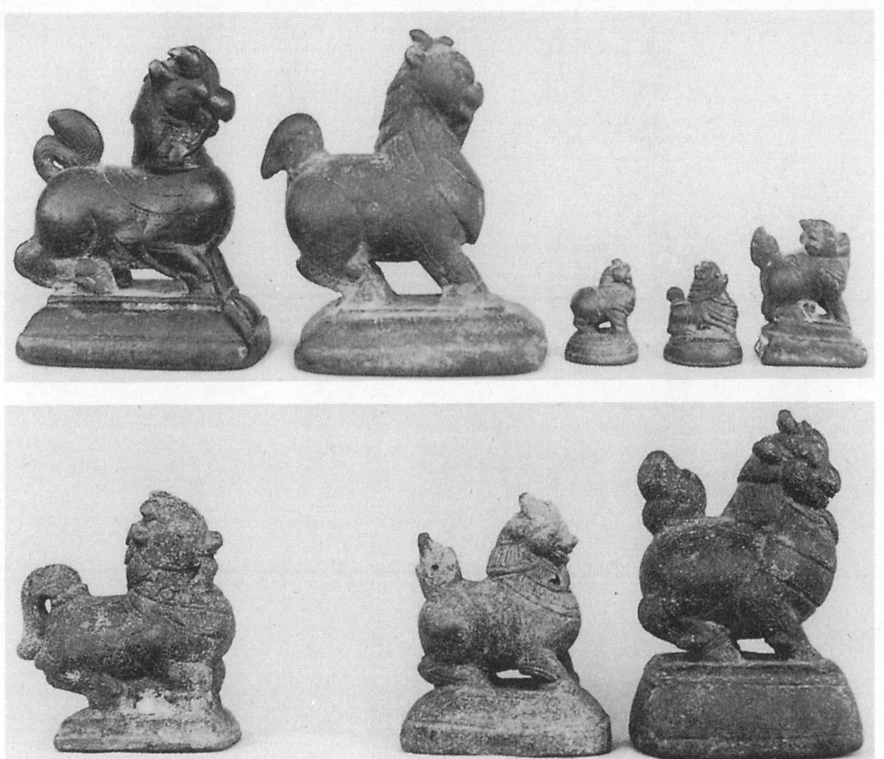

Plate 33: Beast weights, Group 5. Style systems

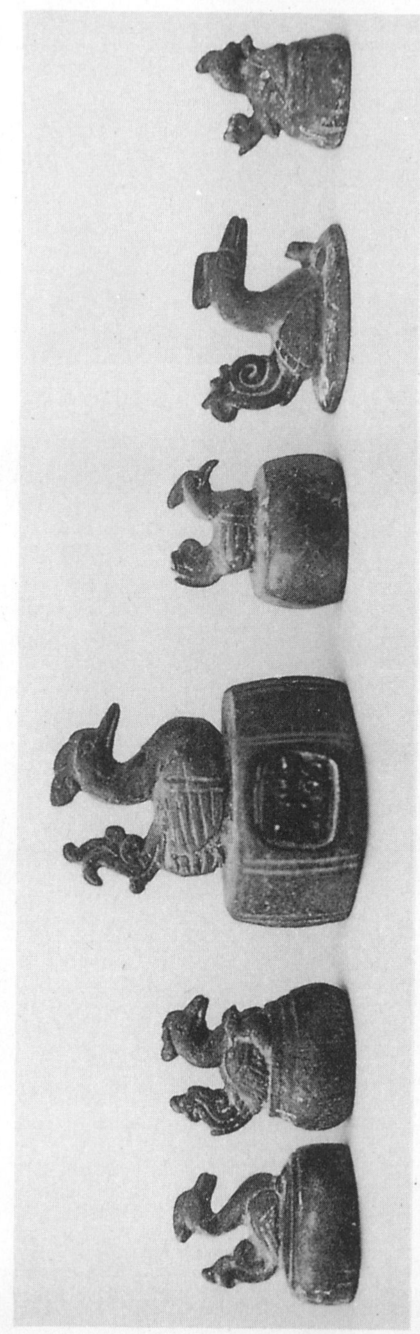

Periods KL and M?

Circular base with 1, 2, 3, birds.
Octagonal base with 1, 2, 3, birds.

Plate 34: Bird weights, Group 6. Style systems

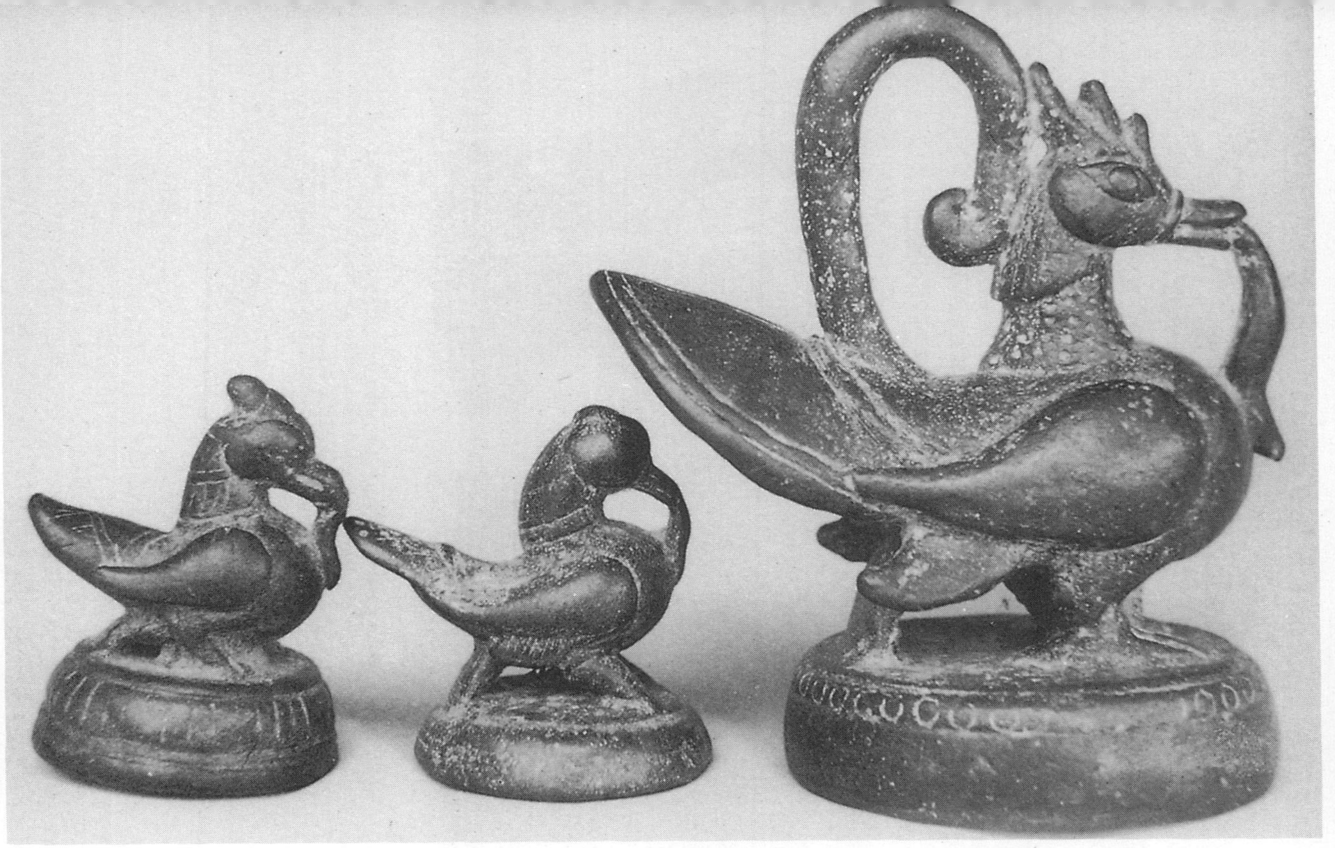

Period L

Octagonal and polygonal base plans with rod mouth appendage

Plate 35: Bird weights, Group 5. Style systems

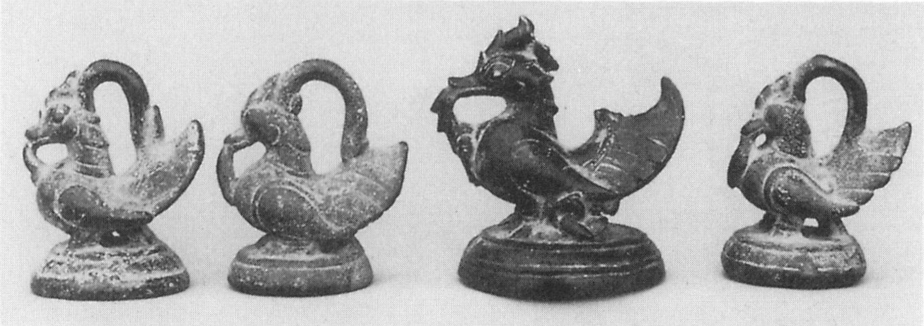

Period K

Circular base plan with rod mouth appendage.
Polygonal base plan with bird mouth appendage.

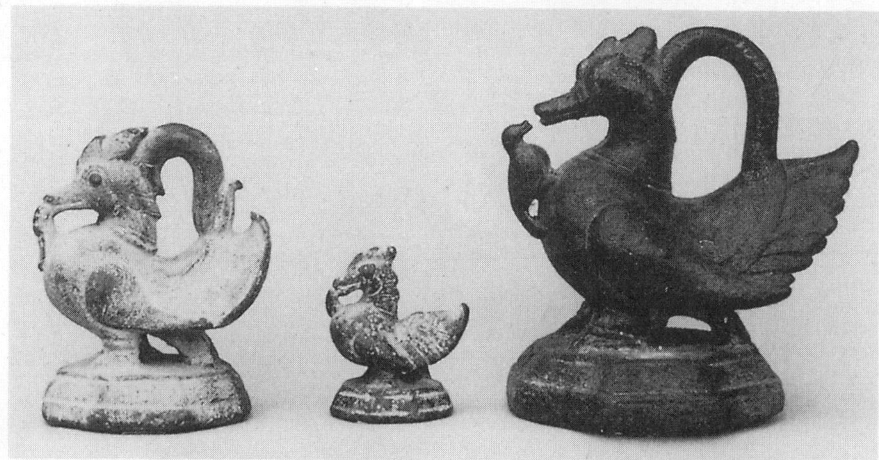

Period K

Octagonal base plan with rod and bird mouth appendage

Plate 36: Bird weights, Group 5. Style systems

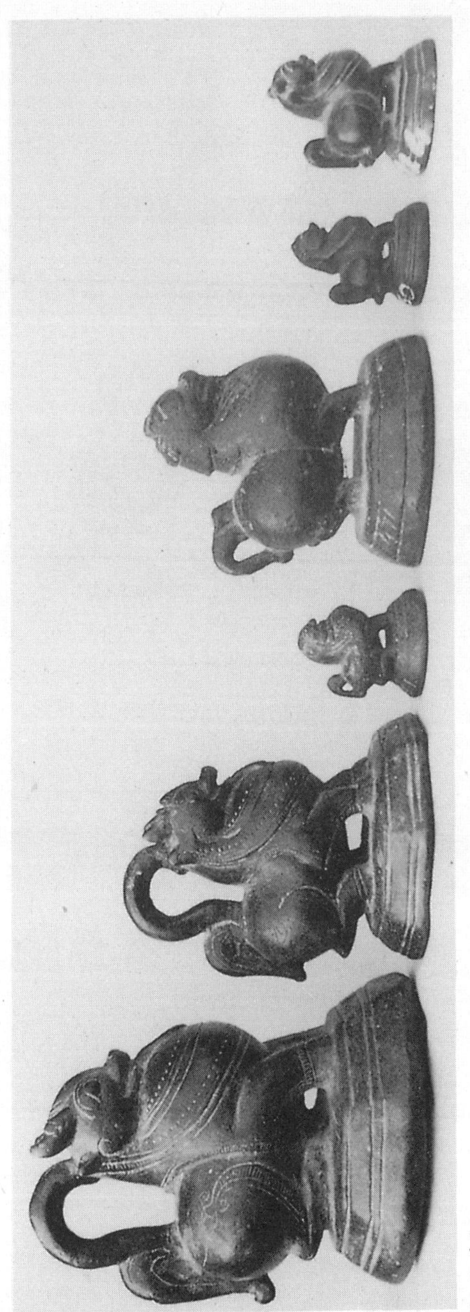

Period J

No and rod mouth appendage on circular base
No and rod mouth appendage on octagonal base

Plate 37: Beast weights, Group 4. Style systems

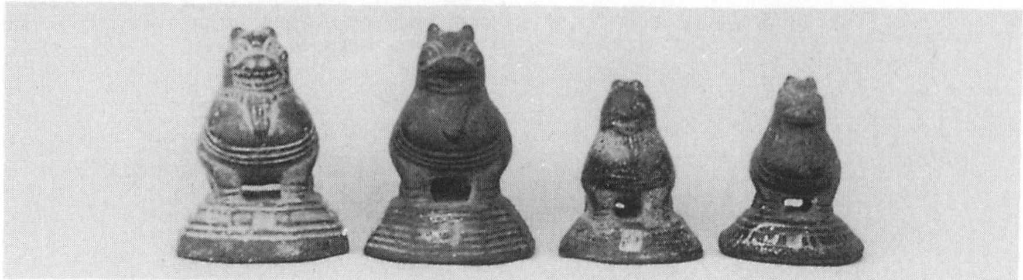

Periods Ib and Ic
Vee mouth appendage on circular and octagonal base

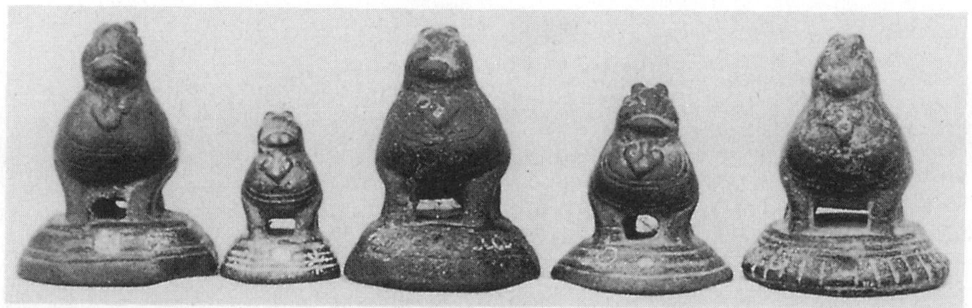

Periods Ib and Ic
Trefoil mouth appendage on circular and octagonal base

Plate 38: Beast weights, Group 4. Style systems

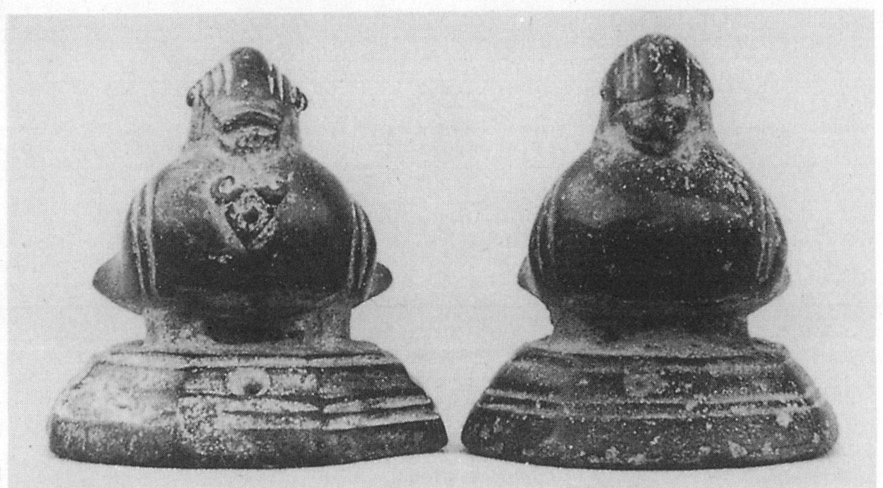

Periods H and Ia (?)

No and trefoil mouth appendage

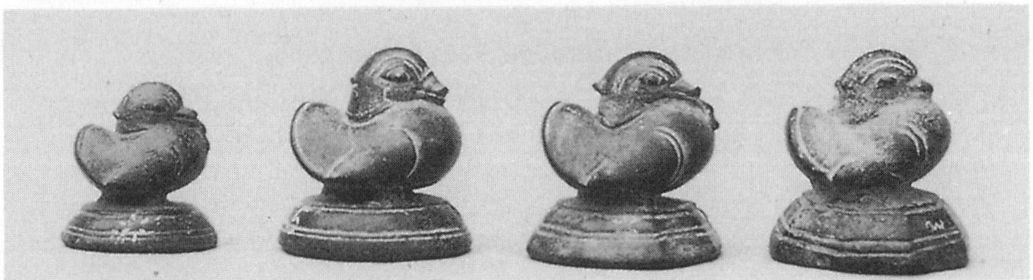

Periods Ib and Ic

No and knobbed spatula mouth appendage on circular base

No and knobbed spatula mouth appendage on octagonal base

Plate 39: Bird weights, Group 4. Style systems

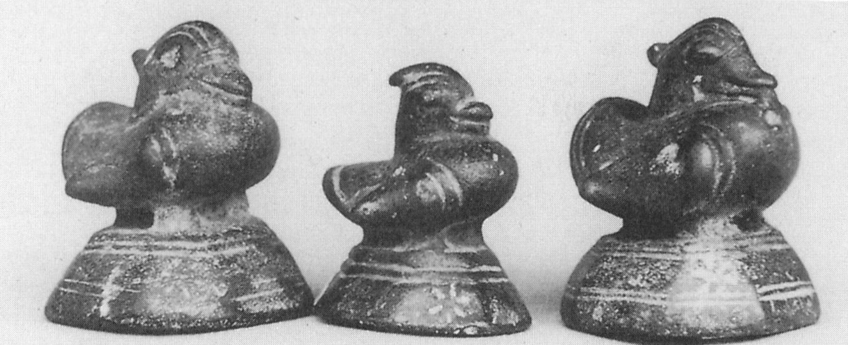

Period E No and trefoil mouth appendage

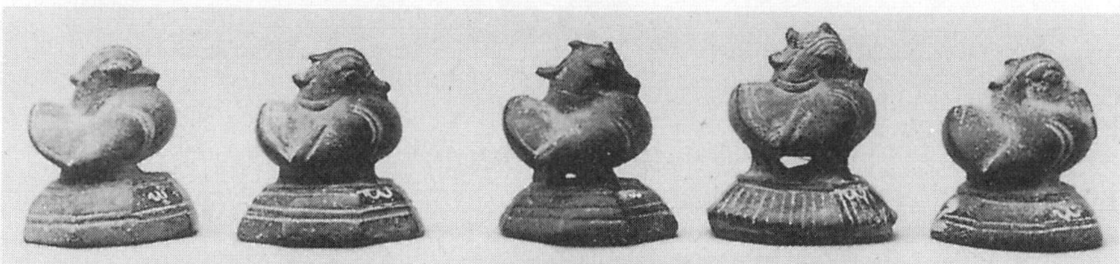

Periods F and G(?) No mouth appendage. 1, 2, 3 head knobs
Trefoil mouth appendage. ? 2 ? head knobs (1 and 3 not found)

Plate 40: Bird weights, Group 4. Style systems

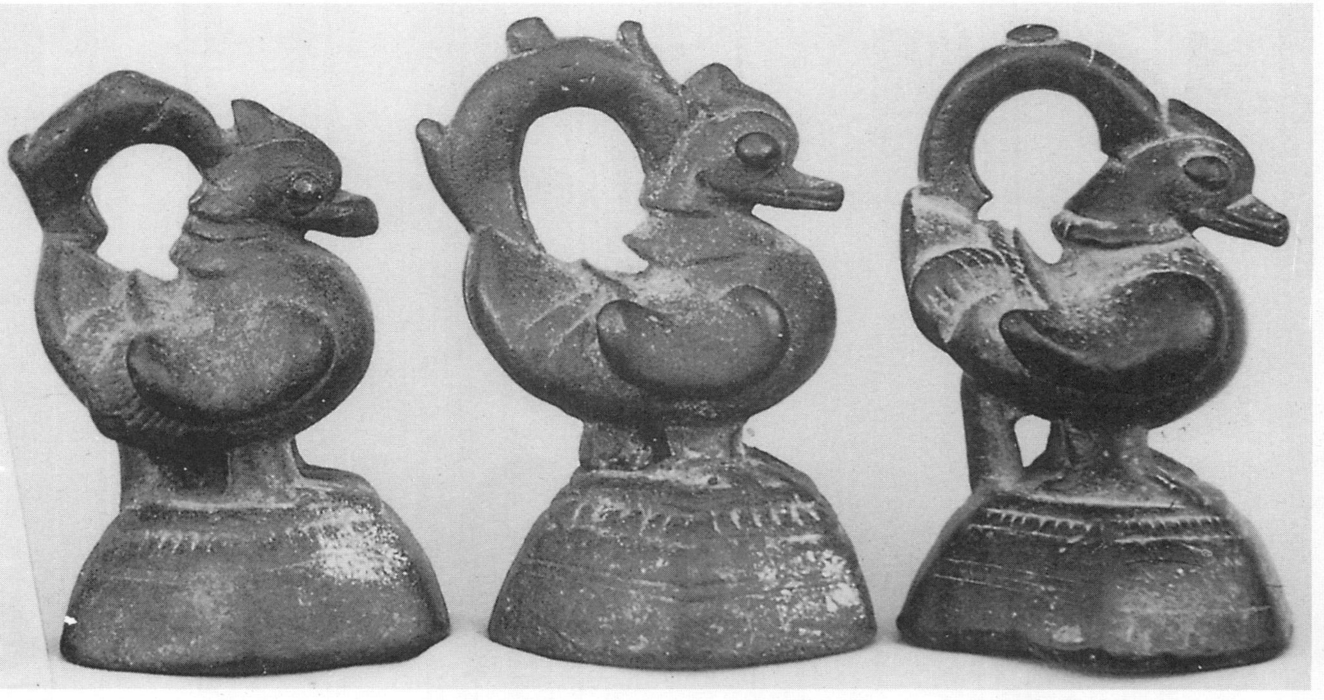

Period C

U-tail. One head knob. 1, 2, 3 manes

Plate 41: Bird weights, Group 2. Style systems

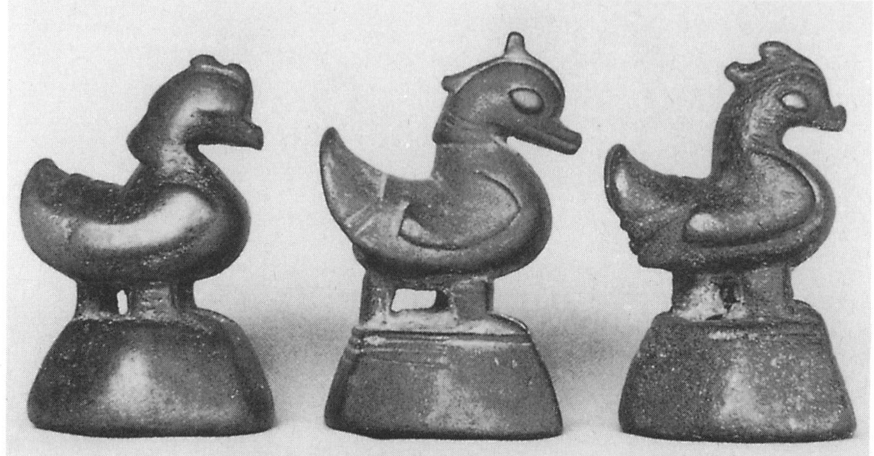

Period B U-tail. Two head knobs. 0, 1, 2 manes

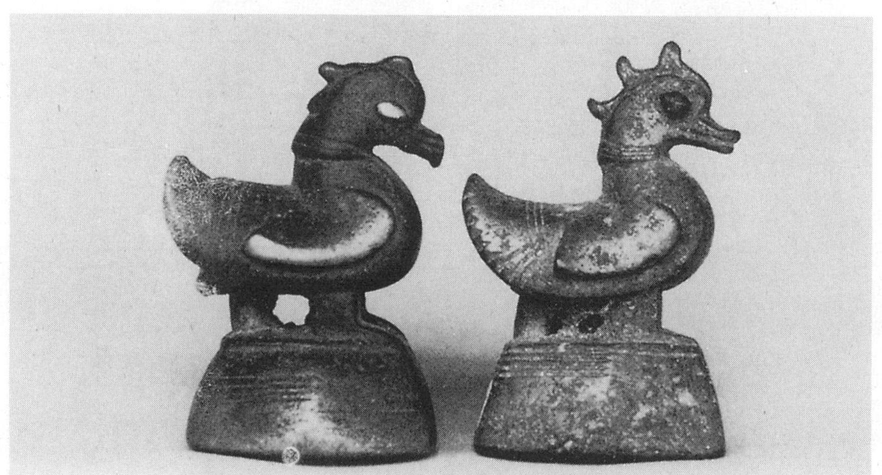

Period B U-tail. Three head knobs. 1, 2 manes

Plate 42: Bird weights, Group 2. Style systems

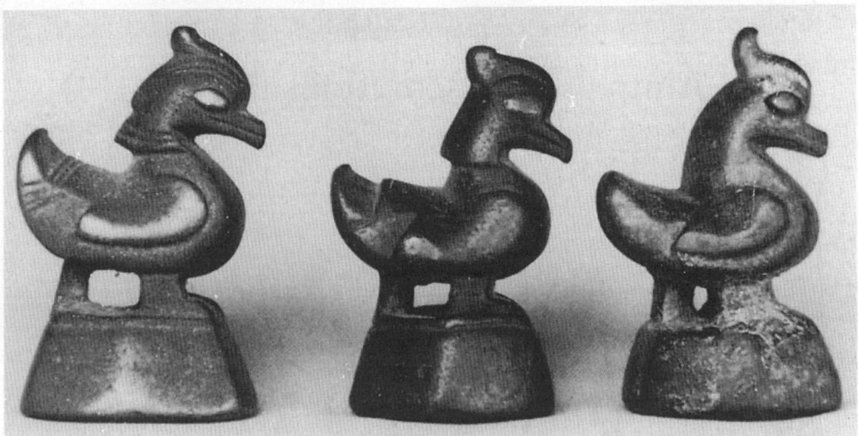

Period B V-tail. One head knob. 0, 1, 2 manes

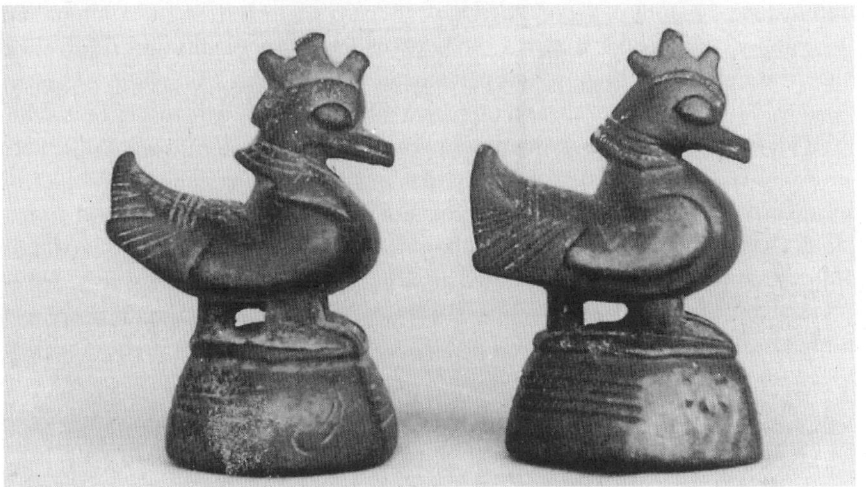

Period B V-tail. Three head knobs. 1, 2 manes

Plate 43: Bird weights, Group 2. Style systems

7
Animal Representations in Asia

Introduction:

This chapter is concerned with identifying the animal shapes on the weights and their ancient pre-Burmese origins. Both the beast and bird shapes are stylized, the former being composed of parts from different animals. In both cases, the Burmese names were examined for indications of the identities of the shapes. The shapes of the whole weight-animals, both bird and beast, suggest possible similar supernatural creatures. Such also were examined. The identification of the most frequent bird shape presented no real problem though there are two infrequent exceptions. The identification of the beast weight presented considerable problems which could be resolved only by first, the study of the mythical beasts of particularly Burma, India, China and the steppes, second, the study of the parts of the beast weights and third, their historical and cultural origins. As experience grew the possible symbolisms suggested other identifications until it became apparent that the beast weight, that is the weight-beast together with its base, symbolised a *chakravartin.* A *chakravartin* is a ruler over an extensive, supposedly universal, domain righteously governing his peoples in accordance with *"dharma"*, an Aryan system of social and religious rights and duties. He also had the power and duty to maintain crop fertility and to keep the cycle (wheel) of life in motion, e.g. the seasons. His capturing of prisoners in war was regarded as essential for maintaining soil fertility. The concept is much older than the peoples of Burma and apparently developed on the steppes but drew its inspiration from ancient Mesopotamia and Persia.

During the investigations it became apparent that the main shape concepts were Buddhist and so came from India. It also became evident that the model for the weight-beast, at least its body, neck and usually its tail, was mostly a maned (and horned) large feline, presumably a lion or perhaps a tiger, to which manes, horns and often a protruding tongue had been added. If it were a lion then its origin presumably had to be in India, especially western India. In fact, however, the motif long pre-dates Indian art. The model for the most common bird shape previously had been identified as the mandarin duck (*aix galericulata)* from China. Neither of these two creatures were indigenous to Burma and so their models had been introduced. It was known that artistic motifs and symbols had migrated across the breadth of Eurasia in ancient times, particularly in the form of animals, notably the animal style art of the steppes and even more particularly the feline-avian motif. Special forms of animal art were the stone and metal weights made in the forms of lions and ducks, "originating" in ancient Assyria and spreading by means of trade and possibly religion throughout pharaohic Egypt, the Near East and the Middle East. The direct descendant of one of these forms, the duck, has continued to be made to this day in modern Persia, i.e. Iran.

With this information, together with that of the associated symbolism, it became evident that the origins of the Burmese weights lay further back in time and further away geographically than the cultures of the present Burmese peoples or those of neighbouring China and India. As a result, this chapter is concerned firstly with Eurasian animal representations, their occurrence and development, secondly with identifying the component parts and entirety of each of the creatures on the weights and thirdly with brief summaries of the origins of the weight creatures. The origins lie either in Mesopotamia or the steppes or both. If the steppe peoples[1] drew their relevant inspiration from Mesopotamia, then Mesopotamia is the origin of the lion-duck motif. If the Mesopotamians and Egyptians drew upon steppe art, then the steppe peoples provided the motif though the former applied them to the weights. However, it seems likely that both motifs originated in a still earlier complex of cultures spread throughout Eurasia which shared many more or less common characteristics, either indigenous or acquired, which subsequently underwent local modification due to different local conditions and events.

Asia, General:

Animal shapes of many kinds have been made using different kinds of material in many styles from wholly realistic to those which are so stylized as to be just recognisable. They vary from sculpture in the round to flat shapes on textiles and paintings and from sphinx-sized structures to carved pin-heads. These forms occur as statues, bas reliefs, mosaics and on many objects of personal use, such as weapons, vessels, horse trappings, belt buckles, toys and jewellery.[2] They have been intended to represent deities or their symbols, incarnations, vehicles, totems, items of magic, guardians of gates, emblems power and legends. They also have been used purely for decoration.

The origin of animal art lies no later than with the Stone Age hunters. In general it was used by all hunting, non-urban and non-literate cultures. Later, crop-growing and community-dwelling cultures replaced animal forms with human figures.[3] In the case of the style of small metal animal representation developed by the mounted, nomadic, predominantly herding cultures of the Eurasian steppes, the period of most widespread use was between about 1000 and 200 BC but it began and ended at different times according to locality.[4] Some of this art appears to have arisen in Mesopotamia in the 4th or 3rd millenium BC along with the development of copper and bronze utilisation. Its motifs, in part, along with bronze-working, seem to have moved east in waves through the Altai mountains of northern Central Asia to Ordos in Mongolia on the Hwang-ho River and to China, the last inflow there of this steppe style probably being between the 5th and 3rd centuries BC. A part of the art may have moved south from Siberia. The spread of the style must have depended on the acquisition of appropriate knowledge, the availability of materials and the existence of the right kind of society. Different styles of animal art continued in Persia, India and China for some centuries longer. In China a different style of animal representation had existed from before 1500 BC. It continues today in suitable 'primitive' cultures and it survives even in Western cultures to a limited extent where it still serves as a kind of status symbol, as in heraldry.[5]

The main interest of this work lies in the animal representations made in Burma and in relation to these, those of Persia, Central Asia and the Eurasian steppes, China, south-east Asia and India, especially after about 200 BC. Those influences from further away which affected the foregoing are also of concern.

Western Asia:

This region comprises Mesopotamia and the Mediterranean region. It includes the animal-like artifacts of the Egyptians, the Minoans, the Sumerians the Babylonians, the Assyrians, the Hittites and the Greeks.[6] The Semites made little use of the animal form.[7]

The region was a principal centre of animal-shaped artifacts from about 3000 to 500 BC and supplied many of its animal motifs to Persia and thence eastwards to Central Asia, to India and beyond. Though of its work was realistic in style, many or most of the mythical composite animals originated from this region. Animals commonly represented were lions, bulls, stags, griffins, sphinxes, snakes and dogs, but there was a considerable variety. They were used as decorations, as temple guardians and, when buried, as protectors from evil.[8]

However, with increasing urbanisation, the animal-like gods became human-like gods. Animal representations then became merely decorative, supporting man as a central theme, their shapes and combinations merely a matter of fancy. Contrariwise, in Asia these shapes and their combinations had serious social purposes as deities or deity symbols and in other ways. It is through their influence on Persia and Central Asia that the animal artifacts of western Eurasia are relevant to this work.

Persia and adjacent regions:

This area is bounded approximately by Mesopotamia, the Caspian Sea and the Pamir mountains. Persia was one of the main nomadic centres of animal-like creations in the 2nd and 1st millenia BC, being influenced at first by the similar work of the Hittites of Asia Minor. The lion and composite monster were common motifs. The mythical and realistic animal representations produced from about the 7th or 6th century BC were partly inspired by influences from Ordos[9] and from Assyria.[10] The Persian (Achaemenid) animal art of this period was predominantly of small, portable, metal, utilitarian objects similar in concept and purpose to those of the steppe nomads. Small bronze models of lions were made similar in shape to those which later appeared on the votive pillars and Asokan columns of India.[11] From Persia the animal motifs spread to India and the Altai mountains[12] and even to Mongolia.[13] To the north of Persia in the 2nd century BC the Bactrian Greeks were still looking to Alexandria in Egypt for their inspiration and so influenced the aesthetics of the region, hastening the growth of anthropomorphism and the decline of animal art.[14] By the 1st century BC Rome and other Mediterranean countries were adversely affecting this field of creative activity along the land and sea routes to the silk road. However, the animal motif remained important and was modified by the Buddhist influence which was dominating the eastern part of the region. Silver, bronze and silk were decorated with mythical animals, e.g. *senmurvs*, Zoroastrian cocks, ducks, griffins, winged lions and many others.[15] The wings and horns on Chinese representations of the tiger and lion derive from the Persian region.[16] Although the Islamic invasion commencing in 637 AD destroyed many Buddhist temples, it did not destroy the considerable output of animal representations which continued in the fields of ceramics, textiles and metal work. By the 13th century AD many Chinese mythical animal were increasingly being incorporated, e.g. *feng-huangs* and *ch'i-lins*,[17] some of which can be matched in style by 19th century Burmese animal representations.

In view of the interconnection of the Persian and steppe animal art, Persia's intimate connection with the 19th to 6th century BC lion-duck weights of Mesopotamia and Egypt, its considerable influence on China, including Yunnan, in many fields of human concern and the long duration of its importance on the sea routes to south-east Asia, it seems likely that Persian influence may have borne on the designers of the weight animal shapes in one or other of the ways mentioned.

India:

Animal representations are abundant and varied in the culture of the Indus valley about 2000 BC.[18] They include infrequent animal combat scenes of the type common in steppe art and are of the same quality as those of Egypt and Mesopotamia[19] from which much of the inspiration for the Indus animal-like creations was derived.[20] When the Aryans from south Russia (?) invaded India around 2000 BC they brought with them little in the way of animal art, animal deities being subordinate to anthropromorphic ones, but frequent Vedic literary references to them exist.

From 1500 to 300 BC both animal representations and Western influences seem to be infrequent,[21] though artifacts are too few to be sure, but what there are appear to be from Persia, especially in north-west India,[22] and Mesopotamia. Persian influences continue to be important until the 7th century AD and probably much later.[23]

Even by the 3rd century BC, in the time of the influential Buddhist monarch Asoka, there are almost no animal representations except for the pillar animals and a few scattered fragments.[24] By the 2nd century BC or a little earlier when stone carving became commonly used in India, the lion, bull, elephant, horse and goose became common motifs,[25] many bearing floral sprays. Shortly after this came the development of the mythical animals such as the *garuda*, the *makara* and the three-headed *naga.*

From the 2nd century BC on, the Buddhist temples were profusely decorated with a variety of animal shapes, reaching a peak which probably has never been surpassed in India. Some include winged and horned lions, griffins and the like. The invasions of the Greeks, Sakas, Parthians and Kushans into the Bactria and north India regions between the 1st century BC and 2nd century AD led to one of the most creative periods in the history of India's art.[26] Another important influence was that of the Romans from about the 1st century BC to the 4th century AD. About the same time many refugees escaped from Christianisation in the west and further stimulated the cultural changes.[27]

By about the 1st century AD representations of animals spouting vegetation became infrequent while the number of animal shapes utilised diminished, though not the quality of craftmanship. During the following centuries, the temples made much use of the Jataka (Buddhist birth tales) for the animal themes used on their structures.

By the 5th century AD the preoccupation was mainly with the human body. The jungle scenes had gone though there was still some notable animal sculpture. The number of animal representations continued to diminish and more and more assumed the role of an attribute or symbol. They tended also to form an integral part of temple architecture.[28] By the 6th century AD the Buddhist animal motifs were infrequent, being replaced by the animals of the myths associated with the Hindu deities Vishnu and Siva.[29] By the 8th century AD Buddhist art was confined to north-east India where lion, griffins and geese were still common while animals in the form of vehicles had become rare. The Jataka animals and those symbolic of events in the Buddha's life continued to be sculpted in large numbers, the bronze effigies being small.[30] In 1162 AD the ousting of the Buddhist dynasty in north-east India by a Hindu one appeared not to halt or hasten the decline of India's animal art. Though the subsequent Mahometan conquest of north-east India in 1200 AD brought destruction of the Buddhist temples and considerable artistic change, nevertheless it also appears not to have hastened the gradual disappearance of India's animal art since it was by this time confined to but a few scattered temples.

Central Asia and the steppes:

The animal art of the Eurasian steppes began in Siberia then, moving south, extended from China across Asia south of the boreal forests and north of 30 degrees latitude[31] into Persia and Mesopotamia and into south Russia and northern Europe.[32] Beginning before 1500 BC, it reached its peak of development between the 7th and 2nd centuries BC,[33] after which it rapidly declined in extent except where its influence still remained, as in south-east Asia, India, Tibet and some other places. By approximately 1000 AD steppe animal art had ceased though its influence persists to this day among the decreasing number of plant and increasing number of human portrayals which have replaced it.

A characteristic of steppe animal art is it appearance on useful articles of bronze and gold of portable size suited to a nomadic life,[34] e.g. belt buckles and harness fittings. The nomadic herdsman's life was one associated with an absence of writing, with an absence of town conditions, with wild animals preying on his herds, with shamanism. Of particular importance in his art were realistic representations (later stylized)[35] of tigers, horses and stags, though many other animals were employed, including the domestic cock and mythical animals, especially the lion-griffin and eagle griffin.[36] Felines, stags, elks, rams and birds are the most typical. Horns and antlers are common in composite animals while animals in procession or in combat are frequent and persistent motifs. Steppe animal art[37] incorporated features from the ancient civilisations of the Near East, Persia, Greece and China.

Of such regions as have been investigated, those of the Persia-south Russia region, the Minussink basin, the Altai mountains and Ordos (north China) have been referred to as important centres of steppe animal art, the last two perhaps being those most important to the transmission of its influence to the animal art of north-west and south-west China. From about 300 until 100 BC the Pazyryk-Tashtyk cultures (Altai-Minussink region) of Sarmatian-Hun origin produced a naturalistic animal art in gold and bronze of griffins, phoenixes, horses disguised as reindeer, monstrous birds and winged felines with a man's head and stag's horns. Hunnish animal art continued to flourished for several centuries longer, though gradually declining.

In Ordos bronzes with animal decoration continued until about the 5th or 6th century AD. This region is sometimes referred to as the last stand of animal art.[38] At some time before the 2nd century BC the Yueh-chieh in present day Kansu, not far from Ordos then at its peak of abundant production,[39] must have contacted the Sakas of near Lake Balkhash and acquired knowledge of its stag art,[40] which, in modified form, could have been transmitted south through Szechwan to the semi-nomadic tribes of Yunnan. The Yueh-chieh subsequently drove out the Sakas (about 160 BC) who then moved into north-west India (Gandhara) while the Yueh-chieh became the Kushans at the west end of the Tarim basin.[41] The Saka retained some of their animal art but the Yueh-chieh abandoned theirs.[42]

From the 1st or 2nd century AD the art forms of Mahayanist Buddhism, which dominated the increasingly literate, urban and commercial Central Asian region, were almost exclusively anthropromorphic while the Islamic invasions of the mid-8th century AD probably gave the coup-de-grace to the surviving animal art of the region.[43] No steppe animal style ornaments have been found in the oasis cities along the old silk route through Central Asia to China[44] while animal representations of any kind have been reported only rarely. They are found further north, e.g. in the Gobi Desert and Mongolia and also in Ordos dating from the 6th century BC. They show marked Persian influence.[45] When the Mongols created their great empire of the 13th century AD, they brought no animal art with them into China and Burma[46] so they could not have initiated the animals on the bronze animal weights. Nor does it seem likely on this evidence that the concept of the animal weight travelled along the old silk route even though there was such a close connection between Mahayanist Buddhism and trade[47] (and so weights). However, at the eastern end of the Tarim basin in the country of the Tangut (Hsi-hsia), Mahayana Buddhism made considerable use of the twelve animals of the solar zodiac, widespread in Central Asia before the beginning of the Christian era. (Objects of this kind in later years are said to have been used in Siam in place of weights.[48]) It is quite possible that the earlier archaeologists investigating the region overlooked or discarded such minor objects as weights.

China:

From about 1500 BC to between 500 and 200 BC was the age of Chinese bronze-working, a skill possibly introduced from the steppes along with jade.[49] The steppe nomads also provided the dominant external influences on the art of north, west and central China.[50] These in turn were affected by the dispersion of animal art from west Asia, notably at the time of the fall of the Hittites.[51] Until about 900 to 800 BC animal representations were used only as decorative themes on vessels but afterwards bronze animal figurines were introduced. As with the nomads of the steppes, small-sized works were characteristic, though the Chinese animals were more stylized, more fantastic and more often mythical in significance.[52] About the 6th to 5th century BC there was a rich flowering in Chinese animal art, possibly influenced by earlier migrations westward from south Russia and the Caucasus[53] and the adoption by the Chinese of many steppe and Siberian animal motifs.[54] At about this time most animal heads became progressively more stylized and animal processions were adopted from Central Asia as a decorative motif.

From the 3rd century BC Indian art forms become noticeable.[55] Animal representations now become more realistic, perhaps due to the impact of the nomad art of the steppes, including Ordos, and of Persia.[56] Tomb guardians in the shape of horned felines with protruding tongues and other funerary animals become evident and animals commenced to become common on mirrors. The use of bronze was going out of favour and iron was replacing it.[57] In addition, animal ornamentation gradually ceased to be emplaced.[58] Animal art continued however and was influenced by the Central Asian cultures during the early centuries of the Christian era.[59] From about the 1st century AD the realistic animal sculpture of the early Chinese Buddhists began to disappear under the pressure of Buddhist anthropomorphism.[60]

Before the 8th century AD much of Chinese art had become dependent on India. The usage of animal motifs continued, animal statuettes being popular. Zodiacal animals were first produced in China early in the 1st millenium AD.[61] By the 9th century AD motifs other than animal representations were becoming dominant. The Mongol invasion of the 13th century AD had no effect on the animal art of China since by this time the Mongols had lost their own. Mammal but not bird motifs continued to decline in popularity and disappeared as the mammals themselves had disappeared from bronzes and ceramics by the 14th century AD.[62] They continued to be used on fabrics[63] while here and there stone or bronze statues of unicorns or lions were used on temples or to line the ways to Ming mausoleums. Even today marked traces of animal art still persist in China.[64]

Tibet, of which no large part was nomadic, has but a little animal art in the steppe animal style and that only among the small groups or nomads of the north and north-east. One possible origin of this nomadic animal art may be from that part of the Yueh-chieh people who lived in the region north-east of Tibet and who moved south when the Hsiung-Nu forced them all from their lands about the 2nd century BC. Among the settled part of the Tibetan population a few winged or horned animal representations, probably of shamanistic (Bon) origin, have been found.[65]

South-west China:

Until 1368 AD south-west China was either mainly independent or fell only under nominal or temporary Chinese suzerainty. As a result, the animal art of these 'barbarians' differs considerably, e.g. in realism,[66] from that part of what is now China which extended north and east of the region. The tribes which lived in the region may have come from the north-west[67] because of the evidence of certain aspects of their culture. Bronze and wood animal themes were popular between the 6th and 1st centuries BC.[68] The bronzes and stone sculptures and other artifacts of Buddhist times from the early centuries of the Christian era onwards were destroyed so completely during the rebellions of the 19th century AD that very little is known about them.[69]

Bronze-working was known in the region about 1000 BC[70] but it was not until after about the 6th century BC that the techniques of high quality bronze work and a strong interest in animal art reached Yunnan. This resulted from the excellent example of the shamanistic Ch'u 'barbarians', who moved into south-west China from its north-east and conquered the Tien (or Dian) living near the lake of that name in east Yunnan.[71] They brought with them also a tradition of wood carving which persisted into the following centuries. Possibly this tradition may have reached south-east Asia, including Burma which has notable wood carving.

Part of the Tien people and those in the adjacent regions of Yunnan appear to have lived like the barbarians on China's north-west borders as semi-nomadic herdsmen of horses[72] for which Yunnan was famous. As a result, their animal art was realistic, differing from that of the urbanised Chinese. Bronze mythical and natural mammals, birds and snakes in a wide variety were made, as were bronze drums bearing animal motifs and gongs.[73] The drums "originated" with the barbarians of south-west China before Chinese influence became effective, an influence which is apparent on the architecture of the region from about the 2nd century BC.[74] Similar drums were made at Dong-S'on in north Vietnam. Since these bronze drums are still valued by the relatively recently-immigrated Karens of east Burma and there may be a connection with the animal figurines on the weights, a listing of their possible distribution follows. From the predecessors of the Yueh-chieh people of Dzungaria (Kansu and Sinkiang in China) to west Yunnan (location of earliest known drum 3rd century BC?),[75] to the Tien people of east Yunnan (ending about the 1st century BC),[76] to the people of the Dong-S'on culture (ending about the 1st century AD)[77] and to Burma, south-east Asia and Indonesia.

Although the culture of the Tien people was probably largely indigenous, nevertheless it appears to have been subject to style influences from the steppes to the west and north of China's borders.[78] Some of these styles show similarities with animal art forms from the Black Sea, from Ordos and from Mongolia.[79]

From about the 1st century BC onwards there is apparently no additional information on the smaller-sized animal art of the region. However, from about the 2nd century BC until the 6th century AD stone winged and horned lions appear in south China,[80] similar in appearance to at least one Burmese animal weight (Group 4, period H). Though native in style, they show the effects of Persian influence.[81] They also show the flat muzzle, rictus and protruding tongue from the Altai (originally from Assyria).[82] These appear on the beast weights and even on modern Burmese wood carvings of *chinthes*.

From about the 8th century AD Indian influences are apparent on Yunnanese architecture.[83] It seems likely that their influence on the smaller artifacts would have been felt long before this.

South-east Asia:

The bronze-working culture of Dong-S'on flourished in north Vietnam from about the 5th century BC to the 1st century AD, during which time it produced, among other artifacts, bronze drums (gongs) ornamented with animal motifs.[84] The movements of the first immigrants, the Mons/Khmers, may have spread the bronze-working techniques and drums of Dong-S'on throughout south-east Asia, though none of its typical ornamentation occurs on the Burmese weights nor did its influence reach Assam. Chinese artistic influence was dominant in south-east Asia at this time and continued so in Vietnam.[85]

After the 2nd century AD until the 9th century AD Indian influences were supreme. At first art forms from southern India are less noticeable in Vietnam and Burma than elsewhere in south-east Asia. Later they are joined by artistic influences from north-east India. By the 13th century AD both Indian and Chinese influences on architecture and Buddhist images were being replaced by indigenous ones.[86] Studies of the animal art of the region are not known to the writers.

Burma:

The main immediate sources of cultural and artistic influences on northern Burma have been north-east India and south-west China. The immigrants of Mongolian blood, the Pyu, the Mons, the Burmese, the Thai and others, brought with them their south Chinese influenced culture.[87]

Though Indian Buddhistic artistic influences may have reached Thaton about 240 BC, it is certain that north-eastern Indian art and architectural influences were exerted on north and central Burma from the 5th century AD[88] or before. It has been mentioned previously that the earliest Burmese representations of the feline/anserine combination have been found in the Pyu ruins of 4th century Beikthano. In the early seventies, the only two of these very small figurines which had been excavated could not be found in the Archaeological Department stores. Nor was there any record of their masses. There is nothing then upon which to found a supposition whether or not they were intended to be weights. But the relative abundance of metal lion-like and duck-like figurines and ornamentation both at Beikthano and Sri Ksetra (Prome)[89] makes it clear that the feline/anserine concept was well established in south and south-central Burma by this time. These Pyu communities were apparently Indian-ruled and it seems probable that the Indians came from eastern India, possibly Amaravati, which is noted for its early Buddhist temples and the animal representations upon them. However, so far as the writers have been able to ascertain, no animal weights have been recorded from anywhere in India except, doubtfully, medieval Indore. On this evidence the lion-duck motif, but not the weight concept, may have been transferred from Buddhist eastern India to the Pyu people of the southern part of Burma before the 4th century AD. It seems likely also that it would have been transferred to those Pyu who settled in the northern central part of Burma before the 9th century AD, e.g. at Halin. Mahayana Buddhism, Hinayana (Theravada) Buddhism and Vishnuism were practised at this time and so influenced the craftsmen's choice of their associated symbolic creatures. In north Burma the influences from north-east India and the Mons continued until the 12th century AD and are visible in the architecture of Pagan and elsewhere.[90] Here the temples built between the 11th and 13th centuries AD have many painting and bas reliefs depicting the animals of the *Jatakas* (tales of Buddha's incarnations, often in animal form). The Mons to the south made less use of these tales.

It seems likely that craftsmen from Bengal were employed on building the temples of early Pagan. This seems evident from the fact that north-east India was the last stronghold of Buddhism in India, from the architectural motifs and from the use of brickwork, plaster and masonry, materials with which the Burmese were less familiar than with wood, as they remained. At about this time it would have been expected that the craftsmen would have been supplemented by those Bengalis fleeing the replacement of Buddhism by Hinduism in 1162 AD and from the Mahometan massacres and destruction of 1200 AD. Mon craftsmen were brought from conquered south Burma to Pagan and their influence was dominant there until about the middle of the Pagan period. After 1287 AD the Mons were once again independent in south Burma until 1531 AD and influential until the mid-18th century. Little is known of their arts and crafts.

Some more obvious possibilities of steppe nomad and Chinese animal art motifs influencing those of the Burmese region before the Nanchao period are summarised here. The Mons, perhaps originating from Ordos, may have been exposed to the bronze animal style art of that region, a style which in Ordos lasted perhaps from between the 15th and 8th centuries BC to the 6th century AD. The Hsiung-Nu are notable for their association with artifacts in the shapes of griffins.[91] Thus the Chiang, close neighbours of the Hsiung-Nu before they moved south and who may have been the proto-Burmese, would have become aware of the griffin.[92] The Tibeto-Burman people of the 3rd to 1st centuries BC who dwelt in Szechwan and Yunnan made bronzes which show features connecting them with the nomads of Ordos, Central Asia and Mongolia.[93] They used the tiger and the mandarin duck as motifs for their bronzes. That the Beikthano Pyu (also Tibeto-Burmans) of the 1st to 4th centuries AD similarly made bronze figurines in the form of a feline and an anserine bird proves only that the proto-Pyu may probably have been familiar with the concept of the combination before they arrived in Burma, either because of the shamanic or the Tien connection. However, the figurines may have been of Indian not Pyu origin.

The effects of the occupation of north and central Burma before and after the 8th century AD by the Nanchao people of Yunnan have not been studied. It is presumed that they must have reinforced the various influences of south-west China on the Mongoloid immigrants to Burma. The conquest of northern and north-eastern Burma by the Mongols in 1287 AD, aided by their Shan allies, finally ended the diminishing Indian artistic influences and led to its replacement by that of the

Mongols, Chinese and Shans, mostly the latter. The Mongols had no animal art but they did have the decimal system. During the 13th century AD more of the animistic and probably Buddhistic Thai (Shan) peoples moved from south China further into Siam, north Burma and north-east India. They dominated north and central Burma until at least 1531 AD. They not only had available to them the culture of Pagan[94] but they also must have had close cultural affinities with the arts and crafts of north Thailand and south-west China, countries they had just left.[95]

So the Bengalis of north-east India, the Shans of south-west China, the Burmese of central Burma and the Pyu and Mons of south Burma could all have contributed to the art, architecture and crafts of Pagan and perhaps introduced their own weight systems.

Treating Burma as one unit, animal forms continued to be employed there until the present time, using, as applicable, bronze, terracotta, wood, especially, stone or brick, to make temple guardians, loom pulleys, weights, pillars of monasteries, prows of boats, betel nut box covers, ornamented cannon, bowls, fabrics, mirror framed and other artifacts.

Notes

[1] The region from the Black Sea to Mongolia, including Central Asia, from before the 8th century BC was occupied by nomadic steppe tribes many culturally and probably ethnically (i.e. they were Indo-Aryans) related to the East Iranians. Many belonged to the Saka group. Those occupying the region north of the Black sea were named Scythians by the 8th century BC Greeks, those west of the Altai mountains were called Saka by the pre-6th century BC Persians and those east of the mountains, for convenience today, are called Saka-Siberian. In Persia related tribes of about the 6th century BC were named Medes and Parthians. On the north-west borders of China and in the Tarim basin region before the 2nd century BC were the Yueh-chieh, also probably an Indo-Aryan people related to the Sakas. They were driven away along with the Sakas by the Hsiung-Nu, a Turki people. Until about the middle of the 1st millenium AD the Tarim basin region region was occupied by the Hsiung-Nu. These were replaced by the Tangut (later Hsi-hsia) Tibetan people, until in the 13th century AD they were overcome by the Mongols, called Tartars by the Europeans.

The Indo-Aryans of about 2000 BC and the Saka group are the most important of the steppe nomads to this work. Firstly, this is because the Indo-Aryan peoples of 2000 BC brought to India in the Vedic religion basic concepts held by the steppe nomads which, together with Indian animism, led to Hinduism and Buddhism. Secondly, it is because tribes of, or related to the Saka group repeatedly invaded India from 2000 BC onwards, so spreading their culture from the kingdoms and republics they established in India and thus leading to the flowering of stone architectural animal art in India from about the 2nd century BC. Something similar may have happened in the small animal metal art field in west China about the 5th to 3rd century BC. The Saka influence on animal art appears to have flowed round both sides of the Tibetan plateau and converged on south-east Asia. This flow of art and people may have led to the foundation of the first Burmese kingdom, Tagaung (See Harvey, 1967, pp. 308, 310: Htin Aung, 1967, pp. 6, 7; Keely and Price, pp. 205, 206, on the foundation of Tagaung by Sakas.) Even today there is a Sak (or Chak) tribe in north-west Burma and neighbouring Bengal. Thirdly, it is because they were related to the 6th century BC Medes and Parthians of Persia, who subsequently became the Persians noted for their land and sea trade, who used the lion and duck weights of west Persia and who had great cultural influence on all the region of interest to this work. Fourthly, it is because the Yueh-chieh (and the Tangut) occupied a region which may have been a place of origin of one of the Burmese peoples. When they were driven away from their homeland by the Hsiung-Nu about the 2nd century BC, part of the tribe moved south towards Burma and part moved west to the northern marches of India, changing their name to Kushan as they did so. There, for about five centuries, they became a great influence on the development of Mahayanist Buddhism and overland trade from the Persian to the Chinese borders. During their migration they drove a part of the Sakas before them and these settled in west India before temporarily extending their sway to the east of India south of the Kushans. Fifthly, it is because important elements of the culture of the steppe nomads can be recognised both in Burma and Central Asia, in the latter case especially in an extensive region around the Altai mountains.

Changing languages, spellings, locations, political situations and destruction or absorption by other tribes have led to confusion in the usage of the names of the steppe nomads. Because of this, the word 'Saka' is found in several forms, e.g. Sakh, Sakka, Sacaea, Sakai, Sai, Saskya, etc., some of which are used in this work.

The persistence of the word 'Saka' in various forms in India and Burma is noteworthy. Sakka is another name of Indra, the Indo-Aryan and Hindu god. Saka is the name of the group of tribes of which the Scythians were one. The Sakas, 'people of the stag', are associated with the animal symbols of the *chakravartin.* Gautama Buddha was the Sakyamuni, the sage of the Sakyas. The Burmese equivalent of Sakka is Thagya who is the chief of the 37 (originally 32 + 1) *nats.* Some scholars have inferred the absorption of those Yueh-chieh, who were driven south from the north-western borders of then China by the Hsiung- Nu about 150 BC, into the Chiang peoples of west China. The Chiang are allied to the Lolo-speaking people who extend into Burma. Other evidences of the connection of the Burmese with the steppe nomads are the used of the numbers 3 and 32, e.g. the three countries of Ramannadesa, each with its thirty-two *myos* (districts); the mythical origins of the kings of Thaton and Pagan from serpent princesses; the legend of the god Indra (Sakka) drawing a circular plan for Pegu instead of the usual square plan of ancient east Asian cities. Apparently there was a similar circular plan for the Pyu city of Sri Ksetra. Several other examples connecting the Burmese peoples and the Sakas occur in the text and notes, e.g. 'The Symbolism of the Animal Weights: Shamanism and the Central Asia Religions'.

[2] Bunker, 1970, p. 105.

[3] Carter, p. 168.

[4] Battisti, p. 939. Yule and Cordier, v. 1, p. 209, quoting Marignolli, 1338-1353 AD, and v. 4, p. 130, quoting Ibn Batuta, circa 1347 AD.

[5] Jettmar, p. 235.

[6] Battisti, p. 946. Elisseeff, p. 1 ff. Bittel, p. 563. Moortgat, pp. 742-755.

[7] Carter, p. 20.

[8] Bussagli, 'Steppe Cultures', p. 375 ff.

[9] Levinson, p. 8. Loehr, pp. 773, 774.

[10] Lemasurier, p. 516.

[11] Irwin, IV, p. 748.

[12] Fagg, p. 815.

[13] Janse, p. 20.

[14] Luce, *The Tan (97-132 AD) and the Ngai-Lao*, p. 202.

[15] Scerrato, p. 721.

[16] Watson, 1975, pp. 39, 52.

[17] Kuhnel, p. 649. Rawson, 1984, pp. 147, 169.

[18] Iyer, pp. 6, 15. Hopkins and Hiltebeitel, p. 217. According to the latter, clay representations of bulls and rams were found most often but also lions and other felines. There were also serpents, tigers, water buffaloes, short-horned bulls, zebu, rhinoceri, elephant, antelope, crocodiles and composites, such as horned tigers and unicorns. The use of clay and the absence of the horse and stag make it evident that in these respects, the effect of the steppes was not yet important.

[19] Iyer, p. 18.

[20] Zimmer (Campbell, ed.), 1982, pp. 94, 95. Funke, pp. 267.

[21] Hallade, 1963, pp. 4, 8.

[22] Battsiti, p. 954.

[23] Fagg, p. 815.

[24] Iyer, pp. 85-87.

[25] Goomaraswamy and Rowlands, pp. 97, 102.

[26] Burrows, p. 139.

[27] Campbell, 1970, p. 289.

[28] Iyer, pp. 87, 88.

[29] Hallade and Lanciotti, p. 488.

[30] Goetz, pp. 24, 25, 32.

[31] Carter, p. 17.

[32] Janse, p. 21. Bussagli, 'Steppe Cultures', pp. 375, 379. Battisti, pp. 951-953.

[33] Jettmar, pp. 140, 219.

[34] Bunker, et al., 1970, pp. 14, 15. Elisseeff, p. 33.

[35] Elisseeff, p. 22 ff.

[36] Griaznov, p. 133. Argan, p. 265.

[37] Elisseeff, pp. 32, 33. Hambis, pp. 815-821. Griaznov displays on page 238 a table which summarizes excellently the history of the animal style art of the steppes between 900 BC and 200 AD.

[38] Elisseeff, p. 33. Hambis, p. 821.

[39] Bussagli, 'Steppe Cultures', p. 394. Bunker, et al., 1970, pp. 105, 106.

[40] Bunker, et al., 1970, pp. 105, 106.

[41] Narain, p. 928.

[42] Bunker, et al., 1970, pp. 83, 108.

[43] Goetz, p. 31.

[44] Carter, p.171.

[45] Janse, p. 20. Rawson, 1984, p. 59.

[46] Carter, p. 171.

[47] Xinru Liu. All the work but especially pp. 88-103.

[48] Kurz, p. 78.

[49] Fontein, p. 487. Elisseeff, pp. 17, 20.

[50] Loehr, 1969, p. 574. Willets, p. 639. Janse, p. 21.

[51] Bussagli, 'Steppe Cultures', p. 381. See also pp. 384-388.

[52] Battisti, p. 954. Watson, 1975, p. 60.

[53] Hallade and Heine-Geldern, p. 45.

[54] Lee, p. 28. Hambis, p. 817. Bunker, et al., 1970, p. 113.

[55] Salmony, p. 52.

[56] Rice, p. 174. Lion-Goldschmidt, p. 120. Watson, 1975, pp. 39, 52.

[57] Gyllensvard, *Metalwork*, p. 820.

[58] Fontein, p. 497.

[59] Shabad, p. 568. Coolidge, p. 652.

[60] Fry, p. 27.

[61] Fontein, p. 514.

[62] Fontein, p. 558.

[63] Schmidt, p. 19.

[64] Bussagli, 'Steppe Cultures', pp. 376, 405.

[65] Denwood, pp. 223-227. Snellgrove, p. 1117.

[66] Rawson, 1983, p. 9.

[67] Bunker, 1967, p. 325.

[68] Mackenzie, p. 96.

[69] Soper, p. 67.

[70] Mackenzie, pp. 90, 91.

[71] Bunker, et al., 1970, p. 113.

[72] Watson, 1975, p. 60. Rawson, 1983, pp. 14, 22, 23, 221.

[73] Bunker, 1967, pp. 299, 301. Rawson, 1983, pp. 17, 23, 238.

[74] Goetz, p. 22. Bunker, 1967, p. 318.

[75] Hallade and Heinz-Geldern, pp. 45-48.

[76] Watson, 1975, p. 63.

[77] Bunker, et al., 1970, p. 113.

[78] Rawson, 1983, pp. 9, 20.

[79] Bunker, 1967, pp. 299, 301, 307, 325. Watson, 1975, p. 61. Janse, p. 20.

[80] Fontein, p. 408.

[81] Watson, 1975, p. 52.

[82] Bussagli, 'Metalwork', p. 820. Fontein, pp. 407, 408. Bussagli, 'Monstrous and Imaginary', p. 268.

[83] Goetz, p. 22.

[84] Hallade and Heinz-Gelder, pp. 45-48. Coedes, 1964, p. 263.

[85] Elisseeff, p. 28. Bussagli, 'Farther India', p. 917. Bezacier, 'Vietnamese Art', p. 779.

[86] Wright, 1979, p. 389.

[87] Bussagli, 'Farther India', p. 917.

[88] Coedes, 1964, pp. 31, 87. Dohanian, p. 91.

[89] Luce, 'The Ancient Pyui', p. 314. Aung Thaw, 1968, p. 55.

[90] Luce, 'The Economic Life...', p. 325. Coedes, 'Burmese Art', pp. 742, 748. Goetz, pp. 29-35.

[91] Elisseeff, p. 33. Hambis, p. 821.

[92] Griaznov, p. 133. Argan, p. 265.

[93] Rawson, 1983, p. 20.

[94] Wyatt, p. 42.

[95] Coedes, 'Burma', pp. 733, 739.

8
The Weight Bird: Identity and Origins

Names:

The bird shapes on the weights have been given the following names by modern writers and dealers. Much the most common are *hamsa*, *hansa* or *hintha*[1] which are Sanskrit and Pali names for geese in general and in particular, the bar-headed goose, *anser indicus*.[2] The Group 4 bird weights are often called Mon ducks. Less common is the name *kalawaik*[3] which refers to a bird shape of Groups 1 and 2, which is similar in all respects to the *hamsa* except that the beak is pointed and curved down and the tail section is in the form of an inverted V. The *hamsa* beak has a rounded tip and curves neither down nor up. The tail section is in the form of a wide-angled, inverted U. A third, very different bird shape is that on the Group 6 weights of which the writers know only the English name, cock.[4] A fourth different bird shape is that of Group 7 which is also called a *hamsa*. See Plates 44 and 45.

Anserine (duck, goose, swan) shapes:

The *hamsa* style of the Group 1 bird weights was recognised in 1917 as a mandarin duck (*aix galericulata*),[5] a bird which occurs in north Vietnam, east and north Chiña, but not in Burma.[6] With minor variations, the same major style is found in Groups 1 to 5 of the bird weights, so that these also may be inferred to be mandarin duck representations. Group 5 with its three overlapping forehead knobs and taller appearance may have been partly modelled on a knob-billed goose. The Group 7 bird weights are of a quite different style. In particular, the tail is vertically spread and is floriated or scrolled. In addition, the head bears a crest and a knob and the wings are convex in cross-section. This style can be matched on Pagan friezes[7] and appeared on Indian stone panels at about the same time as those at Pagan.[8] On the pre-Asokan pillar abaci of north-east India, the bird can be identified as the goose (*anser indicus*).

The *feng huang*:

The *feng huang* is a mythical bird at first portrayed like an Argus pheasant. Later it appears as a hybrid between the pheasant and the peacock. Like the chicken these are gallinaceous birds. There are several varieties, of which one is illustrated on Plate 45, number 4.

The descriptions of the bird indicate that it had the head of a pheasant or cock with the forehead of a crane, the comb of a cock and the neck ruff of a mandarin duck.

Below its beak were beard-like feathery tufts. Its back was domed like that of a tortoise or crescent moon. Its tail had 12 or 13 feathers of five colours.

Illustrations show that the *feng huang* has a cock-like head with a pointed beak which is straight to somewhat curved downwards. It bears a comb with three or more elongated knobs. It may or may not have the near-vertical forehead characterising some species of crane. The neck ruff may be replaced or obscured by wing-like tufts. The beard-like tufts may or may not be present. The back, when visible, is domed but little more than in other birds. The tail usually has between one and seven separated feathers as long or longer than the body and head and often several shorter ones. The long ones often bear eyes like those of a peacock. The tail usually appears like that of a partially spread or fully closed peacock's tail. The wings when closed or open are pointed at the tips.

The cock-like head would be appropriate to either the Group 6 weights or the Groups 1 and 2 V-tailed weights. Lack of a neck ruff, lack of prominent head knobs and the presence of a vertically spread tail in an Indian style on the Group 6 weights suggest that these weights may not have been intended to represent the *feng huang.* Those bird weights which have a markedly abrupt forehead arising from the beak, (crane forehead) three prominent head knobs, a back-cover suggesting the crescent-like back, a mouth appendage and a laterally spread tail, share characteristics which are similar to those of the *feng huang.* The Groups 1 and 2 V-tailed weights, the *kalawaiks,* have all of these characteristics except the mouth appendage. The *feng huang* made delightful music with but five notes. So did the Emperor Asoka's wife who was named Kalavika after the bird of the same name which sang sweetly.[9]

Chicken-like weights:

The Group 6 bird shape looks like a fowl, i.e. a jungle fowl (*gallus gallus*) or domestic fowl (*gallus domesticus*). It has a crest and a floriated (scrolled) tail shape similar to the Group 7 bird shape and the wings have a convex cross-section. Unlike all the other bird shapes, the Group 6 weight is in a broody attitude with its legs invisible, as if sitting on eggs. Sometimes it is accompanied by one or two small birds, possibly its young, or birds of a different kind. It seems that the bird shape m;ay represent a jungle fowl hen or a special form of female *hamsa.* However, no references to ancient art works in the shape of hens have been located nor would they be expected in Buddhist symbolism. It is possible that the shape is that of a jungle fowl cock,[10] sometimes accompanied by one or two other small birds, but no similar representations have been seen from any other country. Moreover, all other representations of fowl-like birds from countries adjacent to Burma do not have the floriated tail. Instead they have upward-arcuate, drooping or trailing tail feathers. Also they are always standing with visible legs, except when small, unlike the Group 6 weights.

On this evidence and knowing that the shape is found in Buddhist countries, i.e. Burma and Thailand, it seems that the bird may be a special style of *hamsa.* Alternatively, it may be a different kind of bird, e.g. a hen, associated with different symbolism. From its crest and tail shapes the bird representation was made after 1200 AD and from its non-mandarin duck style, it was made before the Group 5 weights. From the relative abundance of the fowl-like forms as figurines, on coins, etc., in the southern parts of south-east Asia, it may be of Mon/Khmer origin.[11] However, in the unlikely event that the Group 6 weight is a Burmese or Siamese style of *feng huang,* then it is of Chinese ancestry.

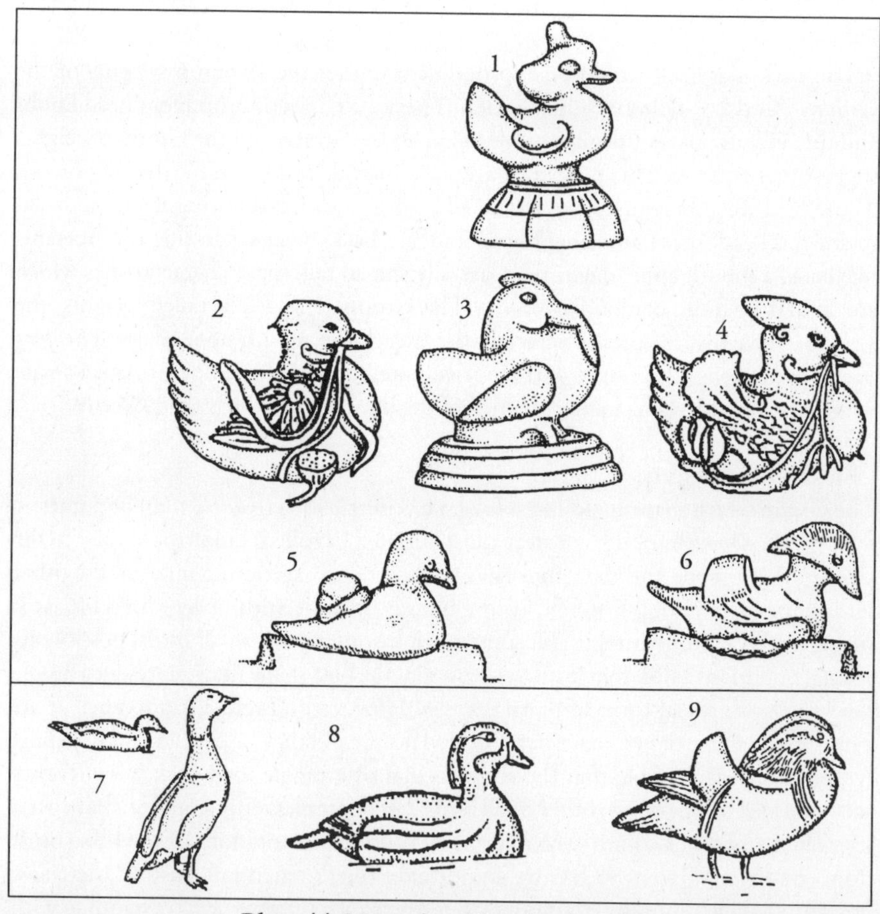

Plate 44: Natural and Mythical Birds

The origins of the weight bird shapes:

The first known anserine bird (swan,goose, duck) figurine in Burma was found during excavations at Beikthano,[12] a Pyu city which existed from about the 1st to 4th century AD. Along with them were sherds decorated with birds including anserines. At Sri Ksetra, near Prome, was another Pyu city dated to between the 5th and 8th century AD, where there were found caskets of gold and silver with Brahminy ducks adorning the covers. At Pagan some friezes on the temples, dated to between the 11th and 13th century AD, also show anserine birds. The Group 7 bird weight is very similar to some of these. All of these bird shapes are Indian in type or influence.[13]

In India anserine birds are referred to in the *Rigveda*, compiled about 500 BC. Perhaps most important to this work is their occurrence on the abaci of several of the lion-capped pillars in India, usually called Asokan (3rd century BC) but probably two or three centuries older. Anserine birds became common architectural motifs in Hindu and Buddhist temples from about the 2nd century BC and continued so, becoming important decorations in north-east India, the last Indian Buddhist region, between about 900 and 1200 AD. From about the 4th century AD the birds begin to display representations of strings of pearls hanging from their beaks, a Persian motif, which later became converted into vegetational representations. These may be predecessors of some of the mouth appendages on the bird weights of Groups 7, 5, 4 and 3. From about the 6th-8th century AD the tail became progressively stylized, a feature displayed on the Groups 7 and 6 bird weights. In approximately the 12th

---

PLATE 44:

Comparison of weight-birds and mandarin 'duck' representations

1. Weight-bird. *Hamsa.* Group 2, U-tail. Period C. Burma. Late 18 - early 19 c. AD. Bronze. Gear.
2. Mandarin duck. Female. China. No date. Painting. Eberhard, p. 177.
3. Weight-bird. *Hamsa.* Group 4. No date. Period J. Burma. Mid-16c. AD. Bronze weight. Gear.
4. Mandarin 'duck'. Male. China. No date. Painting. Eberhard, p. 177.
5. Mandarin duck. Female. China, Yunnan. 1 c. BC. On bronze halberd. Rawson, 1983, plate 127.
6. Mandarin 'duck'. Male. China, Yunnan. 1 c. BC. On bronze halberd. Rawson, 1983, plate 127.

Natural birds supposedly providing the models for the weight birds

7. Ruddy shelduck (*Tadorne* - or *casarca* - *ferruginea*). Male and female. North India, Burma, West China. Natural. Austin, p. 69, and Brown*et al.*, p. 237.
8. Mandarin duck (Aix galericulata). Female. East China. Natural. Austin, p. 297.
9. Mandarin 'duck'. Male. (As for figure 8).

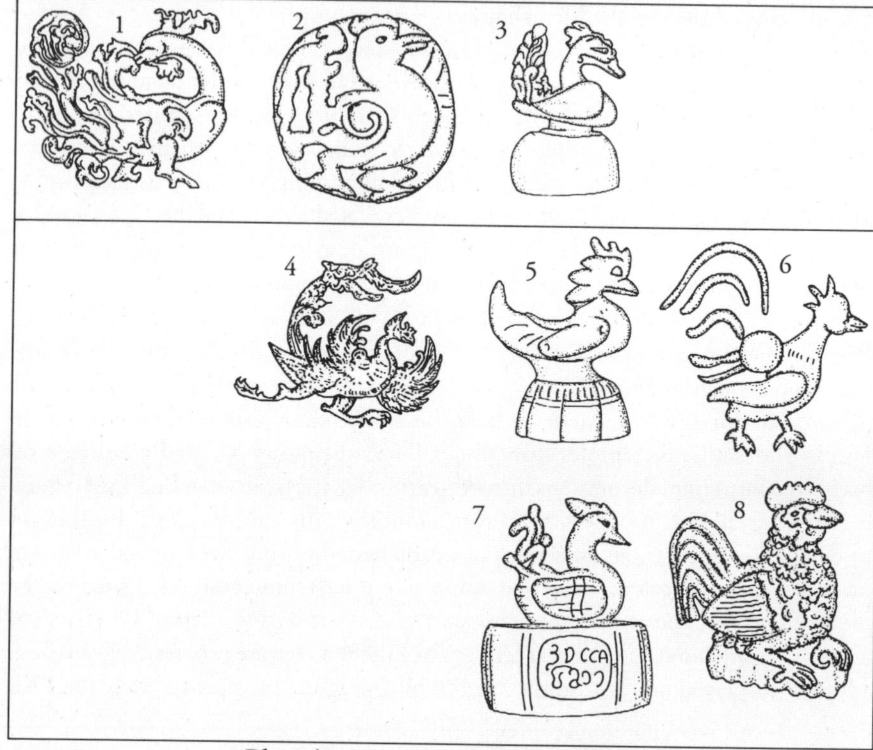

Plate 45: Natural and Mythical Birds

Comparison of older Indian and Burmese representations of the *hamsa*

1. *Hamsa.* India. 8 c. AD. Stone carving. Kramrisch, plate 67.
2. *Hamsa.* Burma. 11 c. AD. Stone carving. Nanpaya temple, Pagan. Professor Cooler, personal communication.
3. Weight-bird. *Hamsa.* Group 7. Period M. 14 c. - 16 c. AD. Bronze. Gear.

Comparison of fowl-like weight-birds and similar shapes

4. *Feng huang.* China. No date. From cover of Percival David Foundation of Chinese Art publications.
5. Weight-bird. *Hamsa* (?). Group 2, V-tail. Period C. Burma. Late 18 c. - early 19 c. AD. Bronze. Gear.
6. Fowl. Male. Tenasserim, Burma. 19 c. AD (?). Coin. Robinson, plate 13.
7. Weight-bird. Fowl? *Hamsa*? Female? Group 6. Period KL. 14 c. - 16 c. AD Bronze. Gear.
8. Fowl. Cock? Java. Pre-7 c. AD. Stone statue. Evans, p. 87.

century AD the first crest on an anserine bird appears. Crests appear on all of the Burmese bird weights. Between 1300 and 1500 AD anserine birds cease to be used because of the increasing Mahometan influence.[14]

In China ducks long have been popular motifs, especially the mandarin duck. The first representations occasionally appear on bronzes[15] but not jade[16] between 1500 and 1100 BC. By 500 BC the Chinese name for the mandarin duck, *yuan-yang*, formed the chapter heading of a book.[17] Bronze ducks with snakes were created by the Yunnanese about the 2nd to 1st century BC and they appear also on their battle axes.[18] By 600 to 900 AD mandarin drakes were common on mirrors[19] and almost every poet of the time wrote about them.[20] They have remained part of Chinese folklore and decoration until today.[21]

The anserine birds are known in Central Asian art from the 5th century BC or before, as at Pazyryk[22] and Zakis,[23] and continued to be fashioned for many centuries thereafter, as for example at Kashgar and Soghdia between the 6th and 8th century AD.[24]

In Mesopotamia and Persia the duck appears on pottery from Susa about 3000 BC.[25] It appears as a stone weight of Sumeria, Assyria and Babylonia from the middle of the 2nd millenium BC until about the 5th century BC. As a weight it also appeared in Egypt and adjacent countries at about the same time and disappeared from them similarly, except in Persia.[26] The duck continued as a popular motif in the art of Persia until modern times. As an example, bronze mandarin duck incense burners were being made in Iran in the post Sassanian period (post 637 AD) which are very similar to those on betel nut box covers made in Burma in the 19th century.These latter may have followed traditional forms. Ivory figurines in the style of the Egyptian and Assyrian duck weights were being made in Persia in the 17th century AD. Even today at Qum marble and alabaster ornaments are carved in the same shapes as the duck weights of ancient Susa.[27] (See Plate 46)

The Chinese *feng huang* appears to be one of a very ancient group of similar mythical birds like the Egyptian phoenix and the Arabian *roc*. Possibly in existence by the 3rd millenium BC, it is mentioned in Chinese inscriptions dated to the end of the 2nd millenium BC. It is referred to by Confucius (551-479 BC) and was an object of worship between 200 BC and 200 AD. Since then representations have become , to be made in quantity even in modern times.

Conclusions:

All the bird shapes represent ducks or drakes with the possible exceptions of Groups 1 and 2v, which may have been modelled on the *feng huang*, Group 5, which could represent gander or a goose, and Group 6. Group 6 may be a hen in a Persian/Indian style or duck in a Mon/Indian style. Groups 1 to 5 clearly show Chinese influence. Groups 6 and 7 are equally clearly Indian influenced.

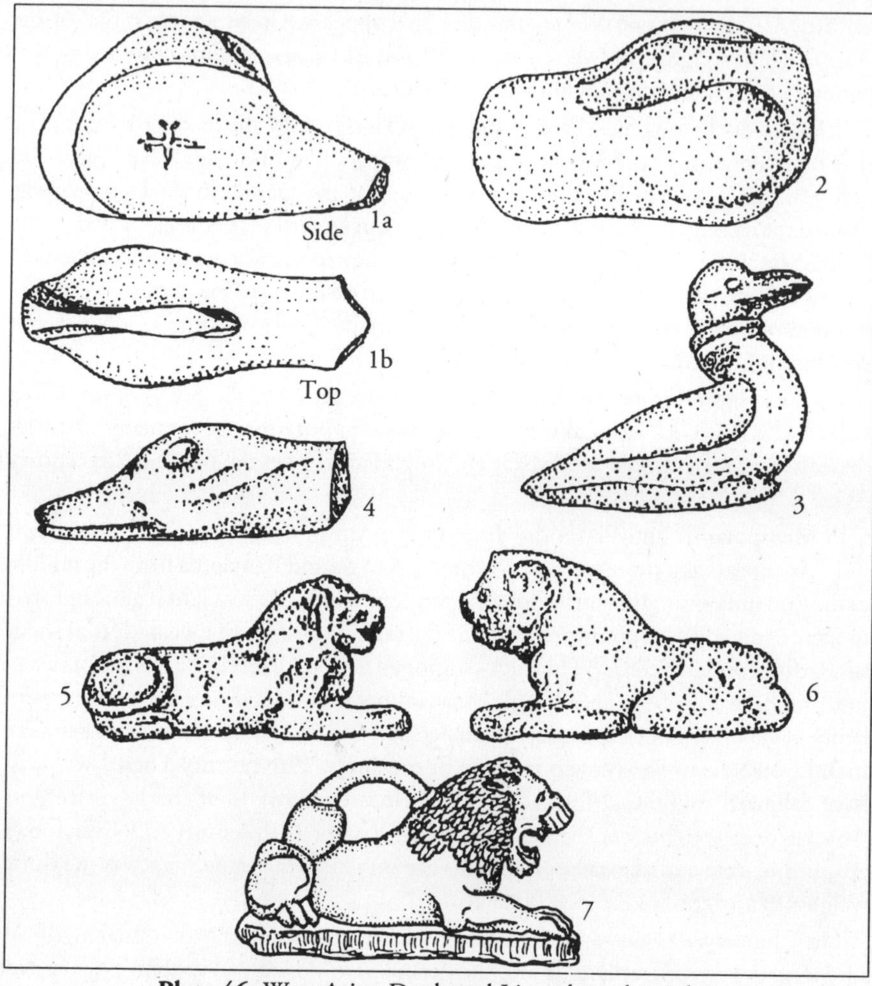

Plate 46: West Asian Duck and Lion-shaped weights

Drawings adapted from:

1a, 1b, and 7	Nineveh and Assyria	Ridgway, p. 245
2	Babylon	Berriman, p. 62
3, 4, 5, 6	Egypt	Petrie, Plate IX

Notes

1 Local dealers and Temple, 1898, pp. 141-143. Temple also quotes the name *ziweso*, the Burmese name of the swallow which builds the edible birds' nests along the coast of south Burma and elsewhere. In Laos Sale, p. 105, named the *hamsa,* the *to fret.* In Siam Gardner, p.1. quoted *suwanna-hong* and *hastalinga* as names for two different styles of *hamsa*, the latter referring to the bird weights of Groups 3 and 5 which have rod-like mouth appendages. In Burma the name *hintha-pungay* was once heard for the *hastalinga*, the *pungay* referring to something being carried in the mouth. Braun referred to the Group 4 weights as sleeping duck weights, perhaps a name used in present day Thailand by the dealers, but also used to describe many of the ancient Mesopotamian duck weights.

2 Salim Ali, p. 108, particularizes the bar-headed goose (*anser indicus*) as the *hamsa.* Smythies, p 476, quotes the name *hintha* for the ruddy shelduck (see note 5) while Salim Ali, p. 109, gives *chakwa*, *chakwi* or *surkhab.*

3 Local dealers and U Thaw Bita, a former monk of Sagaing who studied Burmese history and was interested in the weights. Also U Tun Yee, a collector of Burmese weights at Loikaw.

4 Local dealers.

5 Annandale, p. 197. Temple, 1898, pp. 141-143, incorrectly identified the weight bird as a Brahminy duck, a ruddy shelduck (*tadorne* or *casarca ferruginea*) which is native to and common in India, Burma and Yunnan and which bears the name *hintha* (see note 2).

6 Smythies, p. 470.

7 Professor Cooler, North Illinois University, who supplied a photograph of one taken at Nanpaya Temple, Myinkaba village, Pagan.

8 Gairola, plates 17 and 18.

9 Liebert, p. 118, translates *kalavikarani* as free of limits, i.e. undivided. It is also an epithet of Durga. *Kalavikarnuka* is the name of a propitious goddess. von Manfred translates the Sanskrit word *karavinka* as the Indian cuckoo. Liebert, p. 139, translates *kokila* as the Indian cuckoo. There are several Indian cuckoos but according to Salim Ali, p. 50, only the pied, crested cuckoo (*clamator jacobinus*) has both a crest (one-knobbed) and a down-turned bill like the weight. The bird is supposed to have a sweet note, a musical note. Lessing, p. 118, refers to the connection of the Chinese mythical bird, the *feng huang* , with sweet music. According to Kosambi, p. 128, the Buddhist patron, the Indian emperor Asoka of about 274 to 232 BC, had a wife named Kaluvaki (sweet-voiced). Both the *kalawaik* weight-bird and *feng huang* commonly have three knobs on their crests. Davidson, p. 310, note 19, comments on relationships which connect the archaic Chinese *klwa* with the Mon *klao*, terms which include the meaning 'a large tortoise'. In the Chinese system of direction animals, the tortoise rules the north. Ruling the south is the *feng huang* . If *kalawaik* is connected through *klwa* or *klao* with the meaning 'opposite the north', the name may imply the *feng huang* .

P. Thomas, p. 92, referring to the mythical origins of sparrows and partridges, relates a legend in which a babbling, drunken man was likened to a sparrow, i.e. a *kalavinka.* Williams, 1932 (reprint 1976), p. 442, shows a table of Chinese characters in which opposite the word 'sparrow' is a diagram of a bird closely resembling the bird sign on the base of the *kalawaik* weights. It is possible then that the *kalawaik* weights are so named because of their sparrow-like sign.

Birds, mainly waterbirds, which have similar names to *kalawaik*, are as follows:

Burma	*Kalukwet*	Coot	Scott, 1906, p.489
Burma	*Kalugwet*	Waterhen	Scott, 1906, p. 491
Burma	*Kalagat*	Cotton teal	Smythies, p. 469
India	*Celavaka*	?	Cowell, vol. 5, p. 219*
India	*Cakravaka*	Brahminy duck	Lessing, p. 143†

* *Jataka* 536.An untranslatable bird name from a tale about a royal cuckoo.
† Likened to the mandarin duck symbolism.

The word *kalawaik* may also be associated with speed. Shway Yoe, p. 449, identifies the *kalawaik* as the *garuda* (Burmese *galon*, Cambodian *krut*). He also identifies it as the Burmese crane (p. 163). The *garuda* is Vishnu's vehicle, his messenger, originating from the eagle (Iyer, pp. 51, 55). On p. 365, Scott, 1910, refers to the express messenger boats which carried a representation of the *kalawaik* on their prows. The present writers were told in Burma that the *kalawaik* was a bird which flew so swiftly that if four archers shot their arrows into the sky simultaneously then the bird would fly around and catch them in its beak one by one. Smythies (*Birds of Burma*) does not mention the *kalawaik* and neither the crane nor the eagle can be regarded as being similar to the *kalawaik* or Group 6 bird shapes. Hall, p. 920, refers to a Mon king of 942 AD who bore the name of Karawika. This is also the name of a mountain in Sidantara (Cowell, vol. 6, p. 66). *Jataka* 31, bearing the name Kulavaka, refers to compassionate elephants and a crane.

Some of the bird shapes on the Pagan friezes might provide a clue to the identity of the *kalawaik*, e.g. Luce, 'Smaller Temples of Pagan', plate VIII.

10 Concerning the cock, in the peninsular south-east Asia and Indonesian region, birds of cock-like appearance are found as motifs on the bronze drums of Dong S'on (4th century BC to 1st century AD) of north Vietnam; as pre-15th century Javanese stone statuary (Evans, p. 87) and on temple medallions; on 16th or 18th century(?) coins of Lower Burma; on Cambodian coins of 17th to 19th century AD (where they are known as *hamsas* or *krut* (Panish, plate xxxvi); on the ancient tin gambar weights of Malaya (Pridmore, 1970 and 1972; Temple, 1914, p. 38 and plate III) and on bronze images of the 12 year animal cycle common in Thailand, Laos, Cambodia and perhaps elsewhere (Braun, illustrations 249-401; Sale, pp. 103, 105; Gardner, p. 3; Forien de Rochesnard, pp. 30, 34, 35). Some of these Thai cock-like birds may be of a similar age to the Group 6 weights. Loom pulleys from Burma (Lowry, plate 50) are also similar.

Cock or cock-like representations are common in the metal and other material animal art of Persia from before 600 BC (Bussagli, 'Steppe Cultures', p. 398). They occur in the bronze animal art of the Huns of the eastern steppes after the 4th century BC (Elisseeff, p. 33). They occur in India, whence they probably originated, and the Indianised regions in connection with the worship of Siva, perhaps from the early centuries BC (Herrmann, pp. 830, 831). In China representations appear to be mainly on paintings though they also occur on roof tiles (Ball, p. 225 ff). However, it is in Persian art that the cock appears to be most frequently represented because it was the guardian of Sraosha, the Lord of Obedience, in ancient Persian Zoroastrianism. It symbolised the dawn, the enlightening and the praising of the glorious one. The hen is considered as assisting the cock in his pious tasks (Heissig, p. 402. Goodrich, 1957, p. 64. Gray, pp. 694 – 698). Persians played an important part throughout the East and were employed by the Siamese as governors of ports. It seems possible that the hen-like shape could have been employed on the Group 6 bird weights because of Persian influence and the similarity of the relevant Burmese (?) and Persian symbolisms. It is certain that some Persian 9th century AD mandarin duck effigies closely resemble those of 19th century AD Burma, so that this suggestion is not unrealistic.

11 Apparently fowl-like weights or figurines are more frequent in Thailand than in Burma, judging by the numbers illustrated by other investigators (Braun, Fraser-Lu, Forien de Rochesnard) compared with the numbers found by the writers during their years in Burma. Moreover, the impression is gained that representations of this bird occur most commonly in the southern regions of south-east Asia, i.e. Cambodia, Malaya, Java, Thailand, southern Burma, suggesting that the large bird on the Group 6 weight may be of Mon or Mon /Khmer origin.

12 Aung Thaw, pp. 17, 55 and figure 84.

13 Luce, 'The Ancient Pyu', pp. 308, 314.

14 Derived mainly from Gairola, but see Iyer, pp. 83, 182, and Irwin, I, p. 709 and III, p. 642.

15 Lee, figure 23.

16 Hansford, p. 841.

17 Professor D. Pollard, School of Oriental & African Studies, London and T.C. Chiang, Embassy of Republic of China, Pretoria.

18 Watson, 1973, pp. 177, 181.

19 Lanciotti and Herrmann, p. 833.

20 R. Dunn, Chinese Area Specialist, Library of Congress, Washington, D.C., USA.

21 Rawson, 1984, pp. 140, 175.

22 Bussagli, 'Steppe Cultures', plate 189.

23 Rice, p. 22.

24 Rice, pp. 110, 192.

25 Armstrong, pp. 37, 38.

26 Petrie, 1926, pp. 6, 7.

27 Pope and Ackerman, pp. 304, 306.

9
The Weight Beast: Identity and Origins

Names:

Common modern names for the beast shapes on the weights are *chinthe* and *to*, or occasionally *to-naya.*[1] In 1827 the weight-beast shapes were referred to as griffins and cows.[2] The writers were unable to find a person in Burma who could differentiate between a *to* and a *chinthe*, hence the following definitions are based on observation.[3] The weight-beast shapes are illustrated on the plates "Natural and Mythical Beasts", set against the mythical beast shapes to which they are similar. (See Plates 48 and 49)

The *chinthe:*

The main differences between the *chinthes*[4] and the beasts on the weights are that the *chinthe* sits, the weight-beast crouches. The *chinthe* has no horns, the weight-beast has two. The whole of the *chinthe's* tail is erected vertically or wrapped around its rump on the ground. The weight-beast's tail, after the initial characteristic rise of the tail base, is pendent except for the rare Group 7 weights. The *chinthe* has feet with claws, the weight-beast has hooves. The *chinthe* shape mainly is employed as guardian of temples and palaces, the weight-beast shape is never so employed. Though both creatures may occur on Burma's 19th century AD coins and thrones, the *chinthe* never appears on the weights. *Chinthes* similar to that illustrated occur on carved stone panels from 12th century Pagan. The *chinthe* is clearly a stylized lion.

The *to:*

Two similar descriptions of the *to* are, firstly, a four-legged beast with branching horns, the beast being part horse, part deer and part elephant; secondly, a cross between a deer and a horse.[5] A third description of the *to* is that it is a four-legged dragon with the head of a stag.[6] While the component parts of the *to* seem evident, the manner of their combination is not.

The *naya:*

This creature has been described as a four-legged dragon.[7]

The *to-naya:*

According to one writer the *to-naya* is a fleecy-coated unicorn.[8] Another writer describes it as having the head, four legs and feet of the *to* and a *naga* (snake or dragon[9]) body. It is often depicted in a prancing attitude. Elongated, scaly, horned and maned figures similar to this description sometimes form vase handles and the wooden frames of mirrors. Such have been tentatively identified as *to-nayas.* From photographs of *to-nayas,* with and without wings, on old Burmese monastery pillars, it is evident that the *to-naya* is the Burmese form of the Chinese dragon. Of especial interest is that they bear well-developed two-tined horns.[10] The *to-naya* has also been described as a water elephant.[11] However, this is a likely description for the Indian *makara* and the Burmese *pancama-rupa.* The *pancama-rupa* is a creature with the head of a *to,* an elephant's trunk and tusks, a *naga* body and a fish tail. The *to-naya* appears to be an elongated *to,* i.e. a horned dragon.

Felines:

As a motif the use of the lion has extended from ancient Rome and Greece to Sogdiana, Bactria and Kushania to India. North and east of this region the tiger motif was employed rather than the lion motif until the spread of Buddhism caused its popularity to wane in favour of the lion. Representations of the lion are widely spread in Eurasia, commencing with those of Sumeria and Egypt of about 3000 BC[12]. In Egypt, which had a lion and sun cult, there arose the lion and ball motif that later appeared in China.[13] Lion representations are known from Persia, beginning just after 3000 BC[14] where the solar symbolism is marked. Even today the lion is the emblematic animal of Iran.

The lion is among the figurines created by the people of the Indus valley civilisations about 2000 BC, though it does not become a frequently used motif, at least in durable material, until it was adopted by the Buddhists about the 3rd century BC when it looks distinctly west Asian or perhaps Persian. The adoption of the lion motif instead of the tiger may have been because the Buddhists needed a royal symbol without the ferocious or steppe nomad association of the tiger, the emblem of the warrior caste into which Gautama Buddha was born.[15] In Hinduism, Durga, who represents the sun, is shown with the lion or the tiger for her vehicle.[16] The stone-shafted pillars of India, usually referred to as Asokan pillars, can be separated into two age groups: pre-3rd century BC and later. The early pillars bear, or bore, on their tops copper gilt images of the lion, the bull and the elephant. Of these the lion image is by far the most frequent. It is also the youngest, replacing the bull and elephant images. It occurs in the region formerly occupied by the republican,warlike Licchavis and later by the Nandas. In style the images show the influence of the Anatolian Hittites (20th century - 8th century BC), as do those of the south Chinese lions of the 2nd century BC - 6th century AD.[17] The Indian lion representation gradually changed its form, partly because most of the sculptors probably had never seen a lion, which was rare in India compared with west Asia and which today exists only in west India, and partly because it was intended to represent the broadcasting of a spiritual message.[18] By the 11th century AD its shape had become unrealistic, humanoid and subsequently became increasingly so.[19]

Lion motifs had reached the eastern steppes at the latest by the 4th century BC[20] but probably before 1000 BC. The main feline motif here was the tiger, which was employed long before the lion. It had a generally similar symbolism especially that of ferocity.[21] The lion shape grew in popularity along with Buddhism throughout Central Asia from Bactria to Mongolia and beyond. The steppe region was particularly important in the spread of the lion and other motifs because the main east-west routes traversed it, routes which were followed by migrants, traders, missionaries and armies. Moreover, from earliest times the nomads had been important agents in the distribution of small portable objects and their motifs, these perhaps being the only source of information concerning the lion available to the craftsmen of the steppes and China.[22]

The tiger was the dominant feline motif in China,[23] even though the lion shape reached the country about the 3rd century BC,[24] where it was used on tomb guardians and roofs as protection from demons.[25] The tiger remained dominant in places, as for example, in the case of the bronze tiger figurines which frequently occur on Yunnanese artifacts of the 3rd to 1st century BC.From about the 3rd to 6th century AD the stone, horned and non-horned lion, somewhat tiger-like, statues of south China were created, possibly influenced by the motifs journey along the Sarmatian route north of the Gobi desert. Even after over 1000 years of travel, these creatures still bear a strong resemblance to the Hittite lions of the 13th century BC and the Aramean lions of the 10th century BC. This type of lion continued to be used in south China until the fall of the southern dynasties, about the 6th century AD.

It was about this time that the lion dance originated, in which fanciful bamboo and paper effigies of lions were employed, the dance being supposedly to celebrate the arrival of peace.[26] In Burma it appears to be danced at the Sandawgyi festival. In much of China the lion shape had become more popular as Buddhism increased in popularity but it had been modified according to the type of artifact on which it appeared. The large temple and crop-guarding statues[27] lost their appearance of litheness and activity and became meeker, more passive, tamer.[28] One of this lion's names was 'civilised tiger'. By about the 10th century AD the lion shapes were displaying distinctly Tibetan and sometimes even playful characteristics. The smaller representations often were modified to conform with the *ch'i-lin* since they also commenced to show both a marked spine and flames emerging from the legs. Wood carvings of *chinthes* in Burma often show the latter feature. Moreover, the two sexes were mixed and the lion assumed a dog-like form. After about the 14th century AD three-dimensional lion representations were no longer popular. On paintings however they were often depicted against a background of earth and rock, so symbolising the lion's role as an earth god.[29]

The earliest known lion-like bronze figurine in Burma was excavated from Beikthano (1st - 4th century AD). Couchant lions in stucco were also found.[30] At Sri Ksetra (5th - 8th century AD), further south near Prome, wall panels decorated with the heads of lions were found as were gilded plates with embossed lion's heads.[31] Stylized lions without horns are known from Pagan.[32]

Antlers, horns:

In some classes of Groups 2 and 6 the horns of the weights bear a groove or rarely, a small tine on the lower front. In other groups a marked terminal knob is often present. The horns on the weights which are conical and lack these features could be those of a bovine, in particular, a bull.

There is a variety of deer, the *muntjac* or barking deer,[33] which is characterised by two-branched antlers, both branches occurring above a marked knob or burr which itself caps a long pedicel extending down to the skull. Of the two tines, one is short and directed forwards. One of the two species of this deer is distributed throughout India, south-east Asia and south China and another throughout Tibet and China. Antlers and horns were modelled in China about 3000 to 4000 years ago on the horned creature shapes which formed ritual bronze vessels.[34] Some of these representations, especially those of the dragon and t'ao-tieh masks, have horns tipped with knobs, the 'button-horn' motif. This type is characteristic of an animal known in Mongolia as a muntjac 'sheep' (deer).[35] (See also below, "The to-naya")

The deer was represented over the entirety of Eurasia, from ancient Egypt to ancient China.[36] There were also stag-horned effigies of deities, kings, priests, warriors and composite animals symbolising the sun, tomb guardians and the like.

Stags appear to be particularly important in the art and myths of Central Asia, the steppes and Siberia from 1500 BC and earlier.[37] The stag was especially used by the Altaian Saka people. It was one of the three main animals represented in their art, the others being the horse and the tiger. The use of antlers in the region remained common until about the 2nd century BC.

In the Indus valley civilisations of about 2000 BC stag-horns were emplaced on composite creations. Portrayals show that Agni was horned.Agni was one of the chief gods of the Aryan invasion from the steppes of about 1000 BC.[38] Stag representations are not found again until the last two centuries BC but then seem to be replaced mainly by ruminants with unbranched horns or does. Except in the Upper Punjab, the deer was not worshipped in India. The Punjabi worship was derived from the Sakas, 'the people of the stag', from near Lake Balkash, who ruled various parts of northern India from about the 1st century BC to the 4th century AD.[39]

Stag antlers also were widely used in China well before 1500 BC. The antler and long tongue combination[40] seen on the weights was in use on tomb guardians about the 6th century BC, especially in south China.[41] Antlers were popular motifs between the 7th and 3rd century BC among the Chu in the region centred on Hupeh (Hubei). Horns (or antlers?) appear between the 6th and 2nd centuries BC on the sculpted stone felines of south and south-west China.[42] The horns are directed upwards and though they are all broken off apparently, it seems that they may have been somewhat long and possibly tined. Some similar jade carvings show knobs at the ends of the horns.[43] Branched antlers appear on bronze stags made in Yunnan from about the 3rd to 1st century BC. From the 2nd century BC to the 3rd century AD the sacred deer, the *tian-lu*, was created , which in its early stages bore one or two usually unbranched horns. After about the 5th century AD it bore two horns, became more feline-like and was given the name *bixie.*[44] Many-branched antlers on sculpted felines were mainly abandoned after about the 6th century AD, being replaced by the two-tined horns of the muntjac. These continued in use on tomb guardians and were even employed in the Buddhist Manjusri cult until the 9th to 10th century AD.[45]

From China representations of stags and does were carried south to form important motifs on the bronze drums of Dong S'on of north Vietnam during the period from about the 4th century BC to the 1st century AD. In south-east Asia the ancient Malayans used a barking deer (muntjac) representation on some of their coins.[46] The Burmese would have been familiar with the muntjac and so able to use its antlers as a model.

Weights in the forms of bulls were made in ancient Egypt,[47] and Bactria. Bulls were inscribed on the bag-like weights of ancient Afghanistan. There is a doubtful record of one in medieval India.[48]

The bull formed part of the religious cults of the Egyptians, the Mesopotamians, the Hittites and the ancient Persians. The bull was revered in the Indus valley civilisations of about 3000 to 2000 BC. Nothing remains of Indian bull representations until they appear infrequently capping pillars in north-east India at some time before the 3rd century BC. Similar bull-capped pillars were constructed in Persia about the 5th century BC. There are frequent references to the bull in the Vedic literature dated about 1200 to 600 BC. In fact, no other animal so much concerns the Aryan writers of these works. About the 2nd century BC in the north-west part of India the bull, formerly essentially the symbol of Indra, became associated with Siva. Henceforth the Sivan bull, Nandi, is represented all over India but becomes confined to south India after about the 12th century AD.[49]

Bull representations are rare in Central Asian, Chinese and south-east Asian art.

Since the beast weights, firstly are clearly Chinese in style, secondly, the Chinese rarely portrayed the bull and thirdly the horns are, in some part, modelled on those of the *muntjac* deer, which the Chinese often portrayed, then there is little doubt concerning their Chinese origin. However, the other types of horn may either be poor copies of a *muntjac* horn or perhaps intended to be a bull's horns.

The mythical, horned, leonine creatures:
The main horned leonine creatures are the Mesopotamian and steppe *leogryph* or lion-griffin, the Indian *vyala*, the Chinese *bixie* and some *ch'i-lins*. The sphinx and the *senmurv* are not horned and, like the *chimera*, are too dissimilar from the weights beasts to be considered. Concerning the griffin, originally it had the body of a lion and the head and wings of an eagle.[50] There were several local variations in the kinds of tail, horns and manes but three main styles emerged: the bird-griffin, the snake or dragon-griffin and the lion-griffin.[51] The last was a horned lion with or without wings.

The griffin was in use about 3000 BC in Mesopotamia, Anatolia and Egypt, where it first symbolised deities and then became a temple guardian.[52] It was being used in Babylonia and Susa[53] (the origin of the lion weights) not long after. By 1500 BC the griffin motif was in use all over western Asia. The Assyrians were making much use of it in 900 BC as were the people of Luristan. It became abundant in Persia about 600 BC and continued so until about 700 AD.[54]

The horned lion or lion-griffin reached the Indus valley about 2000 BC.[55] The griffin was utilised by Alexander the Great and he may have introduced or re-introduced it to India about 325 BC.[56] Certainly two griffins of about 3rd century BC occur at Patna.[57] Lion-griffins and eagle griffins were employed in the sculptings on the Indian Buddhist temples of the 2nd to 1st century BC,[58] the motifs possibly having been introduced by the Scythian (Saka) invaders.[59] At that time two types of head appendage appeared: antlers with two or three tines directed upwards and unbranched horns either directed upwards like a goat's or downwards like a ram's. These lion-griffins are not of a fierce appearance as one would expect. They eat grass, are mild and meek-looking like the post 6th century AD lions of China and unlike the pre-6th century AD stone lion-like felines of south China. The vyala[60] was also in existence at about the same time. After the 5th century AD it appears to have replaced the horned lion for the greater part except in central and north-east India, though neither composite creature is common. After the 11th century AD production of leonine and horned leonine creatures greatly diminished.

The griffin and the lion-griffin motifs were distributed by the steppe nomads on the artifacts manufactured in western Asia, just as they transmitted artifacts and motifs from the Orient.[61] The Scythians of east Asia adapted these motifs to their own requirements before the 7th century BC. It is likely that the lion-griffin had reached the Altai, Siberia and China well before 1000 BC.[62] The Scythians preferred combat scenes with griffins as the aggressors and they themselves became especially associated with the griffin by the western world.[63] The western world in fact made the association

Scythia-griffin-gold because at that time, the Altai and adjacent regions produced much of the world's gold. In Burma also the griffin-like *to* is associated with gold. Lion-griffin figurines were present in Bactria in the 4th century BC[64] at a time when colonies of Greeks were working for the Scythians. In the 3rd - 1st century BC lion-griffins were being made at Pazyryk in the Altai.[65] Both the Bactrian and Pazyryk lion-griffins show distinct similarities to the beast weights.

The Persian influenced models[66] for the horned stone felines of South China reached there about the 2nd century BC and the style continued until the 6th century AD when the southern dynasties were destroyed.[67] These creations are similar to one of the beast weights. Elsewhere in China after about the 2nd century AD a two-horned feline shape was referred to as a *bixie*, a term transferred from the sacred deer, (the *tian-lu*). Before the 5th century AD stone effigies of the bixie were placed among those creatures guarding the paths to tombs.[68] In later representations a leonine head and body was combined with the horns and hooves of a deer. The apparent similarity between this description of the *bixie* and the south Chinese horned stone felines of the 2nd century BC to 6th century AD is evident. The latter have been related in style to the jade winged and bearded tiger carvings of about the 5th century BC to the 2nd century AD.

The dragon:

Though their origins are different, the dragon stemming from the snake or *naga*, in China the horned lion has been given some of the characteristics of the dragon. Plate 48, illustrations 1 and 2 show the similarities between the Burmese *to-naya* and the Chinese dragon. The *suan-ni* type of dragon carved on Buddha's throne in China has been identified with the symbolic lion. The similarities of the lion-griffin, the snake-griffin, the horse-dragon and the dragon have been discussed by others. There are several major and many minor types of dragon. They are variously depicted as being with or without horns, wings or scales and with talons varying in number from one to five. The most important to this work is the imperial dragon, the *lung*. This is a serpentine creature with four legs and feet and a scaly skin. It has the horns of a stag (deer), the ears of an ox, a beard and prominent whiskers. Its head is supposed to be like that of a camel and it has five claws on each foot.

The dragon is believed to have first been conceived in Babylon from the body of the snake and components of other animals before being adopted by the Chinese. Though, in its cobra-like form, the *naga*, it may have passed through India to China, it seems to the writers more likely that it reached the Far East by means of the steppe nomads. The animal style art of the steppe nomads and their use of the griffins in metal long before the Indians left permanent records in clay or stone of their animal art, which is almost devoid of dragons, is evidence for this surmise.

Mouth appendages:

The mouth appendage or lip extensions on the weights, which could have been intended to be either protruding elongated tongues or beards, can be matched – to the writers' present knowledge - only on the early stone lion and horned lion statues of south and south-west China and earlier carvings farther north. If the lip extensions are a particular style variation of the long tongue motif, then their origin probably lies in ancient Egypt where it was a protective device against evil spirits. In ancient Babylon (and Tibet) it was a form of greeting. Along the road from Egypt to China long tongues have been found on the Oxus River (5th century BC) and in the Minussink basin of the 7th-10th century AD.[69] In India no illustration of a lip extension on a lion has been seen by the writers. The tongue rarely protrudes and when it does, the protrusion does not extend beyond the lips, as on the Asokan pillars and the lion effigies of west Asia about the 15th century BC. Nevertheless, one of the Hindu goddesses exposed a long striated tongue when the killing lust was upon her.[70] Central China north of Honan (Henan) and adjacent areas, thought to be the homelands of some of the tribes of south-west China and Burma, were occupied by Ti steppe nomads in the 7th century BC. Some appear to have migrated to Yunnan later. They brought with them the motif of the tiger with a protruding tongue in a style clearly not Chinese.[71] Conceivably these may be related to the stone lions of south China. In south China about the 6th to 3rd century BC both animal and human representations often had long tongues which were intended to symbolise a supplication for rain to water the soil.[72]

Combined antlers and long tongues:

Though antlers and long tongues were used separately on images in both China and India, in the latter country they were never combined on one image. This occurred only in the south of China between the 6th and 3rd century BC, probably in the art of some branches of the Thai people. The combination may have arrived there before the 8th century BC. It next occurs, again in south China, on the stone lions of the 3rd to 6th century AD. It is not until the 15th century AD that the combination occurs again, this time on the Burmese beast weights. Because the lip extensions are accompanied by horns, then these motifs on the Burmese weights were not derived from India but from China. In this case it seems that they were derived from, or at least travelled through, those parts of the ancient Chu state centred on the provinces currently named Hubei, Hunan and Anhui, which are located south of the river Yangtze.[73]

Feet:

The feet on the beast weights are like hooves, often marked with lines to suggest toes. It would have been possible for the craftsmen to make lions' feet, just as they made claws on some bird weights even though most weight-bird feet also are hoof-like. That lions' claws were not made suggests that they were not intended. The several toe lines eliminate the horses' hooves and the cloven hooves of the deer as models. The remaining possibility which is also symbolically acceptable, is that the feet of the elephant were intended. Hoof-like feet appear also on the elephant weights. The Chinese lion-like creatures are all shown with normal clawed feet with rare exceptions. However, most of those illustrated are large effigies.

Tail:

The tail is horse-like (flat sided) on two classes of the Group 5 weights and on some 100 and 250 *kyat* beast weights of the other groups. Another class of Group 5 has a short erect tail like that of a deer. All other beast weight tails appear like those of the lion or bull. However, there is a characteristic feature, i.e. the almost always erect tail base, which seems to indicate that the horse was originally intended. This is a rare characteristic among horses but it is found on the Tennessee walking horse.[74] It occurs also on the Yunnanese horse of about the 3rd century BC, and no doubt later, because the Yunnanese bronzes of the time show this feature.[75] It seems that the erect tail base of the beast weights was modelled on that of the Yunnanese horse and it was the horse tail which was intended. On the many Indian and Chinese lion-like representations examined, no tail base was elevated unless the whole tail was raised above the back. When designing the weights, it was necessary to indicate the horse symbolically (see *'Chakravartin'* in 'The Symbolism of the Weights'). In addition it may have been necessary to indicate one of the horned more or less horse-like creatures forming part of the mythologies of adjacent countries. Relevant information on these matters follows.

Representations of the horse abound throughout Eurasia. Along with the tiger and the stag, it was one of the most commonly used art motifs of the steppe pastoral nomads between 1500 and 200 BC. 3rd century BC tombs in the Altai contain horse skeletons fitted with antlers. Even in modern times the horse played an important role in ceremonies in the Altai.[76]

The horned horse and the *ch'i-lins:*

The most well known of the mythical horse-like animals are the centaur, the winged horse and the unicorn. The first does not occur in the Orient. Winged horses of several kinds are commonly represented in China and also on temple friezes in Pagan.[77] The unicorn appeared in the Indus valley from Sumeria and Babylon about 2000 BC[78] and spread to the steppes. While two classes of the Group 5 beast weights could, with imagination, be viewed as much stylised 'unicorns', two horns are present, not one. A single-horned lion-like beast effigy is also present on panels at Sanchi, India, datable to about the 2nd or 1st century BC.[79]

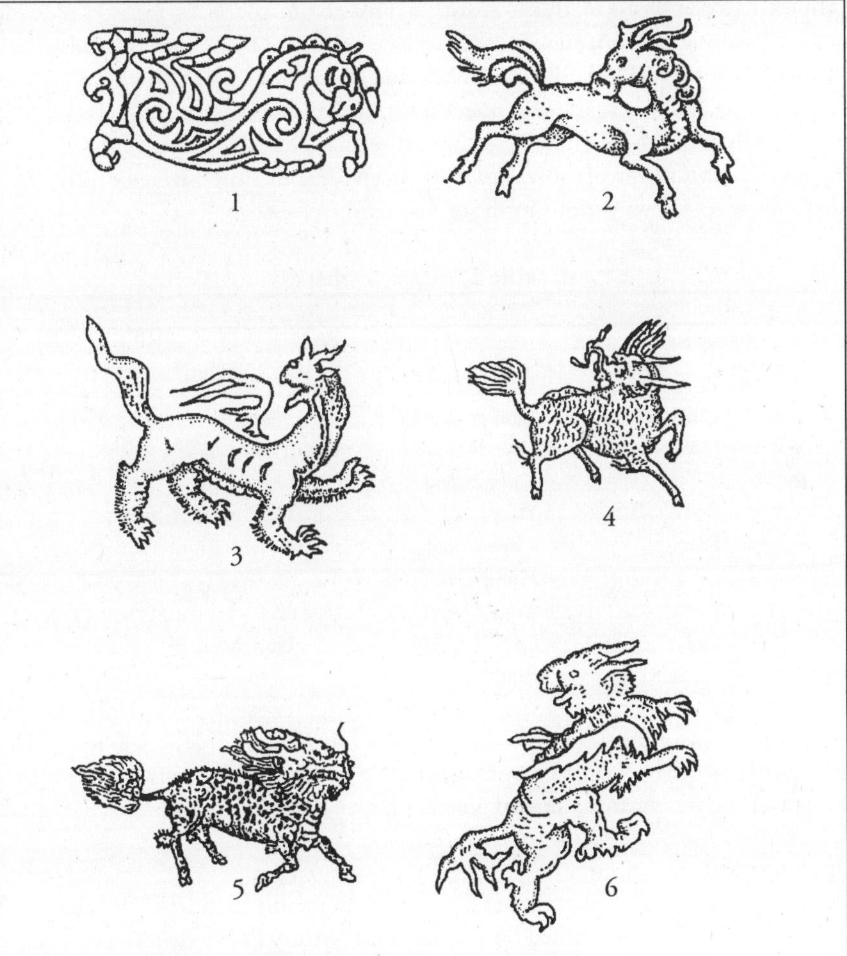

1. Deer. China. About 4 c. BC. Bronze plaque. Loehr, pl. 398.
2. Deer. China. 14 c. AD. Porcelain jar. Rawson, 1984, Fig. 92.
3. *Bixie.* China. 900 – 1400 AD. Porcelain plate. Tsugio Mikami, pl. 70.
4. *Ch'i-lin.* China. 19 c. AD. Chair cover. Christie, p. 125.
5. *Ch'i-lin.* China. Period uncertain. Decorative function uncertain. Lehner, p. 140.
6. *Bixie.* Persia. Approximately 1550 AD. Book binding. Rawson, 1984, pl.7.

The names of the creatures are those given them by the respective writers. It will be evident that creatures of similar appearance have been given different names and vice versa. Most however, show combinations of parts drawn from the lion, the deer and the horse.

Plate 47: Representations of the Mythical *Bixie*, *Ch'i-lin* etc

In China a composite mythical creature named *ch'i-lin* (*ch'hi-lin, kylin, ki-lin, qilin*) or *fu-ba*, among other names, is stated to have appeared first about the 3rd century BC to 3rd century AD, though possibly it may have done so much earlier. It is often called a unicorn because of its frequent possession of but one horn., However, only occasionally is the *ch'i-lin* horse-like, or so it seems. The table below summarizes the reported characteristics of the *ch'i-lin*,[80] of which there are supposed to be six different kinds. Plate 47 shows various forms of *bixie* and *ch'i-lin*.

Table 17: *Ch'i-lin* Shapes

Body Part	*Occurrence*	
	Most frequently	*Occasionally*
Head	Lion or dragon	Horse
Horns	One, fleshy	Two or none
	Unbranched	One-tined
Body	Deer	Lion, horse (± scales)
Feet	Cloven hoof	
	Horse's hoof	
	Tiger's claws	
Tail	Ox	Lion, horse
	Bullock	

The tendency for the shapes of the *ch'i-lin*, the horned lion and the deer to meld has been noted by others.[81] The resemblance of one or more of these kinds of *ch'i-lin* to the *to* and the weight beast animal is evident.

Open mouth:
Open or closed mouths appear to be used equally frequently in India while those south Chinese stone lions and horned lions of the 2nd century BC to the 6th century AD mainly have open mouths. The open mouth is first known from the west, e.g. from ancient Egypt and Assyria.[82] The open-mouthed tiger had become a motif in west China by, at the latest, 700-500 BC.[83] It was employed in the Altai mountains of the 4th to 1st century BC and was in India at the latest by about the 3rd century BC.[84]

Flattened muzzle front:
A flattened snout or muzzle front is characteristic of the pig. It appears on most of the beast weights. The boar was sacred to the ancient peoples of Egypt and Mesopotamia, where both it and the sow were connected with the earth.[85] The flattened muzzle front can be identified also in Hittite art, in Caucasian art of about 1000 BC, on Persian lions, on wolf representations of the Altai from about the 3rd to 1st century BC[86] and

on the stone lion statues of south and south-west China of about the 2nd century BC to 6th century AD. After these it appears to be no longer a popular motif. In Burma however it was still in use by woodcarvers at the end of the last century. The flattened muzzle front has not been seen by the writers on any illustration of an Indian feline representation. Yet in Hindu mythology the drowning earth was rescued from the ocean by the boar-headed avatar of Vishnu.[87] Some of the Chinese stone sculptures may have a rounded muzzle front, though rarely.

Manes:

Manes occur on lion figurines in a wide variety of forms. The most frequent forms on the Burmese animal weights are firstly, that of ridges around the neck, back and sides and secondly, knobs or indentations on the neck back. Of the former, only one illustrated, though doubtful, example has been seen and that on a stone lion from 10th century AD India.[88] Concerning the latter, also only one similar shape has been seen. This was on a lion-griffin from 4th century BC Bactria.[89] One class of weight (period H) from Group 4 carries an animal shape very similar to that of the stone statues of 2nd century BC to 6th century AD south China. Like the horned statues, and unlike most of the non-horned representations and unlike all other beasts on the weights, it has no mane. Maneless lion representations are almost non-existent in the Indian region. Such as occur are known on 3rd century AD coins from Sri Lanka.[90] Elsewhere maneless lions are to be found on some coins of ancient Bactria.[91] The existence of the mane makes it clear that where it occurs, the head of the beast weight was intended to be that of a lion. Lack of a mane suggests a tiger or lioness.

Chest:

Another feature of the Burmese beast weight is the protuberant chest. This is absent from Indian representations of the lion and horned lion but present on the stone sculptures of south and south-west China.

Posture:

The weight-beasts crouch as if ready to spring. The Chinese stone lions and horned lions of south China are lying on their stomachs, sphinx-like, sitting, standing or walking. Other feline representations appear to run or are guardant and rampant. The sitting attitude continues on the antlered humanoid feline figurines of the 7th-9th century AD. Standing and other attitudes are frequent on later representations. In India, the most common attitudes appear to be lying, sitting or guardant. Crouched, ready to spring, is present but is rare and apparently confined to the 11th century AD.

Conclusions:

It is concluded from the foregoing that, with one possible exception, the beast on the weights is a horned, lion-like creature composed of elements of the lion, (tiger) the deer (and bull?), the elephant and the horse. The lion, bull, horse and elephant are represented on the abacus of the 3rd century BC Asokan pillar at Sarnath in India and individually are, or were, to be found capping pillars in north-east India before the 3rd century BC. These elements are similar to those listed in the partial descriptions of the *to*. It has been shown that the elements usually present are arranged as follows:

Lion, (tiger?)-head, maned, with an open mouth and often a lip extension which is probably an elongated tongue.

Deer (muntjac and bull?), antler. These are usually unbranched but occasionally tined, sometimes terminally-knobbed, horns.

Torso. Lion? tiger?

Horse tail with the erect tail base of the Yunnanese horse becoming terminally the tail of a lion or ox.

Elephant (or Sarnath pillar style) feet often marked with lines indicating two to six toes.

The possible exception is the unique weight-beast of Group 4, period H, which lacks a mane, a lip extension and erect tail base. The feline element may be a maneless lion, a lioness or tiger. This weight-beast is similar to the stone horned lions of south-west China.

Plate 48, Notes:

Comparison of Chinese dragon and Burmese to-naya

1. Imperial dragon. China 14 c. - 17 c. AD. Painting. Larousse, p. 405.
2. *To-naya*. Burma. 19c. AD. Vase handle. Lowry, plate 6.

Burmese chinthes

3. *Chinthe*. Burma, Tenasserim. 18 c. AD? Robinson, Coin, plate 13.
4. *Kathayaza chinthe*. Burma, Thaton. 19 c. AD. Wood carving. Gear.

Comparison of deer-like animals

5. Stag. China, Yunnan. 5 - 3 c. BC. Bronze figurine. Rawson, 1983, plate 149.
6. Stag. India, Bharhut. 1 c. BC. Stone carving. Iyer, plate 153.
7. *To* or antelope (?). Burma, Tavoy? 18 c. AD? Tin coin. Robinson, plate 12.
8. *Naya-pyan* (?). Burma, Tavoy. 18 c. AD? Tin coin. Robinson, plate 20, quoting San Bartholomeo.
9. Weight-beast, Group 5. Burma. 15 c. AD? Bronze figurine. Gear.

Observe on 7, 8 and 13 the two tines. 8 may either be one-horned or shown with one horn for convenience. If the former, and intended to represent a 'flying unicorn', then it bears no resemblance to the horse-like unicorns of the steppes, the one-horned, non-natural, *ch'i-lins* of China (so called unicorns) or the rare 'one-horned lion' representations of India.

Muntjac horns

10. Muntjac (Burmese: *gyi*. English: barking deer). China, south-east Asia. Natural. Horns illustrated are fur-covered. Morris, 524.
11. Muntjac skull. To show burr or 'button' and the two tines. Walker, p. 1385.
12. Button motif horns on lion-griffin. China. 18 - 11 c. BC. Bronze vessel. Fry, *Bronzes*, plate 4.
13. Two-tined horns on earth spirit. China. 7 - 10 c. AD Bronze figurine. Argan, vol. 12, plate 497.

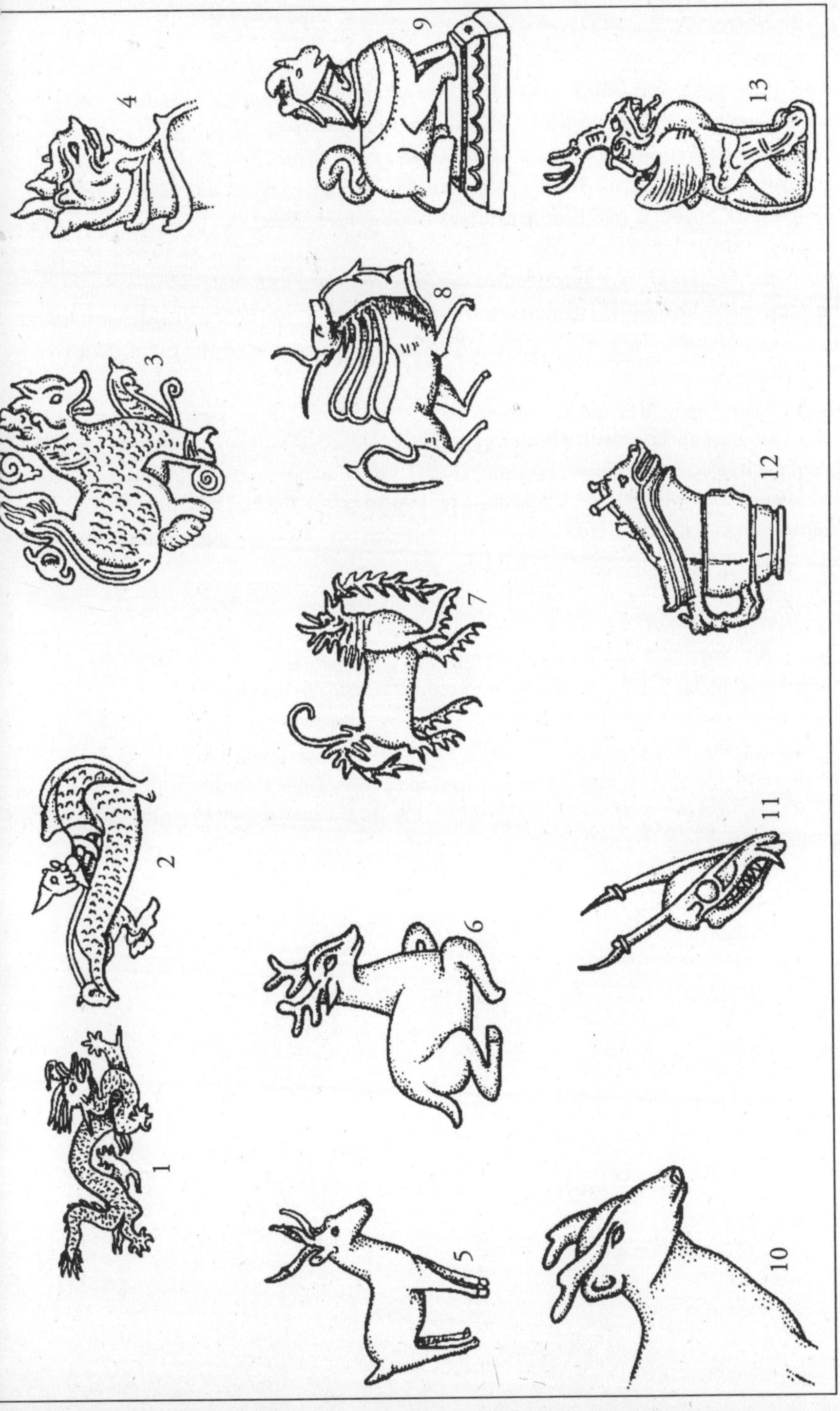

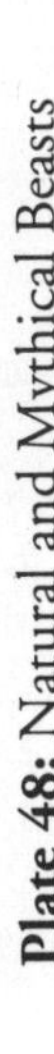

Plate 48: Natural and Mythical Beasts

Plate 49, Notes:

Lion-griffins from the steppes and China

1. Lion-griffin. South-west China. 4 - 3 c AD. Stone statue. Sickman and Soper, p. 17.
2. Lion-griffin. Persia. Gold ornament. Griaznov and Golomshtok, p. 174.
3. Lion-griffin. Bactria. 4 c. BC. Bronze figurine. Rice, 1965, figure 125.
4. Lion-griffin. Pazyryk (Altai). 3 c. BC. Wood carving. Bussagli, *Steppe Cultures*, plate 189.
5. Lion-griffin. China. 2 c. AD. Bronze figurine. Watson, 1975, plate 39.

Weight beasts for comparisonwith the lion-griffins and the Yunnanese horse

The Yunnanese horse (7) is shown in order to compare its tail base attitude with that of the Group 2(r) weight-beast, an attitude which typifies that of the great majority of weight-beasts.

6. Weight-beast. Group 4(Period H). Burma. 16 c. AD. Bronze figurine. Gear. Compare with 2.1. This is the only beast with a horizontal tail base.
7. Raised tail base on horse. China, Yunnan. 3 c. BC. On lid of bronze cowrie container. The raised tail base is typical of Yunnanese horse representations. Rawson, 1983, plate 83.
8. Weight-beast. Group 2(r). Burma. 18 c. AD. Bronze figurine. Gear. Compare with 3, 4 and 5.

Lion-griffins from India

9. *Vyala.* India. Sarnath. 5 c. AD. Stone carving. Iyer, plate 125.
10. Horned lion. India, Sanchi. 1 c. BC. Stone carving. Murthy, figure 34.
11. Horned lion. India, Amaravati. 1 c. BC Stone carving. Murthy, figure 12.
12. Vyalapada. India, Mahabaliparam. 7 c. AD. Stone carving. Iyer, plate 128.
13. Weight-beast. Group 5. Burma. 15 c. AD. Bronze figurine. Gear. Compare with 12. The similarity is slight but it is the closest that a Burmese weight-beast comes to an Indian representation.

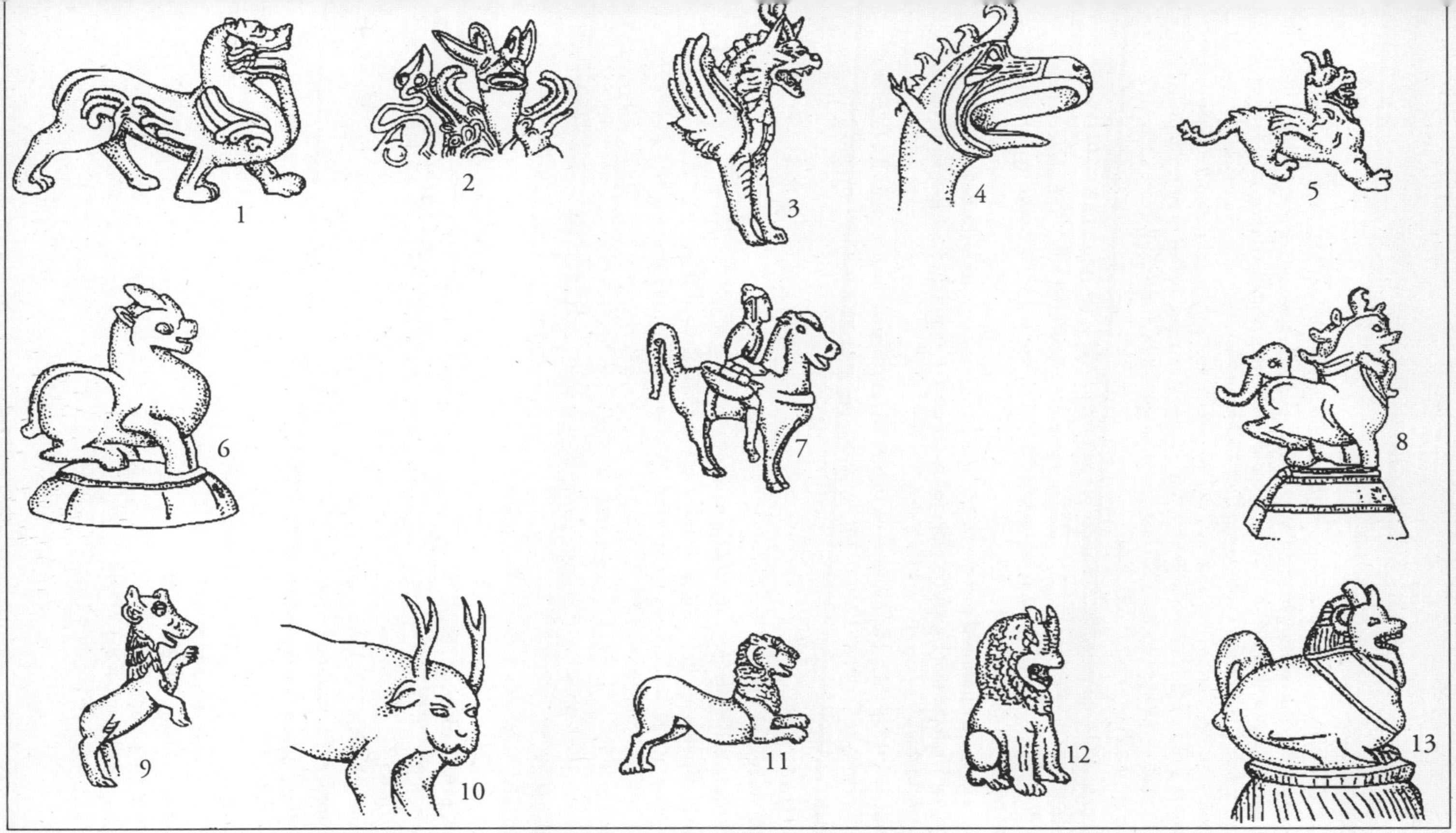

1 2 3 4 5

6 7 8

9 10 11 12 13

Plate 49: Natural and Mythical Beasts

Notes

1 Local dealers usually use *chinthe*. It was used by Temple (1898, pp. 141-143). *To* was used both by local dealers, Temple and Phayre (1882, p. 31). Fraser-Lu uses *to myin* (horse-like *to*) and *to oung* (bull-like *to*) but the writers did not hear these names during their years in Burma. Sale (1910) refers to the *to-sing* (lion-like *to*) in Laos. Gardiner refers to *singha* and *sing-to* as names used in Thailand, while Le May refers to a similar animal on the silver bullet coins as a *rachasi*. It is a *chinthe*. Both *to* and *to-naya* are names applied to animal representations used in the Shan states in celebrations and processions, particularly those concerned with the *sawbwas* or hereditary chiefs.

2 Temple, 1898, quoting J.E. Alexander, *Travels from India to England*. London, 1827.

3 In 1691 the local Siamese people could not name the animals and other symbols occurring on their bullet coins (Le May, p. 25, quoting de la Loubere). In 1882 Phayre (p. 31) could find no one to define the *to*.

4 The writers were informed that the *chinthe* seen outside the temples is the *kathayaza chinthe*. Another kind, the *tina chinthe*, is supposed to eat grass. It is said there are four kinds of *chinthe* in total.

5 Temple, 1913, p. 117. He then goes on to say that, in practice, the *to* assumes the form of a lion. Elsewhere (Temple, 1931. p. 13) he refers to the *to* as being half-deer and half-bird.

6 Phayre, p. 31.

7 Shway Yoe, p. 86.

8 Htin Aung, p. XVII.

9 Anonymous, *Weights of Historical Importance*.

10 Lowry, plates 6, 40. Robinson, p. 6, refers to a two-horned beast shape on a Tenasserim coin described as a *to-naya* (da San Batholomeo, *Systema Brahmanicum*. Rome, 1791). (See O'Connor, facing pp. 38,39)

11 Savill, p. 126.

12 Rawson, 1984, pp. 49, 110. Battisti, p. 945 ff.

13 Ball, pp. 57, 58.

14 Pope and Ackerman, p. 304 ff. F. Anderson, p. 138.

15 Cooper, p. 72.

16 Ball, p. 63. Liebert, p. 261.

17 Irwin, I, p. 710; IV, p. 747.

18 Iyer, p. 85.

19 Kramrisch, plate 125. Iyer, p. 67.

20 Elisseeff, p. 33. Battisti, p. 954.

21 Eberhard, pp. 164, 290. Walters, p. 32. Williams, p. 398.

22 South, pp. 94, 95.

23 Rawson, 1984, pp. 12, 90, 113. Ball, p. 53 ff.

24 Ball, p. 66. Fry, pp. 28, 50. Lessing, p. 21 ff. Hansford, p. 841.

25 Rawson, 1984, p. 90. Ball, p. 61.

26 Ball, p. 60. Lessing, p. 174.

27 Lessing, p. 123. Eberhard, p. 190.

[28] Fry, pp. 30, 50. Ball, pp. 56, 66, 68. Zimmer, 1974, p. 14.

[29] Lessing, p. 121.

[30] Aung Thaw, 1968, pp. 55, 56, figure 84.

[31] Luce, 'The Ancient Pyu', p. 314.

[32] Luce, 'The Smaller Temples of Pagan', plates V, VI.

[33] Kelson, p. 166. Matthews, p. 997.

[34] Fry, plate 4.

[35] Bussagli, 'Chinese Bronzes', p. 89.

[36] Campbell, 1970, p. 402.

[37] Praz, pp. 721, 722. Eberhard, p. 79. Loehr, p. 777.

[38] MacCulloch, p. 792.

[39] Salmony, p. 23. Jettmar, p. 223. Narain, p. 928.

[40] Salmony, pp. 48, 52.

[41] Mackenzie, 1987, pp. 89-92.

[42] Sickman and Soper, plates 15, 17. Watson, 1981, plates 359-363.

[43] Davies, 1975, pp. 13, 67.

[44] Rawson, 1984, p. 108, quoting from the *Han Shu* (*Han History*).

[45] Salmony, pp. 24, 25, 26, 28.

[46] Shaw and Ali, pp. 20-22.

[47] Petrie, 1926, plate IX.

[48] de Rochesnard, p. 71

[49] Iyer, pp. 19-30.

[50] Jobes, p. 691. Perkins, p. 924.

[51] South, p. 87. Lessing, p. 121. There is a gradation between the shapes of the lion-griffin and the dragon.

[52] South, pp. 87, 92.

[53] Fry, p. 29.

[54] Bussagli, 1969, p. 136. South, p. 93.

[55] Iyer, pp. 14, 86, 87. Campbell,, 1970, pp. 90, 102.

[56] South, p. 94. (See Basham, plate LI, for an illustration of a leogryph)

[57] Iyer, p. 85.

[58] Iyer, p. 66.

[59] Murthy, p. 12.

[60] Iyer, pp. 71, 72, plate 125. Liebert, p. 248. A *vyala* is a composite creature. Essentially leonine with parts from the horse, deer, boar, elephant, rhinoceros, wolf, parrot and vulture. Its depictions bear little resemblance to the weight-beast and appear always to be clawed.

[61] Funke, p. 267.

[62] Fagg, p. 314. Praz, p. 722.

[63] South, pp. 86-94.

[64] Rice, 1969, p. 139. Griaznov, p. 180.

[65] Jettmar, p. 100.

[66] Watson, 1975, p. 52. Rice, 1969, p. 182.

[67] Fry, p. 29.

[68] Rawson, 1984, p. 708, quoting the *Hou Han Shu* (*Late Han History*).

[69] Salmony, pp. 33, 52.

[70] Cooper, p. 174. Salmony, p. 41.

[71] Loehr, p. 777.

[72] Salmony, pp. 50, 52.

[73] Salmony, p. 49.

[74] Gorman, p. 705.

[75] Rawson, 1983, pp. 223, 227, plates 5, 26, 31, 36, 31, 44, 46, 48. However, Polo, p. 266, referring to west Yunnan about the mid-13th century AD, remarks that it was the custom to deprive a horse of one joint of its tail in order to prevent the animal from lashing it from side to side and also to cause it to remain pendent. This may have been the cause of the erect tail base. Yunnanese ponies illustrated in 19th century publications do not have this raised tail base, e.g. Gill, pp. 355, 356. Bock, pp. 230 – 232. Crawfurd, Ava. p.192.

[76] Iyer, pp. 37, 38, 39.

[77] Luce, 'The Smaller Temples of Pagan', plate VIII.

[78] Bussagli, 1969, p. 143. Iyer, p. 42, quoting E.G. Pulleyblank. South, p. 9. Liebert, pp. 85, 86. Chinese mythology has several divine horses, often flying without wings.

[79] Christie, p. 130. Yetts, p. 124. South, p. 12. Lessing, p. 114.

[80] South, pp. 9, 13. Ball, pp. 33-40. Temple, (1913, p. 117) believed the *chinthe* and the *to* to have been derived from the Assyrian guardian winged lion and bull respectively. He also links their origin with that of the Chinese *ch'i-lin*. He states that in Japan both the winged lion and winged horse-deer form of *ch'i-lin* are portrayed.

[81] Iyer, p. 87.

[82] Pope and Ackerman, p. 307. Battisti, p. 954.

[83] Loehr, pp. 777, 778.

[84] Iyer, plates 96, 97.

[85] Cooper, p. 22. Iyer, p. 161.

[86] Bussagli, 'Steppe Cultures', p. 386.

[87] Larousse, p. 362.

[88] Iyer, plate 95.

[89] Rice, 1965, figure 125.

[90] Codrington, p. 12.

[91] Rapson, p. 22.

10
The Symbolism of the Animal Weights

Religious beliefs and animal symbolism:
From the first awakening of interest in the weights, it was apparent that the weights had symbolic meanings. Today it may be true to write that few artifacts in the world other than modern communications media, carry so much and so many meanings.

Symbolic meanings of the same shape or representation change with time, natural environment, religious beliefs, culture and human events. Of most importance to the symbolism of the weights appear to be the ancient Eurasian animism of the hunter-gatherers; the earth-god religions of the agriculturalists who developed from them; the sky-gods and shamanism of the herding steppe nomads and the fishing/herding north Siberians; the later, post 500 BC Buddism, especially the Mahayana Buddhism of the eastern half of Asia which, in many ways, paralleled shamanism and was intimately connected with urbanisation, trade and weighing; and finally the events in the continuing histories of the peoples who have made Burma their home. Perhaps most of these symbolic meanings can be traced back ultimately to the desires for beneficial fertility, well-being, protection from harm in this and any after-life and the evidencing of power over people. Coupled with these are the needs to placate harmful beings and to establish a means of communication with the spirits or gods.

With these realisations it became possible to attempt the analysis of the symbolisms of the weights and their parts. Because of the close and sequential nature of the association of the religions, and the tendency for indigenous beliefs to absorb, modify or parallel imported ones, it is often not possible to separate their effects. In other cases the persistence of practices which were abhorred by the Buddhists, e.g. the horse sacrifice or nature spirit propitiation, makes recognition of the differences simple.

In general terms the development of the concept of animal-related gods followed an order in which, at first, everything was a spirit or participated in godship. This was followed by the idea of animals as gods, then came human-beast gods, then humanoid gods symbolised by animals and finally wholly anthropromorphic gods with no or minor animal symbolism. For example, the animistic, earth and sky god symbolisms anciently brought in by the various peoples of Burma from south and west China and the steppes[1] was overlaid by the Hindu and Buddhist symbolism subsequently brought in by the Indians and others. This was what happened to the Christian religion from Israel, which modified its symbolism to accord with the pagan customs of Europe, e.g. Easter.

Similarly, when the animal art and symbolism of the steppes migrated to China it became sinicized and ceased to reflect only the beliefs of the nomad and the hunter.[2] This was the case when the lion and its symbolism was transferred from the India-Persia region to China. The persistence of animal symbolism in Burma was partly due to its widespread persistence in India and south China where the Hindu and Buddhist deities, their preachings and mythologies were represented by animals noted for particular qualities which seemed appropriate,[3] and by animals which remained abundant until modern times. Though in Christianity there is the lamb of God and the lion of Judah, by the time of Jesus Christ wild animals had been mainly destroyed in this region, hence their symbolisms were no longer possible or relevant. In fact, among Semites, there was marked aversion to them, partly because they represented the ferocious steppe dwellers and partly because they were associated with paganism.[4] In the Orient however, animal representations, though no longer the deities they once were, came to assume, and supposedly manifest, appropriate attributes of wholly or partially anthropromorphic deities[5] on a considerable scale. In Buddhism all animals accepted Gautama Buddha as their lord just as they accepted Siva, the Lord of Beasts, in Hinduism. In both religions they appear in many roles as worshippers and servitors. Indeed in every creature there was the potential to become a Buddha.[6] About the 3rd to 2nd century BC, when Buddha was not represented by effigies, he was symbolised by the lion, the elephant or the horse. The *Jatakas* contain many tales of Buddha's kinship with the animals. Another reason for the persistence of animal symbolism was that animal representations could be recognised easily as symbols of power, in particular of divine or royal power.[7] The association of the bird with heaven, light and the gods and the large feline with the earth, darkness and kings was made and symbolised throughout Eurasia.[8] But it was only in Asia that gods and kings alone were symbolised by emblems. Neither states nor towns nor nobles were so distinguished, unlike Europe where emblems were adopted by all of those. Moreover, the concept of combining seemingly appropriate characteristics of different animals, so that the deity or king symbolised by an image of a composite animal supposedly possessed the properties of them all,[9] was widespread in ancient times. These images conveyed, and were expected to convey, messages both religious and political or authoritarian.[10] They were messages which could be understood by the illiterate and by neighbouring foreigners. These practices were widespread in ancient times but persisted longest as a general means of communication in the eastern half of Asia.

There is a specific problem concerning the use of animal effigies on the weights. That is, why were animal effigies adopted when, according to such limited information as is available, they are currently known to have been not used or but little used on the weights of those countries which appear to have most influenced Burma? Moreover, animal representations, though perhaps not their symbolism, had lost much of their importance in those countries by the time the Burmese animal weights are inferred to have been introduced. Was it due to the emphasis on Theravada Buddhism and the *Jatakas* in Burma (and Siam and Sri Lanka)? Was it due to the importance of the influence of the animistic-shamanism of the steppe nomads on the

Burmese? Or was it a combination of both? No doubt these questions could be answered if one could be sure that in those regions most influencing Burma, the relevant evidence had been sought, that it had not been overlooked, that it had not been destroyed by men and that it had not corroded beyond recognition.

Animism:

Animism is an aspect of nature worship which accords life, or good and evil spirits, to unusual or sacred or all natural objects.[11] Such spirits need propitiation to prevent them doing harm. Women may have special power in this respect. Not only does animism underlie the animal art of the steppes, China[12] and of south-east Asia, but it underlies the animal art of Hinduism[13] and Buddhism to this day. In Burma there were the nature *nats*, spirits of air, water, land, trees, animals and the like. These were everywhere at all times and constantly active. To avoid harm, the Burman had to propitiate them constantly. As a result there is a spirit shrine in every house and on the road outside every village, compared with about one monastery in every village. On the strength of these beliefs many writers have commented.[14]

Earth gods:[15]

The Neolithic (?) religions of the Eurasian agriculturalists appear to have been characterised by a belief in earth gods, i.e. those of the earth as a whole, of the soil and of the underworld. This belief was expressed most obviously in the forms of earth mounds, sometimes capped by stones, by megaliths, dolmens, menhirs, small stone pyramids and other structures. Such occur in Central Asia, Tibet, Mongolia, Assam, Orissa, South-east Asia, etc. Later the belief was expressed in the form of pillars in India, Yunnan and Vietnam, as *stupas* in much of the same region and, in South-east Asia, as temples intended to represent mountains e.g. many Khmer temples. Less obviously it was expressed in the form of the earth-crawling snake, around which cults developed in India, South-east Asia, Yunnan and elsewhere.

The natural, or man-made, elevation seems to have been regarded as a substitute for the body of the local earth god (and later for the tribe itself symbolised through the ancestors of the rulers) within which was concentrated the power of the deity, a capping stone often serving to concentrate the power still more effectively. The functions of the elevations were the usual ones described previously. Celestial gods formed part of the cosmology but were less important than the earth gods. A link between the two kinds of god was often made by the use of a pillar or tree on or near the mound, but the elevation itself, especially if high, may have served as the link.

Shamanism also formed a link and means of communciation between, and to, the earth and sky gods, though possibly developing at a later time.

Symbolically the most obvious relation of the earth religions to the weights lies in the pyramid shape of the base, in particular to that with a square plan, a plan closely associated with the Indo-Aryans of 3000 and 4000 years ago, and to that with a stepped profile. Less obvious is the importance attached to the Pole Star and the possible relation of this to the star depicted both on the Dong S'on drums and as a sign on the base of many weights.

Shamanism and the sky gods:

Animism is an intimate part of shamanism. Shamanism is a set of magico-religious beliefs and practices associated with a variety of religions. It originated in Paleolithic cultures over much of the world. It is pre-eminently associated with Inner Asia and Siberia. It was practised by the ancient Indo-Aryans, the steppe nomads of ancient and modern Eurasia, the Bons of pre-8th century AD Tibet and the Chinese of ancient and modern times. The symbols and rituals of shamanism are known also in south-west China, Yunnan, Burma and elsewhere. Shamans were concerned with the achievement of ecstasy. Ecstasy enabled them, among other things, to ascend to heaven, to descend to the underworld, to search for the missing soul of a sick person, and to guide the souls of dead people. A shaman is aided by spirits which he can incarnate and control. Most of these spirits have animal forms but some appear as tree spirits, earth spirits, hearth spirits and the like. The shaman, in his mind, can turn into an animal and speak animal languages, especially those of the birds. His dress symbolises an animal. He is a master of fire. His drum, the most important part of his dress, was used firstly to assist in attaining ecstasy, secondly to summon his spirit helpers and thirdly to transport him on his spiritual journeys. As a result of the last, the drum was usually given the name of a bird, e.g. swan or crane,[16] or called a horse or a boat.[17]

Associated with shamanism in the northern half of east Asia before the 8th century BC and until today, was the belief in anthropomographic gods of nature, each of which was symbolised by that aspect of nature with which it was associated, e.g. animals, plants, storms, etc. The supreme deity was the god of the sky and the sun. He, together with the other nature gods, in time became largely replaced by an ancestor cult. Accompanying these beliefs were cults of fire, gold and the horse. Horse sacrifices, practised by Mongols, Turko-Tartars, Indo-European peoples and others, were always offered to the god of the sun and sky. The horse, especially a white one, symbolised the sun. In the Altai it was the function of the shaman, in a trance, to accompany the soul of a sacrificed horse on its celestial journey and also to offer horseflesh to the ancestors. Horse sacrifices and horse burials formed part of the burial rites of these peoples throughout the entire region, the best known being those of the Indo-Aryan Scythians,[18] Sakas and other invaders of India. These rites and sun worship[19] continued among the Mongols until after the 14th century AD, while Tartar chiefs continued to present thousands of white horses to the Chinese emperor until after the 18th century AD.[20]

To induce ecstasy, both drumming and dancing were employed. Some shamans used narcotics such as hemp and intoxicants such as alcohol. In addition, the Tibetan form of shamanism, Bon, incorporated practices of the Hindu mother goddess Shakti, in which sexual union without pleasurable completion of the sex act was used to attain ecstasy. These practices also included rituals to increase the sexual vigour and fecundity of the community.[21]

Shamanism in eastern Asia was affected by foreign influences. From about 3000 BC the steppe nomads had contacts with south Asia, Mesopotamia and Persia. These contacts influenced their mythologies and cosmologies. Later Buddhism influenced not only the shamanism of the Tibetans[22] but also that of the Mongols further north. In its turn shamanism left its mark on Buddhism.[23] For example, references to shamanic themes occur in early Buddhist texts. Ancient gods were transformed into Buddhist deities.[24] Old beliefs were reformulated to make sense within more modern world views or were retained alongside Buddhist beliefs. But above all, both shamanism and Buddhism, like the other beliefs, had common roots in their prehistoric past.[25]

Shamanism may have existed in Burma from Neolithic times. It appears to have been first recorded in Pagan of the 11th to 13th century AD in the form of the Ari, who were still practising it in 1468 AD. Like the Altai Scythians they were associated with gold and horses. They were also a militant order. [26] In 1948 AD three quarters of the Karen people still were animist/shamanist.[27] The Ari are described as magic-working monks who dressed in dark caftans, had long hair, drank, practised *droit de seigneur* and who withdrew to the solitude of the forests. Their belief in the World Tree may be evident in their use of the *tha-hkut,* or the rose-apple tree, which the Paganese believed grew at the world's end.[28]

The shamans of the Scythians were called En-Aris(Enarees, Enaries).[29] The Medes of ancient Persia, who were both allies and enemies of the Scythians, were formerly known as the Ari-oi.[30] The Indo-Aryan languages are derived from Sanskrit in which the word *arya* first meant 'freedom' or 'of noble character' and later 'lord, noble sir, sir'.[31] It would seem that this is the meaning of the Scythian (and Saka?) Enari and the Burmese Ari.

In ancient India and elsewhere, long hair on shamans, such as the Ari favoured, symbolised the snakes that appear on the costumes and in the beliefs of the Central Asian shamans.[32] The snake played an important role in Central Asian and Siberian mythology and on the shaman's costume. Today the cult of the snake is widespread in the region from Yunnan to Burma and on to north-east India and Tibet. The Indo-Aryans of about 1500 BC, the Scythians of pre-8th century BC and the Sakas of the same stock about the 2nd century BC all invaded India. They drank intoxicants like the Ari and took narcotics (soma, haoma) to attain ecstasy.[33] The *droit de seigneur* may have been part of a rite, either to induce ecstasy or fertility. In Ramree Island off the Arakan coast some 19th century AD sorcerers adopted women's dress and behaved as women like the Scythian Enaris and other shamans. Modern male Burmese shamans seem to be either homosexual, transvestite, effeminate or all three together.[34] Withdrawal to a forest or mountain was one of the premonitory signs of a future Siberian shaman and also part of his initiatory rites.

The Indo-Aryans sacrificed horses to a sky or sun god. Horse sacrifices were made in the Shan regions until 1557 AD. Horses were buried with chiefs in the Karen regions until 1908 AD.[35] This is unlike the situation in India where horse sacrifices were forbidden by the 3rd century BC and rarely occurred thereafter.[36] Nevertheless, there appears still to remain a connection between the horse, Hinduism and the Scythians. Among the Hindu gods who sometimes use the horse as a vehicle is Kubera, the god of wealth. He is also the guardian of the north where the people were believed to have horses' heads.[37] The possible symbolic relationship is obvious with wealth replacing Scythian gold, north standing for the location of the gold and the Scythians and the horse and horse-headed people symbolising the mounted nomads.

The king of Pagan in the 12th century AD selected five widely dispersed pagoda sites, partly by means of the wandering of a white elephant.[38] This is similar to the Indo-Aryan practice of allowing a horse to wander freely for a year before its sacrifice.[39] The shrines of the villages within a large area of Central Burma often contain a wooden effigy of a white horse representing the village *nat*. This *nat* is Myinbyushin, Lord of the White Horse.[40] Similar effigies were used by the animist Kafirs of Nuristan (east Afghanistan) until recently, where they represented the soul-conveying and demon-averting functions of the white horse.[41] The fact that when Myinbyushin was alive he was a blacksmith, supposedly, may connect him with the renowned metallurgical knowledge of the ancient steppe nomads.[42] This Burmese practice is also similar to one in China where at the Feast of Lanterns the ancestral ghosts are believed to visit their living descendants riding on a horse or a bull. To facilitate the entry of the recently dead, small effigies of these two animals are placed at a short distance from the dwelling.[43]

The Indo-Aryan Assaka people of about 5th century BC Deccan in India were called 'the horse people'.[44] The Saka have been referred to as 'the people of the stag'. Since Saka horses were frequently buried bearing stag antlers on their heads, the expression 'horse people' would be equally apt. The Manipuris refer to the horse as 'the Burmese (Shan?) animal'.[45] There may be a significance in the association of both Sakas and Burmese (Shans?) with the horse, particularly since the dominantly Mongoloid Manipuri and the neighbouring Naga peoples include an Indo-Aryan component.

The Sakas had ancestor cults. Ancestor worship persisted until recently among the remoter Burmese peoples but even the Burmese cult of the 37 *nats* is a form of deification of hero-ancestors who guarded both the kingdom and the religion. The first royal list of *nats* in 1089 AD contained only 33 *nats*, including the chief, the Hindu-derived Buddhist god Sakka (Sakra in Pali, Thagya in Burmese).[46] In Hindu-Buddhist cosmogony there are 33 gods who reside on the summit of Mount Meru, among whom Indra (Sakka)[47] is king. In Central Asia the people of the Altai mountains have a belief in 33 heavens.[48] In the 7th century BC Saka-Siberian burial site of Tuva, just east of the Altai mountains, the chief's tomb comprises a large central mound surrounded by a stone wall 44 meters away. The annulus so-formed is separated into sections by 32 radial spokes built of stones bearing incised depictions of horses. This is a temple of the sun.[49] The number 32 appears also as that of the number of bodily marks characterising the *chakravartin.* There seems to be good reason to think that the 32 fiefdoms of the *chakravartin* are derived from the solar cult of the Indo-Aryans and reached Burma through the Sakas. In this respect the nomadic influence appears to be evident also in China where the Taoist Western Paradise included 33 pavilions, some of which were sealed with the teeth of the 'Celestial Stag' (the lung dragon).[50] Particular variants of the *feng huang* also lived there, suggesting its connection with the peacock, first recorded as a symbol among the Scythian sun worshippers,[51] and closely associated with the Burmese and Shans.

Drums were an essential part of the shaman's equipment. The earliest known bronze drums which possibly were made for shamanic purposes, originating in motif perhaps from the Lake Baikal region,[52] were made in 3rd century BC Yunnan. Similar drums were still in use by the Karens in Burma in the 20th century AD. The frogs which appear on the upper surface of many of the Karen and south-east Asian metal drums may refer to the Altaic myth which relates how a frog induced the supreme god Ulgen to permit the creation of the first man and woman. It may also refer to a belief of the Yakuts that the most outstanding shamans originate from a toad.[53]

Turning to the Chinese region north of Burma, many of the peoples of today's Yunnan are shamanist/animist just as they have been since before the Nanchao (Yunnan) king's adoption of Buddhism in the 9th century AD. The Lolo (Yi) people of south Yunnan are shamanists.[54] The Tien people of 2nd century BC Yunnan had shamanist ceremonies at which surrounding clans gathered, erected an altar and a pillar (akin to Vedic rites and these of the earth-god religions?), made sacrifices to the earth god and vowed to respect each other's boundaries.[55] Those practices and their symbolism are identical with those of some Siberian tribes. The Na-khi people of south-west China practise a pure Bon, pre-Tibetan-Buddhist form of shamanism. They are, in fact, nomads originally from the north-east of Tibet. They preserve the faith of the herdsmen of Central Asia. In the cases of both Lolo and Na-khi, part of their shamanic rites are similar to those of China. In China itself, shamans were the real priests before the Christian era.[56] Confucianism, Taoism, Buddhism and shamanism were mutually adaptable and so permitted the old beliefs to continue in hybridized forms.[57]

It is clear from the foregoing that there is a considerable likelihood of the practices, beliefs and symbolism associated with the ancient religions and shamanism of the people of Tibet, Central Asia and Siberia having been transmitted to the peoples of Burma. It is possible that the Aris of Pagan came from north-east India, where Tantrism developed, perhaps deriving from the Sakan invaders of about the 2nd century BC, particularly since horse sacrifices are known but rarely from post-3rd century BC India nor from the central Burma region nor are they recorded as an Ari practice. It seems certain however that the shamanism of the Shans was brought with them from the Chinese who may have obtained it from the steppe nomads, particularly since the horse burials and sacrifices continued in the Shan states for about 1300 years after they had almost ceased in India.

Buddhism:

The two major divisions of Buddhism of concern here are Mahayanaism (Great Vehicle) and Theravadaism, formerly known as Hinayanaism (Little Vehicle). Mahayanaist scriptures are mainly in Sanskrit and those who follow the Mahayana creed believe that Gautama Buddha was but one of many Buddhas. It emphasizes the divinity of a Buddha and the importance of donations to shorten the period of merit accumulation. It is a more worldly-wise religion and formerly used both trade and the practice of medicine (cf. the Group 2 beast weights and an important use of the Chinese small liang) for proselytisation. It is associated with animistic views. Those who practise Theravadaism believe that salvation is to be attained by adhering to the ways supposedly laid down by Gautama Buddha. He is regarded not as a deity but as an exemplar. It is believed by its followers to be a purer, austerer form of Buddhism. Its scriptures are written mainly in Pali.

The Aryan invasion of Persia and India of about 2000 BC introduced the Vedic religion. The descendants of the Aryans and the Saka/Scythians (sun and snake worshippers) became allies perhaps before 700 BC, the Sakas becoming known as 'the serpent' or *naga* race, while the *naga* itself became one of the most important associates of the Brahmanic, Hindu and Buddhist pantheons. According to the Indian Puranas, Gautama Buddha originated from the solar race of Iskshvahu and at the commencement of his ascetic life, he was protected by the *naga* king.[58] Tombs of Gautama's own Sakya tribe, excavated in the 19th century AD, each contained an effigy of a *naga*. This is one of the several pieces of evidence linking Gautama Buddha with the steppe nomads. Vedism and indigenous animism merged to form Hinduism about the middle of the 1st millenium BC, about the same time as Buddhism emerged.[59] In the section entitled 'Trade' the voyages of the allied Sakas and Indo-Aryans as sea traders are mentioned and it is these which may have given rise to some of the Hindu legends, such as the 'Churning of the Ocean', and the association of the *naga* with water.[60] By about 240 BC Buddhism was dominant in north India[61] and by the 1st century AD two major sects had arisen,one being Mahayanaism, which was associated with India, Central Asia and north China, and the other Theravadaism, associated with Ceylon (Sri Lanka).[62] In India Buddhism reached its peak in the 1st to 2nd centuries AD and commenced to decline thereafter.[63] Its Theravada form had disappeared by 500 AD though it still flourished in Sri Lanka. In its Mahayana form it continued in north-east India, i.e. Orissa, Bihar and Bengal, whence it exerted considerable influence on early Burmese and Yunnanese religion and architecture.[64] Bihar, the cradle of Buddhism, fell to Hinduism in 1161 AD and then fell to Islam in 1200 AD.[65]

Beginning not later than the 2nd century BC and increasing during the 1st to 3rd centuries AD, Buddhism spread to Central Asia under the dominant influence of the fervently Buddhist and commercially-conscious Kushans (the former Yueh-chieh), whose empire stretched from north India to Bactria and from the Parthian empire of Persia on the west to that of the Chinese Han empire on the east.[66] Eastern Persia, a home of bronze animal art, and much of Afghanistan, also adopted Buddhism, parts retaining it until after the Moslem invasions.

In China, Mahayana Buddhism may have arrived before the 2nd century BC. Perhaps from the 2nd century BC onwards, especially from the 2nd century AD when the silk route was fully operational and the Kushans were actively supporting Buddhism, the religion became well-established in this region. By 380 AD the region had become important as a centre of Buddhist learning.[67] Elsewhere Buddhism made little headway until the To-pa steppe nomads established themselves in northern China about 386 AD. Buddhism reached its maximum extent between about 300-500 AD and 900-1100 AD[68] when most Chinese were Buddhists. By this time it no longer needed the foreign missionaries from India and Central Asia who had taught and spread Buddhism and by 845 AD, a reaction against non-Chinese religions and their exponents led to their persecution.[69] Although it remained important, Buddhism in China, as in India, became a much modified variant of its former form.[70] By the 11th century AD Chinese pilgrim traffic to the former Buddhist centres of India appears to have diminished considerably.[71] In Tibet, Buddhism was not strong compared with the shamanic Bon religion until after 750 AD. By 794 AD it had become the state religion and many Bon supporters moved east and south. Repression of the Buddhists began about 840 AD. These also fled east and south and it was not until about 200 years later that Buddhism had recovered.[72] Mahayana Buddhism had reached Yunnan probably from Mongolia by the 1st century AD,[73] possibly before,[74] and Buddhist monks were travelling the India-Burma-Yunnan road by the 2nd century AD.[75] In 880 AD Buddhism became the state religion of Nanchao (Yunnan).[76] From this time onwards Buddhism was increasingly adopted throughout Yunnan.

Thus those migrants who travelled from China and Tibet to Burma from about the beginning of the Christian era would have been exposed to both Mahayana Buddhism and shamanism, according to the prevailing conditions. Animism may have remained their dominant belief while their Buddhism may have been influenced by the adjacent Tibetan and Cambodian forms,[77] a mixture of Mahayanaism, Theravadaism and Tantrism.[78] Possibly this was also similar to the practices in the Burmese region and to the beliefs of the immigrating Burmese and Shans.[79] So all the region of north-east India, north Burma, north Siam and south-west China may have professed much the same beliefs (compare the mass scales). In 1253 AD Yunnan fell to the Mongols and about 20 years later, it received a Moslem governor who encouraged the immigration of people, including Persians no doubt and especially Moslems, from far parts of the Mongol empire.[80] Persian Buddhist influence may have been important in the choice of the lion-duck motifs for the weights.

Theravada Buddhism may have reached Thaton in south Burma about 240 BC. With the arrival of increasing numbers of southern Indians after the 2nd century AD, the Mon peoples of south Burma and Siam were introduced also to Hinduism and Mahayana Buddhism by Indians from the Amaravati region in India.[81] By the 4th century AD a mixed form of Buddhism was present in south, west and central Burma, e.g. among the Pyu people.[82] In the south the Theravada Mons of the Thaton region cooperated closely with the Theravada Mons of Siam as far north as present day Chiengmai for almost a thousand years.[83] Meantime, during the period of the 7th to 10th centuries AD, it appears that Shans of the eastern side of Burma and Yunnan may have been partly converted from animism to Mahayana Buddhism by Mongolian and Chinese missionaries. North Burma had close connections with Nepal and before 1000 AD was receiving Indian culture and Mahayana Buddhism through Assam and Manipur.[84] In the 11th and 12th centuries AD a revival of Theravada Buddhism took place at Pagan and from this time on, it became the dominant form of Buddhism in Burma and south-east Asia. Its ruler captured Thaton and took its monks and craftsmen north to Pagan where, together with monks from Sri Lanka and craftsmen from Bengal and Bihar, they played an important role in shaping Burmese Buddhism, culture, architecture and art.[85] In the 13th and 14th centuries AD there was a Theravada revival in Thailand, followed in the 15th century AD[86] by another in Ramannadesa,[87] which again called upon Sri Lanka for religious guidance. In the 1550s, when the Burmese conquered both the Burmese and Chinese Shans, it seems that the dominantly shamanistic and Mahayanaistic Shans were more or less converted to Theravada Buddhism.

Tantrism:

Tantrism was a form of worship associated with mysticism, magic utterances, gestures and diagrams which first began to become popular in India about the 5th century AD in both Mahayana Buddhism and Hinduism. Its sacred writings were the Sanskrit tantras which probably were written between 500 BC and 400 AD. Its practices were intended to obtain spiritual experience, satisfy worldly desires, control and purify bodily functions. It is popular in India and, most of all, in Tibet. About the eighth century AD Tibet adopted that variety of Mahayana Buddhism which incorporated Tantrism. Because of the close similarity of this variety to Bon shamanism most of the latter was absorbed into Tibetan Buddhism.

It has been inferred that Hindu, Mahayana Buddhist and Tantric influences from India, China and Tibet and native animism all played a part in local beliefs in northern Burma before the 11th century AD. Because of the relatively late date of the introduction of Tantrism in India and the liklihood of its effects being localised in Burma before the 11th century AD, the absence of Buddhism and Hinduism (?) from 7th century AD Assam, and the presence of dragon *(naga)* worship in Burma in the 4th century AD, perhaps associated with the Ari, it seems that the shamanistic sky-god beliefs derived from the early Mongol invaders from the north-east and/or from the 1st Century BC Saka invaders from the west, or the Tibetan Bons of pre-8th century AD together with the residue of the Neolithic earth god religions, may have been more important than the effect of Tantrism.

Notes

[1] Backus, p. 128. Harrison, p. 7. Page, p. 10. Salmony, p. 27.

[2] Watson, 1975, p. 60.

[3] Battisti, pp. 938, 945. Vajda, p. 793.

[4] Carter, pp. 18, 20.

[5] Battisti, p. 955. Ball, p. 240.

[6] Iyer, p. 9.

[7] Praz, p. 716.

[8] Higgins, p. 15.

[9] Iyer, pp. 2, 50.

[10] Rawson, 1984, p. 21.

[11] Noss, pp. 984-985. Wyatt, p. 4.

[12] Harrison, p. 7.

[13] Rapson, p. 35.

[14] E.g. Bock pp. 309,341.

[15] Most of this section has been adapted from Quaritch Wales.

[16] Dunnigan, p. 188.

[17] Siikala, pp. 212-213. Eliade, 1972, pp. 89, 93, 457, 499.

[18] Rice, 1969, p. 118.

[19] Yule (Cordier, ed.), quoting Ibn Batuta, v. 4, p. 130.

[20] Yule (Cordier, ed.), quoting Odoric, v. 2, p. 239.

[21] Eliade, 1972, pp. 12, 67, 80, 190, 198, 506.

22 Siikala, pp. 206, 208.

23 Eliade, 1972, pp. 406, 408, 428-441, 501.

24 Heissig, p. 405.

25 Keyes, pp. 512, 518.

26 Harvey, pp. 17, 313. Htin Aung, 1975, p. 130. Duroiselle, pp. 79 – 93.

27 Htin Aung, 1967, p. 314.

28 Harvey, pp. 18, 48.

29 Meserve, p. 241. Raevskii, p. 146.

30 Cameron, pp. 68, 69.

31 Kosambi, p. 72. Htin Aung, 1975, p. 129.

32 Eliade, 1972, p. 407.

33 Siikala, p. 214.

34 Eliade, 1972, p. 351. Spiro, p. 220.

35 Harvey, pp. 47, 166, 343.

36 Kosambi, pp. 102, 162, 168.

37 Ball, p. 107.

38 Harvey, p. 48.

39 Kosambi, p. 87.

40 Spiro, pp. 85, 95.

41 Litvinskii, p. 519.

42 Meserve, p. 241. There is another connection with a blacksmith. The *nat* lords of the great mountain Min Mahagiri are the spirits of a brother and sister who were killed, perhaps in 638 AD, by the king of Burma's first recorded kingdom, Tagaung, traditionally said to have been founded by an Indian prince driven out of India. Possibly he was a Sakan. Originally the highest in status of the *nats*, about 400 years later they were replaced by Thagya (Sakka or Indra). According to the legend, the brother was a mighty blacksmith. He was burned to death along with his sister under a 'Saga' (Saka?) tree. All horses and catle which subsequently sheltered under the tree died until it was cut down and floated downstream to nearby Pagan, where the statues of the brother and sister were carved from the tree and transported to Mount Popa (Mahagiri), where ever since the spirits of the two have been revered. Thenceforth each year their memory was recalled in a festival of lights, the sacrifice of many white horses and cattle and dances in the skins of animals. The association of the blacksmith with the steppe metal-working traditions, of the festival of lights with the Scythian fire cult, of the sacrifices of the white horses with the Scythian solar and horse cults and the 'animal dances' with shamanism seems evident (reading Sakan for the more specific Scythian). The name by which the sister is remembered, Lady Golden Face, together with all the other references to gold in the legend, suggest the Scythian cult of gold. See Htin Aung, 1975, pp. 61-67 and Harvey, 1925, p. 16.

It is possible that these events are commemorated by the beginning of the Burmese era in 638 AD, an era which was used also in Siam until 1887 AD, where it was known as the Chula-Saka-Raj(note the Saka) era. One of the Indian calendars (since 1957 the official one) is known as the Saka era and commences from 78 AD (Fillozat, pp. 721-722). It was used by the Pyu from 80 AD (Htin Aung, 1967, p. 16). The name Saka in all the above occurs too frequently for coincidence. One can conjecture that the myth of the Min Mahagiri really commemorates an adverse happening to the Sakan descendants of the founder of Tagaung.

[43] Ball, p. 91.

[44] Kosambi, pp. 112, 190.

[45] Htin Aung, 1967, p. 55.

[46] Spiro, p. 43. Htin Aung, 1975, p. 83.

[47] Spiro, pp. 133, 248, quoting Heine-Geldern.

[48] Eliade, 1972, pp. 276, 277.

[49] Litvinskii, p. 519. Shorto (1963) provides information showing that the concept of 32 *myos*, districts or countries, was also employed in Martaban in 1766 AD, in Burma as a whole in the middle of the 16th century AD, in Martaban in 1287 AD, in Thaton in 1087 AD and in the country of the Pyu about the 4th century AD.

[50] Ball, pp. 28, 30.

[51] Ball, p. 219

[52] Eliade,1972, p. 497.

[53] Waida, 'Frogs and Toads', pp. 443. Eliade 1972, p.68.

[54] Backus, p. 8, 21, 28.

[55] Rawson, 1983, pp. 15, 219.

[56] Eliade, 1972, pp. 442-445, 455.

[57] Zurcher, pp. 415, 416, 418.

[58] Ball, p. 166.

[59] Basham, p. 508.

[60] Ball, pp. 169, 170.

[61] Hamilton, p. 36

[62] Hamilton, p. 356

[63] Scott, 1910, pp. 37, 38.

[64] Goetz, p. 22.

[65] Goetz, p. 22.

[66] Burrow, p. 138.

[67] Goodrich, p. 90.

[68] Harrison, p. 20.

[69] de Groot, p. 555.

[70] Christie, p. 93.

[71] Goodrich, p. 160.

[72] Wylie, pp. 1110C, 1110D.

[73] Bagchi, p. 6.

[74] Sainson, p. 28. Terrien de Lacouperie, p. 6.

[75] Goodrich, p. 62. Sainson, p. 28.

[76] Backus, pp. 128, 129, 159, 160.

[77] Milne and Cochrane, p. 27.

[78] Cochrane, p. 152.

[79] Cochrane, p. 156.

[80] Goodrich, p. 174.

81 Taw Sein Ko, p. 184. Harvey, 1969, p. 440. Lester, p. 72.

82 Coedes, 1968, p. 63. Parker, pp. 15, 16. Hamilton, p. 358. Scott, "Buddhism in Burma", p. 39. Temple, "Burma", pp. 20, 21. Luce and Pe Maung Tin, pp. 290-292.

83 Wyatt, p. 76.

84 Coomaraswamy, pp. 169, 172.

85 Temple, 1910, p. 20.

86 Wyatt, pp. 38, 39.

87 Lester, p. 78.

11
The Symbolism of the Bird Weights

Continuity:
One of the most interesting of the several symbolisms of the bird weights is their continuity from the 15th century AD. This continuity reflects that of the Buddhist Burmese monarchies. It was important to the legitimacy of each new dynasty of united Burma to show its descent from the rulers of Pagan. In this way they could justify the monarch's right to rule Burma according to the *chakravartin* ideal of rightfulness. On the other hand, the beast weights symbolising the *chakravartin* were produced to show justification for the rightfulness primarily, but not solely, of his non-Burmese conquests. These do not show the continuity of the bird weights.

The identity of the weight-bird and its symbolism:
The bird shapes on the weights have been identified as anserine (gander, drake) on Groups 1 to 5 and 7. Other possible bird shapes and associated symbolisms examined were those of the fowl (*gallus* species), the peacock, the eagle, the *garuda* and *feng huang*. None fitted the available information as did the anserine bird but the *feng huang* appears to have influenced the shapes, most markedly in Groups 1 and 2(v).

The shaman is closely associated with birds and has been since ancient times, as evidence from artifacts shows. In the Altai mountains, Siberia and elsewhere the goose is the preferred form of animal in which the shaman chooses to manifest himself (see 'Religious Beliefs: Shamanism').[1] Images of waterbirds, especially the swan, symbolise the shaman's immersion in the aqueous underworld. Swans and geese, like the horse, are the animals most closely associated with the shaman's journeys to the underworld and heavens.[2] A similar association appears in Burmese history according to which, in 1555 AD, the town of Ava was attacked, the journey by river to the town being made in war canoes shaped like ducks and horses and some other animals.[3] In pre-Buddhistic Vedic symbolism the gander is closely associated with Indra (Sakka) and appears on the pre-Asokan pillars. The bird forms a link between the monsoon rains which follow just after its migrations from the parched plains and the abundant waters which are present on its return. Thus it forms a link between heaven and earth and symbolises the flight of the soul. In Vedic mythology and ritual the pillar linking heaven and earth was founded in the waters below the earth where all life originated. A similarity between the Vedism of the Aryans and the shamanism of the steppe nomads is apparent.[4] The anserine bird was associated in Buddhism with spiritual purity and gentleness.[5] Unlike all the other bird symbolisms examined it was associated with the purity of water and with the purity of the lotus flower. Because of its purity it could repel dross and attract the pure and good, as for example, differentiating between the more impure and purer silver alloys and differentiating between false and correct weights. In Hinduism, Brahma, the creator, rides on a gander (*hamsa*), symbolising the god's creative principle. The epithet 'gander' is applied to Hindu ascetics and teachers to symbolise their intellectual creativity, the term *parahamsa* being applied to those who have been spiritually awakened.[6]

This symbolism is translated into folklore terms in India and China, where both the Brahminy duck (or *chakravaka*, i.e. *anser casarca*)[7] and the mandarin duck (*aix galericulata*) are associated with marital fidelity, connubial bliss and "the perfect union towards which even the celestial beings fly".[8] In China, about 1500 BC the wild goose symbolised ancestors and hence the sovereign.[9] See also 'The Symbolism of the Anserine-Feline Combination'.

Turning to the *feng huang*, which appears to have influenced the weights in varying degrees, the qualities of this bird would appeal to Buddhists because it abhors the taking of any kind of life and is particularly associated with righteousness (e.g. the crest), gentleness and all the virtues. Its vaulted tortoise back (back cover?) may have symbolised oracular wisdom or right behaviour. To the Chinese animist it symbolised the sun and its life-giving warmth, sovereignty over the birds, the superiority of its status being indicated by its beauty and by both the number and length of its tail feathers. From before 600 BC its appearance had come to indicate the rule of a wise and just emperor over his subjects. By about 100 BC it had come to symbolise the empress. In more modern times its main symbolisation was that of feminine beauty.[10] It was noted for its intelligence and its erudition. The last two qualities were shared by the mandarin duck. With the arrival of Buddhism the choice of the mandarin duck as a model for the *hamsa* would have been understandable. There are in China many legends connecting Gautama Buddha with the mandarin duck.

Concerning the symbolism of the Groups 5 and 6 weights, in the case of the latter it is considered that the bird representation may be a special style of *hamsa* or an unknown bird kind from Siam (Martaban). However, there remains the problem of the weights with either two small birds by the sides of the large one or one small one in front. There appear to be four alternatives to explain these occurrences. Firstly, the large bird, if it be the *hamsa*, could represent the Buddha. The two small birds then could represent the two disciples often accompanying him, i.e. Moggallana and Saraputta. But then what is the explanation for the one small bird in front of the large one that occurs on some weights? Secondly, Lower Burma once had a queen ruler, Shinsawbu. At first, being the sister of the then king of Lower Burma, she was married to the king of Ava in 1425 AD. She left him in 1429 AD and returned to Lower Burma along with two monks from her own country. In 1452 AD, in the absence of male heirs to the previous ruler, she became the ruler of Lower Burma. Later, one of the two monks relinquished his religious status at the queen's request and was made heir-apparent while the other was imprisoned and later executed. The queen ruled for another seven years and then decided to retire from her duties but not her rank. The remaining former monk assumed the duties of government under her rule until the death of the queen in 1472 AD, when he became king.[11] He died in 1492 AD as a *chakravartin.*

Both the queen and her successor were highly regarded for centuries after their deaths. Their qualities and reputations suggest an association with the spiritual meaning of the *hamsa.* The old name for Lower Burma recorded by foreigners was Ramanya or Ramannadesa (country of the Mons).[12] It was used in connection with the queen's successor in 1475 AD. This name was superseded by that of Hamsavati,[13] apparently first recorded by Europeans in 1608 AD.[14] That is, it seems that the name by which the country of Lower Burma was known to foreigners changed to one incorporating the name *hamsa* at some time between 1475 and 1608 AD, thus establishing a particular association with the *hamsa.* This in turn could be associated with the religious revival in the region, culminating in 1475 AD and so with the two sovereigns, Queen Shinsawbu and King Dammazedi. At the Shwemawdaw pagoda in Pegu, the former capital of Lower Burma, there is an effigy of two *hamsas* standing one above the other in order of decreasing size upwards. Related to this there is a legend concerning the foundation of Pegu, the site of which was supposed to have been located by a male and female *hamsa* looking for a place to rest. The site was in a lake and was so small that only the male could land and the female had to alight on his back. In passing there is a saying in Pegu that one does not marry a Pegu girl because she will forever be riding on one's back, i.e. she would rule one.

The significance of the foregoing may be as follows. The queen, because of her virtue, was regarded as a *hamsa.* The *hamsa* is a gander.[15] Therefore, as one way to suggest her femininity on the Shwemawdaw effigies, she could be shown in a smaller size than the *hamsa* representing one of the monks, but to indicate the hierarchical relationship, the small effigy would be shown standing on the back of the large one. On the Group 6 weights, however, where strength to resist handling was required, Queen Shinsawbu could be shown as the large bird in the appropriate brooding position of a hen or duck, with her feet usually hidden as female propriety required. On the Group 5 weights the femininity could be indicated by a broad (laterally-spread) tail indicative of the duck, as compared with the more pointed tail of the drake, and a generally feminine, non-predatory appearance indicated partly by rounded facial contours. In both cases the two monks could be most suitably shown as small birds, one close by each of the two sides of the large one. With the removal of one monk upon her retirement, these weight effigies would show but one smaller bird, which would be positioned in front, presumably for symmetry. The one large bird alone on its base would represent the monk who became king upon her death.

The third alternative concerns the only reference to a hen, and a sitting hen at that,[16] which has been found. Crossing the ruined walls of Thaton there is a gap nowadays referred to as 'the sitting hen' gap. This marks the place where King Anawratha of Pagan entered the city during his final and successful assault on the town in 1057 AD. Nearby are statues of the two heroes of the attack. These heroes were Indian Mahometan brothers who had been shipwrecked at Thaton. For various reasons subsequently one was killed by the Thaton king's men, forcing the other to flee to Pagan where he joined the Burmese army. When Pagan attacked Thaton he was sent out as a scout. While he was reconnoitring the walls the spirit of his executed brother approached him and told him that the town walls were impossible to cross except at one place where there was a gap, though the gap was only as wide as a sitting hen. The gap sufficed for the successful assault, the second brother being killed during the attack. The symbolism is evident, the sitting hen with the two brothers, then with one brother, then with none. The difficulty here is that, providing the weight is a Mon weight of the 15th century, then it seems unlikely that the Mons would commemorate a defeat by an enemy of four hundred years previously.

The fourth alternative is that the three kinds of Group 6 weight, i.e. two small birds, one small bird and no small bird, represent the three countries of Ramannadesa. (See "The horse, the horse tail and the *ch'i-lin*"). The problems of the script and the manner of manufacture may be explainable if the weights were made in Siam to the order of the Mons or the same manufacturing techniques were used in Ramannadesa as in Siam and the script be an old local form of Mon or Thai.

Turning to the Group 4 weights the introduction of this new style was associated with the introduction of the new dynasty. The squat form may have been chosen for a mechanical reason rather than a symbolic one, the squat sturdy Group 4 style being ideal for weights as compared with the easily-broken design of the Group 6 style and the space occupied by the tall Group 5 style.

The subsequent replacement of the Group 4 style by the Group 2 style, however, probably had not only a dynastic significance but also a political one. The early rulers of re-united Burma from 1531 until 1635 AD preferred to retain the Mon capital as the capital of united Burma, there being, initially, but little animosity between the Burmans and Mons. The style of bird weight introduced in the mid-16th century by the new dynasty was still known to local dealers in 1972 AD as a Mon duck. After the Mon rebeliion of 1740 to 1757 AD the Burmese victors began a genocidic process of destroying the Mon people, their language, their literature and their culture. This probably included the cessation of manufacture of the Group 4 weights, which, like the weights before them, were associated with the Mons.

The symbolisms of specific parts of the bird shape:

The bird shapes on the weights of Groups 1, 2 and 4 may have one, two or three prominent head knobs. On Groups 3 and 5 there are three knobs. When discussing the weight beast tails later, the likelihood of the tail-shapes of the Group 5 beast weights symbolising the three countries of Ramannadesa is considered. It may be that the three head knobs of the Group 5 bird weights had the same significance. The fact that the three head knobs overlap may have symbolised the unity of the three countries within Ramannadesa. Henceforth, when Ramannadesa had become part of united Burma, there is only head knob on the weight-bird until periods F and G when two and three head knobs are also found. These introductions are inferred to have been made at about the time, between 1740 and 1757 AD, when Lower Burma regained the independence it had formerly enjoyed as Ramannadesa. All this suggests that the former single head knob may have symbolised united Burma. The stylistically contemporaneous occurrence of one, two and three head knobs suggests a revival of the symbolism of the three countries of Ramannadesa just as the silver medallion/ amulets bearing one, two and three hamsas produced at about the same time, or a little later, are inferred to do. (See Plates 50 and 51.)

The vertical position of the tail on the Groups 7 and 6 weights is distinctly different from the lateral disposition on the bird weights of Groups 5 to 1. The vertical position is accompanied by convolutions on the tail-sides. This style occurs at Pagan and was common in India in the later centuries of the 1st millenium AD. It seems to be decorative in function and without meaning. With the Group 5 weights the Indian influence on the styles disappeared, being replaced by a Chinese influence, a change probably due to the greater use of Burma for Chinese trade at this time. (See the sections Trade and Routes). The lateral (duck-like) tail attitude occurs first on the Group 5 weights where, with the rounded feminine contours and the occasional accompanying small bird, it may have been intended to signify the feminity of Queen Shinsawbu. This lateral U-tail shape continued in Group 4 with the "Mon ducks" and into Group 2. After the quelling of the 1740-1757 Mon rebellion by the Burmese, the drake-like lateral V-tail shape of Groups 3, 2 and 1 came into being coupled with a facial appearance that can be called predatory, masculine, aggressive or even garuda-like. It may be that this style was introduced to symbolise the Burmese conquest of Queen Shinsawbu's people, parts of whom still practise a snake-cult (cf. the garuda versus the snake legends). Since the V-tailed Group 1 was the last of the weight styles, appearing after the annexation of Lower Burma by the British in 1852 AD, it may be that this style was associated particularly with Upper Burma while the U-tailed weight may have been associated particularly with Lower Burma, the country of the Mons.

Unlike the mouth appendages of the beast weights, those of the birds can be seen to emerge from between the mandibles in Groups 3, 5 and 7. The mouth appendages

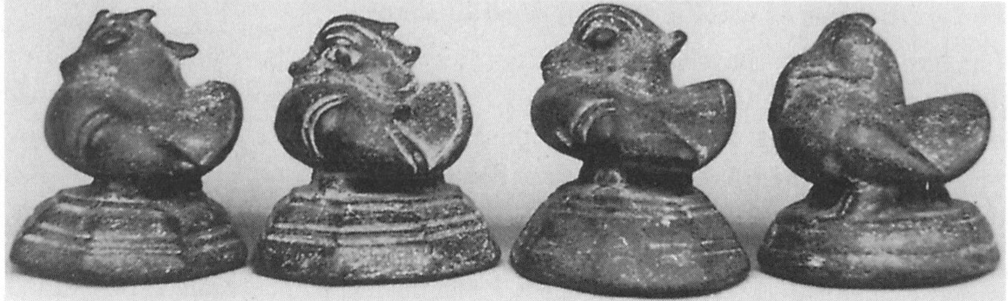

The one, two, three head knobs on Group 4 bird weights of ca. mid-18 c. AD. Possibly introduced by the newly re-independent Mons to symbolise Ramannadesa

One, two and three head knobs also occur on the bird weights of Groups 1 and 2, though the same symbolism may not apply.

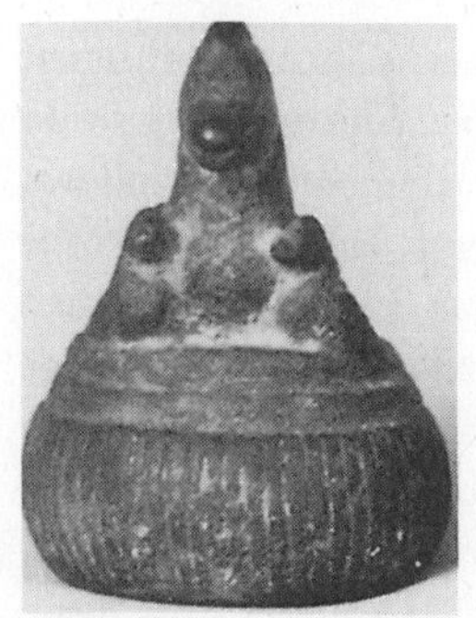

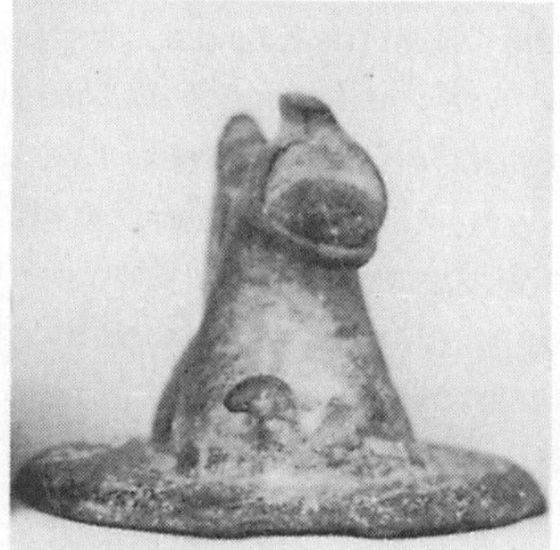

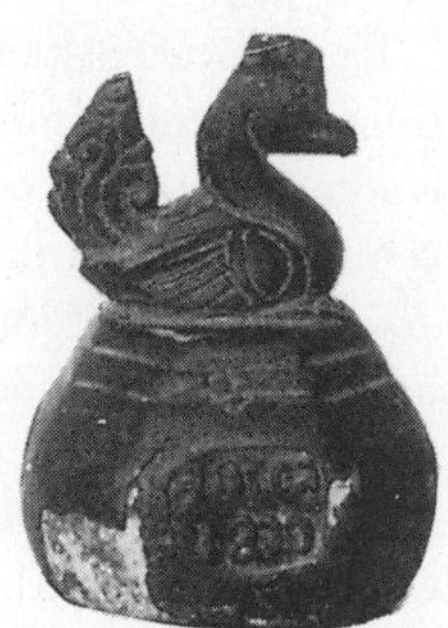

The one, two, three birds on Group 6 weights of the second half of 15 c. AD.

Plate 50: Possible Symbols of the Three Countries of Ramannadesa

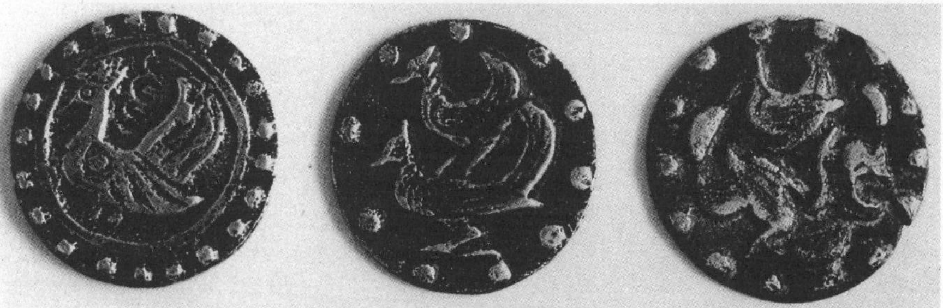

The one, two, three hamsas on silver amulets possibly used by the Mon resistance movement in the Tavoy region in the second half of 18c. AD

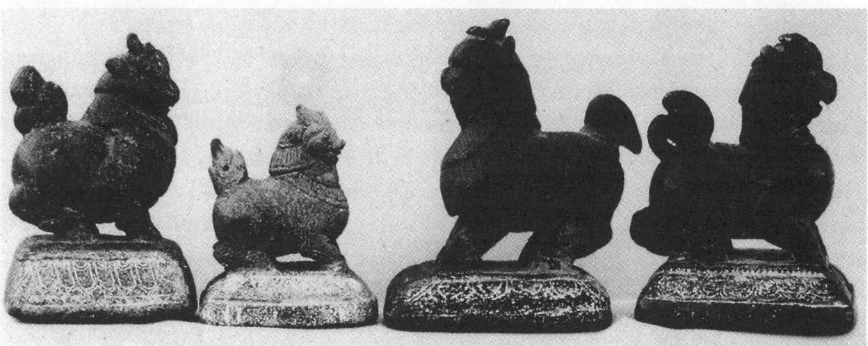

The three tail shapes of the weights probably of the "Lion King" Razadarit, 1383 – 1423 AD

Plate 51: Possible Symbols of the Three Countries of Ramannadesa (Pegu, Myaungmya, Bassein)

in the shapes of spatulas and trefoils on the bird weights of Group 4 emerge from below the mandibles and may represent lip extensions or just a style change. In India and elsewhere, the spouting from the mouth of fruit, flowers, sprays, pearls and the like, symbolises an abundance of good things.[17] In China, mandarin drakes bearing lotus vegetation between their mandibles symbolise not only married happiness but also the hope of sons.[18] If the vegetational emanations are intended to represent the lotus, as in India they often are, then purity may be intended, so emphasizing this quality in the *hamsa*. Support for this view is to be found in the commonness on the mouth appandages of the trefoil, presumably symbolising the *triratna* (Buddha, the Law and the Community) though the Scythians used a similar shape for some of their arrowheads. The mouth appendage on the Group 5 bird weights may also be associated with the following interpretations.

From 1423 until 1539 AD when Pegu, the capital of Rammanadesa, fell to the army of Toungoo, the region may have enjoyed the happiest and most prosperous years of its existence.[19] It was this prosperity which attracted the attention of their conquerors, the Toungoo army of the Nyaungyan dynasty in 1531 AD. In 1475 AD King Dammazedi brought to its peak a renewal of Theravada Buddhism in Burma, a renewal in which he was joined by the king of Lan-Na in northern Siam. This renewal extended over mainland south-east Asia so that once again part of present-day Burma was a spiritual centre. The combination of spiritual renewal and material prosperity could be represented by the Group 5 *hamsa* symbolising prosperity. Another interpretation is that the Group 5 weight represented, possibly in addition to the first interpretation, the hope for a noble son. The ruler of Burma before Dammazedi was Queen Shinsawbu, who had no live male members in her family. Her daughter she considered too gentle to rule by herself so the daughter was married to the former monk Dammazedi, perhaps in the hope of begetting a son, a hope which was fulfilled. This symbolism could have been incorporated into the symbolism of the Group 5 weights by the means of the 'emanation' from the mouth.

In each of the Groups 3, 4, 5, 6 and 7 there occurs, very rarely, a small bird attached to the chest of the larger one, the beaks both touching or almost so. The rarity of these weights and their occurrence in so many bird weight groups suggest that they may have been weights kept for standardising purposes. The small bird 'mouth appendage' is dove-like in appearance and would seem to symbolise the Sibi *Jataka*, just as it does on balances in China.[20] Buddhism prohibits the use of incorrect weights[21] in the same way that it prohibits all forms of lying and stealing. This injunction is illustrated by the legend in which King Sibi, an incarnation of the Buddha-to-be (i.e. a *bodhisattva*), sheltered a dove from a pursuing hawk. The hawk, who was Indra, protested, saying that if it did not eat the dove as nature intended, it would die. The king offered to the hawk an equal amount of his own flesh. The dove, however, kept increasing in weight so that the king, to keep his word, had to increase the amount of his flesh placed on the balance which was weighing it. Finally, he placed all of his body in the pan and so paid what was due and kept his promise.[22]

The symbolism of the bases:

Concerning the hexagonal base, the foetus which became the Gautama Buddha is commonly said to have been a six-tusked elephant which entered the side of his mother, Queen Maya.[23] An alternative account states that it was a six-rayed star.[24] The six-rayed star is a common sign, mostly on the bird weights. Every Burmese boy at some time enters a monastery where he is taught the six excellences of the law[25] which in turn may be related to the six refulgent rays which emanate from the Buddha's body and possibly symbolise the six virtues. Since the overlying animal shape is a hamsa, it seems likely that the appropriate six-sided base symbolisms relate to heavenly rather than earthly matters.

The symbolism of the eight sides of the octagon also may be heavenly rather than earthly. It may symbolise the Tusita heaven where the *bodhisattvas* dwell.[26] It was also an ancient and widespread custom to place a chief deity in the centre of eight minor ones,[27] as in the case of the nine planets. This is the symbolism of the eight sides of the second main layer of the Burmese pagoda.[28] There is another possible symbolism in the case of the Group 6 bird weights. The octagon is an intermediate form between the square and the circle, that is, between earth and heaven. Some of the Group 6 bird and Group 5 beast weights have eight pairs of vertical grooves, one on each corner of the octagon. These may represent the eight pillars of heaven which in Chinese Buddhist mythology support heaven above the earth.[29] These in turn would be related in their symbolism to the number eight and the eight pairs of heavenly elephants created by Brahma.

The base in the shape of the circular truncated cone on the bird weights may have a different symbolism from that of the beast weights. The circle represents the highest of the heavens and this may be the meaning.[30] It also may represent the Wheel of the Law, the *dhammachakra*, which Gautama Buddha set in motion with his first sermon. A somewhat similar symbolism of the circle existed in China where the perforated jade disc, the *pi*, was a symbol of the heavens.[31] The circular base also occurs as part of a truncated near-hemisphere, the overall shape being of an inverted basin or umbrella. Typically this is symbolic of the heavens.[32] The Chinese viewed the heavens, the sky, as an inverted bowl rotating on an axis through the Pole Star and the earth's centre.[33] There is also a cylindrical base with vertical or oblique striations on its sides. It occurs mostly on the bird weights of Groups 7, 6 and 5 and very rarely on the beast weights of Group 5 with the lotus bud tail. Clearly it has a special symbolism. Speculating, it could represent the pillar of the Tien people of Yunnan, which was associated with the worshipping of the earth god. Alternatively, it could be connected with the sky-supporting pillars of the shamans of Siberia.

Seven-sided, nine-sided and ten-sided bases have been seen on the bird weights but these are rare. In the shaman's ecstatic techniques throughout central and north Asia, the number seven plays an important role, one which is due ultimately to influences from Babylon.[34] On his costume, a Yurak shaman may have seven balls representing the seven celestial maidens. There are also the common beliefs in seven or nine each of celestial and infernal levels, though rarely, up to 33 occur. In his rites the Altaic shaman climbs a tree or a post notched with seven or nine steps to symbolise his ascent to the most powerful one. The seven steps are similar to the Buddha's seven steps mentioned below, a concept derived from Buddhism's parent, Brahmanism.[35] In legend Buddha could walk immediately after his birth. He took seven steps in the direction of each of the cardinal points and claimed possession of the world. Seven days after Gautama Buddha's birth his mother died.[36] After Buddha's enlightenment he meditated for three periods each of seven days. After these he was wrapped in seven coils of the serpent king, Mucalinda, and endured continuous rain for seven days.[37] Seven also symbolises the horse, one of the seven treasures of the *chakravartin.*[38] This shape is associated in the Group 7 bird weights with the double pyramid, one being inverted upon the other. The cross-section is the same as that of the altars which support images of the Buddha and form the seat of the Lion throne of King Mindon of Burma.[39] There the shape represents two lotuses, one inverted on top of the other. Because the lotus appears to spring from pure water and not from the underlying mud, it has become a symbol of purity and divine rebirth. Hence the symbolism of the cross-section is that of the divine, the spiritual and the pure. The shape has other symbolisms. It can represent the royalty of Gautama Buddha. It could also represent Mount Meru or have an older but related symbolism, that of the pillar linking earth and heaven.

In Burma an essential form of instruction received by every Burmese boy when he enters the monastery for his obligatory period is that of the nine excellences of the Buddha. For more information on the importance of "nine", see "The Sign and Style Symbolism: The nine-rayed star". Concerning the ten-sided base, a candidate for Buddhahood must possess the ten virtues in their most perfect form. The number ten is associated with the ten powers of the Dasabala, which is one of the titles of the Lord Buddha.[40]

The symbolism of the base decoration

The horizontal grooves, flutings and lines on the bases of both bird and beast weights, like the steps on the bases of some elephant weights, may have had their ultimate origin in the megalithic stepped pyramids. The semi-circles, ogives, inverted V's and U's most probably represent petals of the lotus. The significance of the short verticals is not known. The ring of small bosses or impressed circles occasionally found on older weights may have stemmed from the coins of the ancient Pyu who could have obtained the motif from the Indian Gupta or Pallava dynasties, who in turn, obtained it from the earlier Sakan rulers.[41]

Notes:

1 Eliade, 1972, pp. 89, 149-155. Siikala, p. 213. Rice, 1969, p. 118.

2 Dunnigan, p. 188. Eliade, 1972, pp. 153, 191, 404.

3 Harvey, p. 164.

4 Irwin, 1, p. 709; III, p. 642; IV, pp. 741, 742.

5 Iyer, pp. 82, 83. (See also all of Vogel, 1962)

6 Zimmer, pp. 47-50.

7 Armstrong, p. 440. Liebert, p. 53.

8 Ball, p. 239.

9 Waterbury, pp. 118, 119.

10 Ball, p. 27 ff. Cooper, pp. 130, 175. Eberhard, p. 235.

11 Page, pp. 25, 26. Harvey, pp. 117-118. Taw Sein Ko, pp. 125-128.

12 Page, pp. 9, 14. Hall, p. 143. Luce and Pe Maung Tin, pp. 389, 393. The name for the region of Lower Burma in the old language of the Mons was Rmen whence the classical name Rammanadesa. Rmen is found in Javanese inscriptions of 1021 AD.

13 Zimmer, p. 104. He connects the word Irrawaddy, the name of Burma's principal river, with Iravati, the name of a great river in the Punjab. In Sanskrit, rivers and waters are feminine, maternal. So he translates Iravati as follows: Ira (of fluid), vati (she who is possessed). By analogy the region and river Hanthawaddy would become Hamsavati and be translated as: Hamsa (of the [divine spirit of the] *hamsa*), vati (she who is possessed). Literally this would mean 'the river of the *hamsa*'. Symbolically it could refer to Queen Shinsawbu, i.e. she who is possessed of the qualities of the *hamsa*. This would be a way of overcoming the possible problem of her sex in representing her as the customary male *hamsa*.

14 Yule and Burnell, p. 184.

15 In Buddhism women are a primary source of desire and temptation and so are regarded as little more than a necessary evil on the path of salvation. The *Jatakas* have many tales illustrating how men have been led astray by their wives. This attitude was in opposition to the matriarchal antecedents most evident in Mon, Shan, Yunnanese and Szechwanese cultures. Even in the mid-14th century AD it was reported that the Mongol men treated the women with great consideration as if they were of a higher rank than they. (Ibn Batuta, quoted by Yule and Cordier, v. 4, p. 6). When in Burma, the writers were told by a Burman lawyer that the Burmese legal system was more enlightened in relation to women than that in Britain. This opposition may have presented the weight designers with a problem in symbolism. There was yet another problem since, as a female, Queen Shinsawbu could not be a *chakravartin* so that she could not be represented by a beast weight. Other possibly relevant matters are that the Lolo people of Burma and Yunnan have a creation myth, according to which a white hen produced good people and a black hen, bad people. In China a hen with its chicks symbolises good relations between parents and children. In this case it would symbolise good relations between ruler and ruled (Eberhard, p. 144). The hen can also drive away evil spirits (Williams, p. 24) and in north China it was the object of a special cult. The cock is a symbol of lust and so not highly regarded by Buddhists. Presumably a hen would be

even less well regarded. It is possible that the symbolism was mainly political while the legend of the male and female *hamsas* and the new country name Hamsavati may have been derived subsequently to provide religious justification for the situation.

[16] Myo Chit, p 82.

[17] Herrmann, p. 143. Jobes, p. 677.

[18] Lessing, p. 143. Eberhard, p. 177.

[19] Harvey, p. 121. Htin Aung, pp. 101-103.

[20] Needham, 1962, vol. 4, pp. 24-27. Equal-armed balances were common in China, especially for the smaller weights, from about the 4th century BC., The two arms were suspended from a bar supported on two posts. On the bar there was often the representation of a bird. It was the dove waiting for the flesh of King Sibi (of the Sibi *Jatakas*) to be weighed accurately.

[21] de Groot, p. 554.

[22] Larousse, p. 356. Hallade and Lanciotti, 'Myth and Fable', p. 437.

[23] Ball, p. 79.

[24] Coomaraswamy, p. 34.

[25] Scott, 1910, p. 40.

[26] Scott, 1910, p. 43.

[27] Lowry, plate 12.

[28] Htin Aung, p. 7 ff.

[29] Eberhard, p. 93.

[30] Taw Sein Ko, p. 178.

[31] Anderson, p. 138.

[32] Cooper, p. 82.

[33] Christie, pp. 56, 57.

[34] Eliade, 1972, pp. 274-279.

[35] Eliade, 1972, pp. 405-406.

[36] Larousse, p. 348.

[37] Zimmer, p. 67.

[38] Liebert, p. 255.

[39] Anonymous, *Mandalay Palace*, pp. 24, 32, 33.

[40] Jobes, p. 1544.

[41] Gutman, p.12.

12
The Symbolism of the Beast Weights

The identity of the weight-beast and its symbolism:
The weight-beast has been shown to consist of elements of a large feline, probably the lion, the horse, the stag and the elephant. The horned equivalents of the foregoing are the lion-griffin and the composite creature, the *ch'i-lin.*

Certain parts of the weight-beast carry particular denotations essential to the symbolism of the whole weight but which are not essential parts of the symbolism of the animal representation itself. The symbolisms of the base shapes, for example, are intimately related to the weight-beast shapes just as they are to the weight-bird shapes, though sometimes differently. In the totality of animal and base the whole beast weight is a symbolic representation of a combined *bodhisattva* and a *chakravartin* i.e. a divine universal monarch. The likely origin of the elephant on the north Siamese weights is also derived from the symbolisms of the lion-griffin and the *chakravartin.*

The tiger and the lion:
In order to avoid repetition and confusion, the symbolism of the feline part of the beast weights is here described under the concepts of lion, horns and horned lion/lion-griffin. However, initial reference to the tiger is required, partly because of the much greater familiarity of the steppe nomads, the Chinese and the south-east Asians with the tiger rather than with the lion, and partly because of its similar symbolism, e.g. of courage, royalty and especially power.

In south-east Asia, north Asia and elsewhere the tiger was important in the shamanic rites practised by Scythians, Sakas, Medes and others. It was invoked by shamans both when healing and when a neophyte was being initiated.[1] In Vedism it was the emblem of the *kshatriyas,* the warrior caste of Gautama Buddha,[2] because of its fierceness. The Persian equivalent of the word 'Kshatriya' was in use by the Medes about 600 BC.[3] The tiger remained important in Chinese symbolism. For example, about 800 AD in Nanchao (Yunnan) where the people were still animistic, they preferred the tiger, status being indicated by the amount of tiger skin a person was permitted to wear.[4]

Turning to the lion, though in Egypt it symbolised the sun god-king, as in Asia Minor,[5] it also symbolised the god of the underworld and of the dead when coupled with the crescent moon.[6] Here then began the association of the lion with the earth and its role as a tomb guardian, a role much used in China. Here too arose the practice of placing one of a pair of lion shapes on each side of a temple portal in order to guard the interior against the entry of evil spirits. This practice spread to China and Burma, among others, the figures being known as *chinthes* in the latter country. It is in ancient Babylon and Sumeria that animals first appear as steeds (vehicles) to transport and symbolise deities. That of the earth goddess was a lion, that of the rain and thunder god was a bull and that of the sun was a horse.[7] Here the beginnings of the symbolism of the *chakravartin* and the Burmese weight-beast become evident.

In China after about 800 AD lion shapes became protectors of the portals of wealthy people, the number of bumps on the lion's head, along with the size and posture of the creature, displaying the status of the owners of the mansion. Though the lion symbolised Gautama Buddha it also symbolised Manjusri,[8] one of his disciples who had a cult of his own. It may have been this cult which led to the practice common in China and south-east Asia of placing lion images before the household shrine.[9] One of the Burmese Group 4 weights lacks a mane (the only beast weight without one), so it could be a maneless lion or lioness. The lioness also symbolises the earth and is particularly associated with the female consort of one of the Mahayanist *bodhisattvas*.

The symbolism and mythology attached to the lion in Burma is mainly Buddhistic. As in India, Gautama Buddha is referred to as Sakyasimha, 'the lion of the Sakya clan'. His postures are known as lion postures, his doctrines are the lion's laws, his sermons are known as the lion's roars[10] and Burma's kings sat on lion thrones.

Horns:

Horns were associated with power[11] which might be physical, divine, magical, anti-demonic, intellectual, virile, regal, solar or lunar.[12] Antlers, which re-grow annually, were symbols of fertility, rebirth and were employed in funeral processions.[13] On composite creatures they are either relics of the Paleolithic/Neolithic animal gods or additions made to convey one of the meanings mentioned above.[14] They help to ward off evil and evil spirits and facilitate communion with the earth spirits.[15]

In central and north Asia horns are indicative of the ecstatic flight of a shaman, mounted on a stag or reindeer, into the mystical realms. One of the three main types of costume worn by a central Asian shaman is that of a stag, using its skin, and a cap bearing its horns. He uses these to facilitate communication with the earth spirits.[16] His drum is often called a 'black stag' since it is believed to transport him. Some of his spirit familiars appear as stags.[17] In the Altai about 600 BC stag horns were being employed on effigies as tomb guardians. In the Pazyryk tombs of the 3rd to 1st century BC antlers converted the horse to the sun-animal of the nomadic creation myth.[18] In pre-Scythian times the stag was a totem and continued to play a mythical role in Sakan culture, thereafter becoming a being which may have been deified.[19]

The stag was worshipped as a god in Yunnan[20] and on the Eurasian steppes,[21] especially by the Scythians and Sakas.[22] It was similarly worshipped by the ancient Sumerians. The deer in ancient Egypt was sacred to Isis. It was a holy animal in Anatolia. It was the steed of the Hittite god of the animals and other deities.[23] In China non-Buddhist mythology referred to the dragon as the celestial stag but the stag motif was not extensively used.[24] Shen-Nung, one of the three legendary god-emperors of ancient China, wore two horns on his head rather like the shamans. Like them too he practised healing. He was god of nature and animal life.[25] In Chinese Buddhist mythology the deer (doe) was sometimes regarded as one of the three senseless animals.[26] In India the deer was not worshipped, its holiness in Hinduism being due to its symbolisation of Siva as lord of the animals, because it was the vehicle of the wind god[27] and because it was associated with meditating holy men. Hindu myths also include those of a *chakravartin* who temporarily entered the body of a golden stag in order to gain identity with the sun.[28] Representations of the stag rarely if ever appear in Indian art until the advent of the Sakas during the 2nd century BC. In Buddhism almost everywhere the deer symbolises Gautama Buddha's first sermon in the deer park at Sarnath and it is frequently used allegorically in the *Jatakas.* The most important qualities which it symbolises are gentleness and swiftness.

At Sarnath stands one of Asoka's edict pillars. On its abacus appear decorations in the form of the bull, the lion, the elephant and the horse. All of these are Buddhist and pre-Buddhist Indo-Aryan symbols. They are also the animals parts of which appear on the weights and symbolise the *chakravartin..* The bull was revered throughout Mesopotamia, Persia, the Indus valley and by the Aryan invaders of India. This was not so with the Chinese. For the Indo-Aryans it was one of the symbols of royalty, of Sakka (Indra), of Yama, who was later adopted by the Buddhists as god of the dead, and by the Hindus of Siva. It symbolised especially strength and aggressiveness, virility, fertility, rain and storms.

The griffin:

The lion and the griffin both contribute to the symbolism of the lion-griffin, the horned lion with or without wings. Characteristically they all symbolise courage, strength, swiftness, endurance, vigilance and guardianship, especially of mines and treasure in the case of the griffin. They also signify enlightenment and wisdom. They mean royalty, the power and light of the sun. The griffin frequently implies aggressiveness, rapacity and lust. In China the griffin symbolises Tou, the name of the constellation of six stars in the shoulder and bow of Sagittarius. There is another association between the earth and the lion-griffin. It is a legend that China's flood water control ditches were dug by a winged and horned dragon with its tail because the earth was being submerged by flood waters. This took place in the time of the legendary emperor Huang Ti, supposedly about 2700 BC. A 6th-5th century BC representation of this 'dragon' is a lion-griffin, complete with mane and wings, though the lion motif is supposed not to have reached China before the 3rd century BC. The griffin is frequently symbolised, like the lion, as the enemy of the horse or unicorn, a significance that might stem from inter-nomadic conflict.

The dragon:

Since the connection between the griffins and the dragons has been referred to when discussing the identity of the weight beasts, the symbolism of the dragon is mentioned here. In China the dragon is a good-natured creature. It, like the *naga* in India, is associated with water in all its forms, especially water necessary to plant growth. The five-clawed dragon, the *lung*, symbolised the emperor. The azure dragon symbolised new life, rain and the East. The dragon is particularly associated with the number nine (recalling the nine-rayed star sign) which symbolises male vigour. Chinese Buddhism abounds with legends of dragons. A primitive form of dragon, a *k'uei* – dragon, is believed to drive away the sin of greed, a belief appropriate to its use in commerce. Most kinds of dragon symbolise watchfulness and protectiveness, e.g. dragons guard Chinese temples against evil spirits just as the *naga* does in Burma.

In Burma the dragon dance, with dragons up to 30 metres in length, was used at the Tawadeintha feast of October/November. This marked the Buddha's ascent to the second of the *nat* (nature spirit) heavens, the Tawadeintha, to preach to his mother.[33] This festival was followed shortly afterwards by the Sandawgyi feast to celebrate Buddha's enlightenment. In this case it appears to have been the lion dance (or, using another name, tame *to-naya* dance?) which was used, the lion being a creation with one man or two men inside.[34] The writers were informed that somewhat similar animal shapes which were known as *to-nayas* or *tos* were used at funerals of monks and of Shan chiefs.

The forehead protuberance:

The protuberance on the forehead between the weight-beast horns represents a third eye, the all-seeing eye, the eye which sees all things in their entirety, the eye of transcendent wisdom.[35] All Buddhas and *bodhisattvas* have a usually lens-shaped mark on their forehead, symbolising the third eye.[36] It is associated particularly with Mahayana Buddhism and also with the Indian god Siva and the Chinese goddess Tou-Mu. It was because the kings of Burma were potential Buddhas that their representations on the weights bore them. The titular names of the kings often indicated their existence as *bodhisattvas. Paya* or *hpaya* means a god or Buddha in Burmese. Kings Bodawpaya and Alaungpaya are best known by these titles.[37]

The snout, the open mouth, long tongue and antler combination:

The flattened snout, if it were intended to suggest the pig, would symbolise delusion to a Buddhist, i.e. the delusions of a sensual, desirous life.[38] However, it may be a style feature brought from the steppes. It is present on a lion-griffin representation of 4th century BC Bactria. In China and Siberia the boar was associated with all the heroic warrior qualities. This was true also in ancient India where, in addition, it was part of a creation myth in which it rescued the drowning earth. It was a sacred animal in ancient Egypt and Babylonia where it too was symbolic of the earth.[39]

The open mouth of the weight-beast signifies the Buddha's preaching of the *dhamma* (ideal justice). However, if the muzzle of the weight beast was intended to be that of a horse and not that of a pig (see below), then the open mouth may also carry the older Indo-Aryan meaning of ruling, taming and subjection. The horse's mouth was where it was bridled and curbed and so subjected to the domination of its master.[40]

It has been concluded previously that the lower lip extension represented a long protruding tongue, at least originally. A long tongue was one of the major signs of a Buddha. It symbolised the worldwide spread of the word, i.e. the tongue covers the world.[41] However, it is an ancient and widespread symbol and it may have been a style feature adopted from south China or further north. The long tongue could also have been intended to indicate the monarch as subservient to Buddha, the protruding tongue being a sign made by an inferior to a superior in the East.[42] The Buddhistic symbolism may be preferable because the trefoil shape of so many of these lip extensions probably symbolises the *triratna*, i.e. the Buddha, the Law and the Community.[43] Other Buddhist symbolisms would also be relevant and some local symbolisms, e.g. the three 'countries' of Rammanadesa (Lower Burma).

The combination of antlers and protruding tongue on the beast weights symbolises the enlightenment of the Buddha and the dissemination of the Buddhist doctrines.[44] It has an older symbolism which could have come through south China, which is that of sun and water and so crop growth.[45] In Burma as elsewhere in the East the king, a *chakravartin*, was regarded as the protector of his people's welfare and their crops and it is likely that, traditionally, he ploughed the first furrows of the season as in Siam of pre-1688 AD.[46]

The mane and the chest decoration:

The mane conveys the idea of masculinity and strength. It can also symbolise intellect and wisdom.[47]

The chest decoration symbolises a *bodhisattva*. One is usually portrayed in Buddhist princely dress. This includes a necklace, a garland to the navel and a garland to the thighs. The beast weights have at least one set of chest decoration and often two or three. Such sets do not occur on the bird weights because the bird weights do not symbolise the monarch in his role as a *bodhisattva*.

The horse, the horse tail and the *ch'i-lin:*

The horse is most clearly indicated on the beast weights by the tail. In Sumeria and Babylon of 5000 years ago the horse symbolised all or some of the sun, the sky, the wind and water, speed, endurance and virility.[48] The horse was important in the religion of the Turki, Mongolians and Indo-Europeans of the central Asian and north Asian areas[49] who practised various ceremonies in which the horse was an essential part. Such ceremonies were brought to Persia and India by the Aryan invaders of 4000 years ago,[50] so establishing the cult of the *chakravartin* in these regions. The horse was associated with the sky, sun and fire cults of the Sakas, Persians and Indians, being known as the 'heavenly animal'. These attributes were especially true of a white horse.[51] To the Indo-European nomad, a stallion symbolised freedom, kingship, the power of the warrior and fertility and abundance of the virile male.[52] The horse was associated with the burial of chiefs throughout the entire Scythian and Saka regions, large numbers of horses and people being killed for the occasion.[53] Similar customs were practised by the Shans, certainly until the middle of the 16th century AD.[54] In its funerary role the horse symbolised death.

Shamans, both in the central and north Asian regions and pre-Buddhist India, required the souls of horses to guide them or carry them to the celestial regions. For this purpose a horse would be sacrificed ritually to the sky god or the wind god.[55] In Burma the Myinbyushin *nat* (spirit of a heroic ancestor) was carried on the back of the White Horse.[56] See also the section entitled 'Shamanism'. In India representations of the horse and its solar symbolism, like the stag, do not become prominent until after the arrival of the Śakas and the Yueh-chieh, about the 2nd century BC.[57] The horse is one of the marks of the *chakravartin.*[58] It appears on the abacus of the pillar at Sarnath erected in the 3rd century BC by the emperor Asoka, later supposed to be a *chakravartin* himself and the first monarch of note to patronize Buddhism. Gautama Buddha was also later called a *chakravartin.*[59] To the Buddhist the horse had special symbolisms. Along with the elephant and the hare, the horse was one of the three animals which attained Nirvana on its own merits.[60] When Gautama Buddha left his family to become an ascetic he travelled on a horse. The riderless horse became an early symbol for the Buddha.[61] The horse is often shown carrying the flaming pearl and the sacred books of the Buddhist law on its back.

Among the Turki people of Central Asia in the 14th century AD the horse tail was used as a standard to denote supreme military command, just as the yak tail was used in China.[62] In north Burma and Siam of the early 15th century AD horse or yak tail fans were used by kings while the cavalry carried them at the heads of their lances as a mark of "high nobility".[63] The Altaian-Scythians of the 8th and 7th centuries BC strewed the floors of the burial chambers of their notables with horse tails, perhaps to denote similar meanings.

The three different tail shapes on the beast weights of Group 5 could have represented three different *chakravartins.* In 1548 AD there were three different *chakravartins* in the region. These were the rulers of Burma, the combined kingdoms of Lan-Na/Lan-Chang and Ayudhya.[64] The last two were conquered by Burma shortly afterwards. A second possibility is that they represented the three 'countries' of Rammanadesa.[65] King Razadarit, the father of Queen Shinsawbu (who is associated with the Group 6 weights) ascended the throne of Rammanadesa in 1385 AD and reigned until 1423 AD. He is credited with the division of Rammanadesa into the three countries of Myaungmya, Bassein and Pegu. Each of these countries was composed of 32 provinces,[66] the ideal number to be ruled over by a *chakravartin*. Hence each of these three countries could have been represented by one of the three Group 5 beast weights. Possibly the lotus bud tail represented Pegu because of the religious associations. After the re-uniting of Burma between 1539 AD and 1559 AD, the three kinds of tail were replaced by the single style of rod-like (rarely horse-like) tail. Was this intended to symbolise Burmese unity like the one head knob on the bird weights?

The combination of lion, deer and horse in China to form a variety of composite creatures known collectively to Europeans as *ch'i-lins* has been mentioned previously. They are similar in concept and sometimes in appearance to the weight-beasts. The best known is also perhaps the most horse-like. It symbolised wise administration, unbiased justice, benevolence and peace.[67] It appeared only to a ruler of supreme virtue. The similarity of the *ch'i-lin's* symbolisation to that of the *chakravartin* as represented on the beast weights is close.

The elephant and its feet:

The elephant's feet on the beast weights symbolise both a *bodhisattva* and a *chakravartin*.[68] On the pre-3rd century BC pillars in north-east India, the elephant symbolised Indra and later, perhaps, royalty. The white elephant is a symbol of both Gautama Buddha and his birth.[69] A white elephant with six tusks appeared to his mother just before his birth to announce the arrival of a *chakravartin*.[70] It was an elephant which the Buddha saw first after his enlightenment.[71] The elephant symbolises might, wisdom, mercy and longevity.[72] It also symbolises, with the white elephant in particular, water, rain, clouds and so crop growth. As king of a people who lived mainly in the dry zone of Burma, a people who adhered to its old animistic beliefs, it was the last symbolism which, in practical terms, made possession of a white elephant so important. These symbolisms were accepted throughout India, south-east Asia and China, with local modifications to suit local customs.[73] In addition, the elephant's feet on the beast weights symbolise the role of the eight pairs of elephants, which were born from Brahma's breaking of the cosmic golden egg, in supporting the universe[74] and so make evident the relation with the universal monarch, the *chakravartin*. A similar symbolism appears as eight pairs of vertical grooves on some bases. The number of marks indicating the toes on the weight-beast's feet varies from none to six, perhaps as a result of craftmen's choice or their forgetting or ignorance of the symbolism.

The four animals included on the beast on the weights:

Each of the four animals included in the beast on the weights at first symbolised life-giving rain and fecundity and became guardians of the four quarters of the universe. Later they also symbolised the ruler. Some time after the arrival of Buddhism they added other symbolisms. The lion was chosen to symbolise Buddha, 'the lion of the Sakyas'. The elephant symbolised the late legend of the conception of the Buddha; the horse now symbolised Buddha's departure; the bull came to symbolise the constellation under which he was born. Together they symbolised Buddhism itself and its doctrines.[75] When the weight beast was combined with its base in the correct position it symbolised a universal monarch.

The symbolism of the bases:

In India, the number four or the square is associated with the idea of totality, universality and perfection.[76] The square symbolises the earth with its four cardinal directions: north, south, east and west. A fifth direction joins the centre of the square earth to the Pole Star. A similar cosmic symbolism existed in China where the four boundaries of the square base represented the bounding mountains separating the central mountain from the four oceans and the four continents.[77] The pyramidal shape symbolises the divine mountain, like, for example, the ziggurat, the church spire, the *stupa* (*hpaya*) and many Indian tumuli of Aryan times.[78] These, like the sacred trees and pillars, symbolise the axis of the earth joining together the watery underworld, the centre of the earth's surface and, at the sun's zenith, the heavens. It is at this centre where is placed the throne of each universal monarch and it is through the association of the 'axis mundi' with the throne that the ruler acquires power.[79] This is the importance of the location and orientation of the weight beast on the square base and its relation to the sides. This is why the Burmese weight bird never occurs on a square or rectangular base unless false.

The square plan combined with the pyramid came to represent the four-sided Mount Meru of Indian belief, the central peak of the world, with the apex symbolising the highest spiritual or hierarchical attainment. Each of the sides is the abode of a guardian spirit.[80] In 8th century BC India the burial tumuli of Aryan kings had a square base plan while those of non-Aryans had a circular plan.[81] The towns of Central Asia were built to a square plan. All northern Hindu and Buddhist temples, royal palaces and cities followed this same general square plan. In Indian cosmogony the universal monarch sat on a throne representing Mount Meru at the centre of the universe, while the walls of his palace running north, south, east and west in the four directions represented the four continents which bounded the earth.

So with the Group 2(r) beast weights where the beast shape on the square pyramidal base symbolises the semi-divine universal monarch at the centre of the world facing due east. Much the same significance attaches to the beast weights with the rectangular

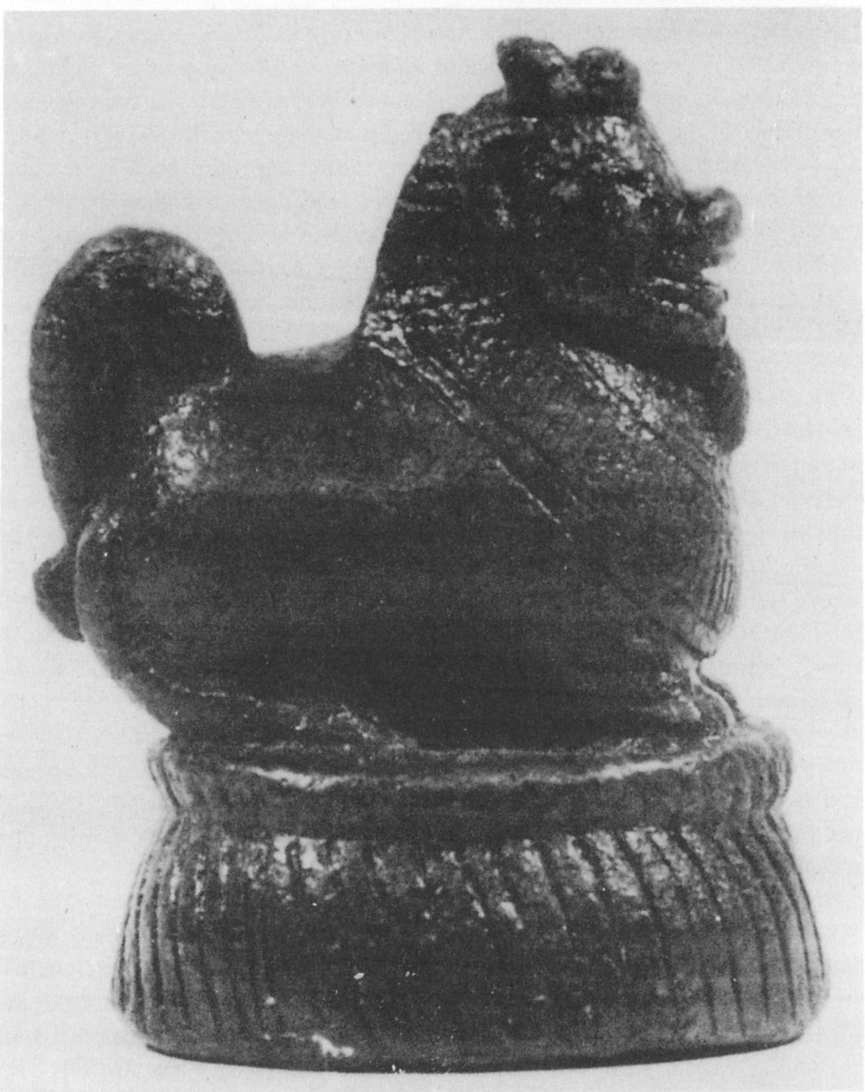

This combination of animal shape and oblique grooves is present in bird weight Groups 4 and 5 (both rare) and more frequently in bird Groups 6 and 7. It is present, as shown, in the beast weights of Group 5, though rarely. The symbolism of the base cross-section is that of the altar (see Group 7 bird weights) but that of the grooving is not apparent.

Plate 52: Circular Base Plan with Oblique Striations on the Base Sides.

base and possibly to the rectangular base without corners (lozenge shape). Returning to the Group 2 beast weights with the square base, there may be an additional significance. In the section on the symbolism of the nine-rayed star sign, the significance of this sign on the Group 2 weights in relation to Shwebo and Halin is discussed. At Shwebo occurs the square-based Shwebawgyun pagoda, with which King Bodawpaya would have been familiar since he was raised there when Shwebo was the Burmese capital. At a much later date, when he was king, he remembered the region well enough to recall the Pyu coins sometimes found at Halin and have copies made in 1797 for use as a coinage. It is possible that his remembrance of the square base of this pagoda may have influenced the return to the square base on the Group 2(r) beast weight which is associated with him.

The octagon, combined with the pyramid, may represent the Chinese Tushita (Burmese Tawadeintha) heaven, the abode of the Bodhisattvas, showing eight instead of four major points of the compass. Each of these sides would be guarded by one of eight deities. Above the Tushita heaven is the highest heaven of all, the whole weight thus yielding a symbolism similar to that of the square truncated pyramid.[82] The elephant weights often have an octagonal base with a stepped cross-section, each step symbolising a cosmic level which has to be surmounted to attain the highest spiritual state, the highest heaven. There are other symbolisms depending on the number of steps.[83] Ultimately the steps are probably associated with the somewhat similar symbolism of the megalithic stepped pyramid. However, in Burma the octagon may also be associated with a pre-Buddhist cult of nine planets (one having been deleted for an unknown reason). Each side of the octagon is associated with a particular planet and each planet is represented by a particular animal. At a pagoda a Burmese Buddhist will pray either at the side of his own birthday planet or at a side associated with a specific planet.[84]

In the case of the beast weights the circle, to the Buddhists, probably symbolises the Wheel of the Law.[85] By an older symbolism it could signify the solar origin of the Burmese monarchy. It was believed that the Burmese monarchy descended from two Indian dynasties, one claiming its origin from the sun god and the other from the moon god.[86] The symbolism of the truncated cone with its tapering shape is that of the similar tapering shape of the pyramid. A few of the Group 5 beast weights have cylindrical bases with vertical striae. This combination occurs mostly on the bird weights of groups 6 and 7. The symbolism has not been identified. (See Chapter 11, section "The symbolism of the bases" and Plate 52.)

The universal sovereign and the *chakravartin:*

At first the expression 'universal sovereign' possibly meant one who supervised on behalf of the gods of nature the meeting of their requirements on earth in order to ensure fertility. He applied laws concerning his people's behaviour to ensure cosmic order through the use of his power. Later the notion appears to have become associated also with the idea of territorial sovereignty. This conception may have arisen among the ancient steppe nomads. Powerful and cruel, they were able to roam and plunder in whatsoever direction they willed because of the horse and their riding and archery talents. This ability, together with their caste system, possibly convinced them both of their right to behave in such a fashion and of the importance of the horse. The association of the wheel and the *chakravartin* perhaps arose from the ox carts used to carry their families and possessions. The shape of the wheel with 32 spokes (symbolising the royal chariot and the sun's motion) appears in the shape of the Altai-Sakan tombs of the 8th century BC at Tuva. The association of the horse and wheel appeared also in the burial rites of the 14th century AD Mongols.[87] They had continued to use the term 'universal monarch'. In fact the name Chingis Khan means just that.[88]

The earliest known use of the *chakravartin* concept is in Mesopotamia, firstly by the Accadians, then by the Assyrians (about 13th century BC) and then by the Babylonians. It seems to have entered India along with the Aryan invaders of about 1700 BC. Several centuries after Gautama Buddha's life, the term *chakravartin* is known to have been used by the Mauryans to describe a universal sovereign, a world teacher, one who turned (vartin) the wheel (chakra), i.e. one whose journeys everywhere were unopposed because his domains were so extensive and his power so great.[89] However, while there are representations which. show the *chakravartin* associated with the wheel so there are others which show a discus.[91] The discus is a symbol of Vishnu, who is associated with the sun and day. He is closely allied with the Indo-Ayran god Indra (Sakka) who was born of Heaven and Earth and who was introduced into India by the Indo-Aryans. The discus preceded the *chakravartin* in his tours of his empire. There is clearly a solar symbolism here and, in fact, there was a sun cult both in north India and in the empire of the Kushans. The solar symbolism is evident also in the clockwise circumambulation of the animal representations on the abacus of the Sarnath pillar.[92]

Theravada Buddhism demanded of the Burmese monarch, as from all Buddhist kings, that he be the religious exemplar and the guardian of social order among his people.[93] The former was the more important and symbolically was represented by the weight-bird, i.e. the spiritual *hamsa.* The latter was symbolised by the weight-beast which portrayed the earthly *chakravartin.* The willing acceptance by his people of a monarch's authority to rule depended upon its being legitimised partly by Buddhist theology and partly by its continuity with the past. In Burma it was necessary for every monarch to trace his descent from the Pagan monarchy, which itself had looked to the Indian king Asoka, patron of Theravada in the 3rd century BC. The territorial concept of the *chakravartin* was particularly important in the conquests of neighbouring kingdoms because they could be justified as *dhammavijaya,* righteous conquest by a *chakravartin* who was also a *dhammaraja*, a king of righteousness.[94] This particular symbolism facilitates the dating of the beast weights.

A *chakravartin* was supposed to be recognisable by several special body characteristics.[95] There were other characteristics required also, the first of the two main ones being that he had to rule over a sufficiently extensive region. Before 1200 BC this area was decided by that over which a wild stallion would freely roam during a year of freedom.[96] Later it was supposed to be a monarch who ruled over 32 kingdoms.[97] The second main Buddhist characteristic was that he owned a *saptaratna*, i.e. seven jewels or treasures.[98] These commonly were the wheel, elephant, horse, wife,. minister, general, thought or jewel. Of these the first three were the outwardly apparent symbols of a *chakravartin*. The concept of a universal monarch could imply a degree of divinity and Burmese monarchs who claimed to be *chakravartins* also claimed to be *bodhisattvas*. In Burma, however, the notions of kingship, though similar, were drawn not just from Indian but also from Chinese and indigenous sources. It is because of the widespread recognition of the association of the conquering universal sovereign with the inseparable symbols of his rule, the symbols which gave him his power, that the beast weights were made as they are found today. The latter conveyed to illiterate peasants or conquered peoples speaking different languages meanings clear as the written word. Even in the late 1950s many people of central and northern Burma in the small towns and villages attached real power and authority to the royal animal weights. Some said that they had been known to have had these powers "since the earliest days". Dealers in hardware in such places still stocked these weights when they could get them, for that reason. One expression of this power was that it was thought that such a weight had the power of healing, this being especially true of the King Bodawpaya beast weight (Group 2(r)).

When establishing their rule over conquered peoples, the Burmese did so over regions noted for the diversity of their languages. There were also diversities of currencies and, perhaps to a lesser extent, of weight systems. Colonial rule had to be paid for by revenues in kind obtained from the conquered peoples. To the Burmese in particular it was obviously simpler to use their own weight system in order to measure the revenues, so avoiding difficulties caused by the differences in language and weight systems. In addition there had to be a way by which the conquered peoples could recognise the Burmese weights and accept them. Since states did not have emblems and since animism and Buddhism were the beliefs of the great majority, these then provided a source of widely understood symbols of relatively fixed meaning. Fixed because from before the 11th century AD the Buddhist symbols, if not the animist, were documented. So arose a system of Burmese weights for use in countries occupied by them, shaped in forms which, while conveying the Buddhist message of a powerful universal monarch legitimately ruling over his extensive domains, also conveyed to animists the message of the god-protector of soil and crops.

It seems clear that each time there was a new righteous conquest of sufficient importance the Burmese ruler became a *chakravartin* unless he was one already. The cost of his wars necessitated the rebuilding of destroyed or disturbed revenue systems. Since payments were in kind and by bullion, not a coinage, this included the rebuilding of the weight system. When the conquered territory regained its independence there was no further need for the Burmese weights. Thus there was a series of more or less connected events on the occasion of each conquest. For convenience these events of conquest and defeat are tabulated in relation to the weight groups, the conquering kings and the regions gained or lost. See the table 'Approximate Dates of Events Accompanying New Beast Weight Issues'.

Table 18: Approximate Dates of Events Accompanying New Beast Weight Issues

Weight Group	King and DYNASTY	Re-unification and Conquest		WS	RI	Disunification and Loss	
		Year	Territory	Year		Year	Territory
					1803	1796	N. Siam
	Bodawpaya	1785	N. Siam				
		1784	Arakan		1784		
	Hsinbyushin					1776	N. Siam
2	Hsinbyushin	1767	N. Siam				
	Alaungpaya	1757	All Burma				
						1740	Burma disunited
	KONBAUNG	1752		Dynasty begins			
	Taninganwe					1717	N. Siam
					1638		
	Anaukpatlun	1620	N. Siam				
	Anaukpatlun	1615	All Burma				
						1595	Burma disunited
						1595	N. Siam
4				1575			
		1557	N. Siam	to			
		1555	All Burma	1553			
	Tabinshweti	1539	Toungoo and Lower Burma				
	NYAUNGYAN	1531		Dynasty begins			
5	Dammazedi	1472	?	?		?	?
5?	Razadarit	1417	L. Burma	?		?	?
	WARERU	1287		Dynasty begins			
				1150			
			All Burma				
	Anawrahta	1057	Empire				
	PAGAN	1044		Dynasty begins			

WS = Weight Standardisation RI = Revenue Inquest

The meanings of the Chinese words *tu* and *tou*:

The Burmese name, *to*, appears to have originated from Chinese words. The anglicized forms of these words are *tou*, *tu* and *t'u*.[99] They have several meanings, of which the relevant ones are earth, a measure and beginning.[100] They are applied also to the ladle shape of the six stars (the Measure) in the shoulder and bow of the constellation Sagittarius. They appear in the names of the supernatural beings.

Tu Bo is the lord of the underworld. He has a horned, squarish tiger's head with three eyes, a long protruding tongue and a body like a bull's with a humped back.[101] *Tu kuei*[102] means 'earth spirit'. Between about 600 and 900 AD effigies of two such spirits were placed in the tombs of high officials along with two other tomb guardians, similar in appearance to *lokapalas*.[103] An illustration of a *tu kuei* appears on 'Illustrations of Natural Beasts and Beast Representations'. The *Ch'i-tou*, perhaps identical in purpose with the *tu kuei*, is a tomb guardian which originated in the devil-abolishing ceremony of the New Year dragon dance. It appears in a sitting posture, is beast-like, horned, with large ears and flaming shoulders and with a terrifying face intended to drive away demons.[104] The group of stars called the Measure is represented by the griffin.[105] In the constellation of Sagittarius lives Tou Mu, the goddess of destiny, who is associated with earth and the idea of measuring, e.g. human life, rice.[106] She has a third all-seeing eye, like Tu Bo, in the middle of her forehead. She has nine sons who are the stars which help in determining the dates of birth and death. One of the *ch'i-lin* family which has a lion-like head and body has the name *tou-jon-sheu*.[107]

The similarities of Tu Bo's, the *tu kuei's*, the griffin's and the *ch'i-tou's* appearance to those of the beast weights are evident. The association of the *tou*, Tou Mu's sons and the *ch'i-tou* with beginning, birth, New Year and hence the lion (dragon) dance are clear. The idea of measuring is common to *tou* and the weighing implied by the animal weights. The nine-rayed star on the weights is similar in concept to the nine star sons of Tou Mu. The six-rayed star sign is similar in concept to the six stars of the Measure. Both the six-rayed star sign and the nine-rayed star sign were used by the Shans as 'measures' of the fineness of the silver alloy used for ornaments. There is then abundant evidence to connect the Chinese word *tou* and its symbolism to the Burmese or Shan word *to* and its symbolism. There is an even closer connection between *t'u* and the *chakravartin*, symbolised by the beast weight. About 794 AD. which was a time when animism appears to have been important, Nanchao signed a treaty with China. The ritual associated with the signature required the symbolic sanction of earth-sky-water.[108] About 880 AD the country adopted Buddhism as its state religion. Its king aspired to be a *chakravartin* and adopted the title 'Earth-Wheel King' (*t'u-lun wang*).[109] The association of *t'u* with earth and the *chakravartin* is clear. It also seems clear that the symbolism of Buddhism with its lion/earth and gander/water/sky symbolism easily would have fitted the animist symbolism of earth-sky-water.

The *bodhisattva*

A *bodhisattva* is a Buddha-to-be. He is one who has understood the idea of, and is on the way to, Enlightenment. In Mahayana Buddhism he is also one who has postponed his attainment of the ultimate blessing, Nirvana, in order to help others. Because of this, he or she, has been elevated to the status of a divinity, a god or goddess. The presence of the garland-like chest decoration on the beast weights makes it clear that the bodhisattva was intended to be symbolised. The all-seeing eye on the beast weights, characteristic of many sky-gods, in which the *bodhisattvas* can be included, may have been intended to further indicate a *bodhisattva.* The horns, most probably signifying in this case, divine power, may have been similarly intended. However, both in the case of the all-seeing eye and the horns, the symbolism may equally well suit the older beliefs. It is the symbolism of the *bodhisattva* on the beast weights which makes the appearance of similar garlands on the bird weights symbolically impossible.

The god-king

In south-east Asia, as in India and elsewhere, a king having the characteristics of a *chakravartin* was regarded as a divinity. This had come about because, anciently, the chief symbolised his ancestors and, through them, the often divine tribal ancestors. (Hence, in later times, the importance to his people of the continuity of the ruler's descent from the ancestors). God, king and ancestors all played their parts in ensuring the fertility and well-being of their peoples and, aided by their various symbolic representations, e.g. the mounds of the earth religions, the discus and horses of the sky religions, ensured the essential communication with the earth and sky gods. (See "Religious Beliefs")

On the beast weights the symbolisms of the divine *bodhisattva* and the earthly *chakravartin* are combined. That is, the beast weight symbolises the god-king[110]. This combination of the powers of the *bodhisattva* with those of the *chakravartin* probably explains in large part why these weights are, even today (1972), regarded by many of the people of Burma as having "power" and as having had this power "from the earliest times".

Notes

1 Eliade, 1972, pp. 334-346.

2 Cooper, pp. 172, 173.

3 Cameron, p. 68.

4 Rawson, 1983, p. 238.

5 Iyer, pp. 63-65.

6 South, p. 355. Cooper, p. 99.

7 Jobes, p. 999.

8 Ball, pp. 53, 54. Liebert, p. 26.

9 Schmidt, p. 18.

[10] Ball, pp. 53, 62. Jobes, p. 1000.

[11] Salmony, p. 50.

[12] Jobes, p. 104. Salmony, p. 24. Coudert, p. 30.

[13] Iyer, pp. 11, 79. Larousse, p. 386.

[14] MacCulloch, pp. 791-794.

[15] Salmony, p. 30. Rawson, 1984, p. 108.

[16] Cooper, pp. 50, 158. Salmony, pp. 29, 30.

[17] Eliade, pp. 89, 147, 155, 174.

[18] Salmony, p. 49.

[19] Rice, 1969, p. 119. Iyer, p. 79.

[20] Rawson, 1983, p. 238.

[21] Eberhard, p. 79. Iyer, p. 79.

[22] Rice, 1969, pp. 118, 119.

[23] Cooper, p. 158. Iyer, p. 79.

[24] Eberhard, p. 78.

[25] Veith, p. 313.

[26] Cooper, pp. 50, 172. The deer was associated with love sickness. The other two animals were the monkey (too greedy) and the tiger (too ferocious). All of these were associated with immoderation and desire. The inclusion of the deer and the tiger also suggests a disapproval of the nomads whose art, mainly used as personal ornaments and on harness and weapons, made much use of the tiger and the stag. A similar inference of disapproval may be drawn by the appearance on artifacts, e.g. bronze mirrors, of the (Chinese) dragon of the east in directional opposition to the (steppe nomad) tiger of the west. This example is one of the many mentioned in these pages which connect Yunnan and hence the Burmese people with the steppe nomads, because in barbarian Yunnan of the 3rd century BC, both the tiger and the deer were worshipped as gods.

[27] Iyer, pp. 78, 79. Liebert, 180.

[28] Salmony, p. 23.

[29] Jobes, pp. 690, 1000. Praz, p. 722. Salmony, pp. 23, 49, 50. South, pp. 94, 357.

[30] Christie, pp. 87, 88. Williams, p. 443.

[31] Funke, p. 466.

[32] Ball, p. 165. Eberhard, pp. 83-86. Williams, pp. 132-141. Cooper, pp. 55-56.

[33] Scott, 'The Burman..', pp. 328-333.

[34] Anonymous, *Weights of Historical Importance.* Scott, 'The Burman...', p. 339.

[35] Cooper, p. 170.

[36] Larousse, p. 356. Herrmann, p. 828.

[37] Harvey, p. 276. Htin Aung, p. 17.

[38] Herrmann, p. 830.

[39] Iyer, pp. 80, 81. Williams, pp. 326, 425. Cooper, pp. 22, 166. Eberhard, p. 236.

[40] Jobes, p. 1132. Cooper, p. 99. Liebert, p. 270. O'Flaherty, 'Horses', p. 467.

[41] Salmony, p. 37. Cooper, p. 174.

[42] Jobes, p. 1587. Rockhill, p. 125, referring to the 14th century Sumatra.

[43] Rapson, p. 66. Lessing, pp. 36, 71.

[44] Salmony, p. 38.

[45] Salmony, p. 52, quoting Mus.

[46] de la Loubere, pp. 20, 43.

[47] Zimmer, p. 157. Jobes, p. 1057. Cooper, p. 77.

[48] Cooper, p. 85. Walters, p. 32. South, p. 358. Jobes, pp. 789, 790.

[49] Eliade, 1972, pp. 11, 183.

[50] Larousse, p. 329.

[51] Litvinskii, pp. 319, 519, 520. Eliade, p. 467.

[52] Gimbutas, p. 269. Rice, 1969, p. 118.

[53] O'Flaherty, 'Horses', pp. 363, 364. Eliade, 1972, p. 198.

[54] Harvey, pp. 166, 343, 344.

[55] Eliade, 1972, pp. 79, 80, 182, 198.

[56] Harvey, pp. 53, 220.

[57] Litvinskii, p. 518.

[58] Liebert, p. 28.

[59] Rowland, pp. 704, 705.

[60] Walters, p. 297.

[61] Iyer, pp. 40, 41. Cooper, p. 85. Jobes, p. 791.

[62] Yule (Cordier, ed.), v. 1, p. 223.

[63] Major, p. 14, quoting Conti.

[64] Wyatt, p. 96.

[65] Harvey, p. 115.

[66] See the comments on '32' in 'Religions: Shamanism'. Shorto (1963) also discusses in detail the three countries of Rammanadesa, each with its 32 *myos* or districts. At this time, when a country was conquered, it did not lose its identity. Its ruler, even though he might be a nominee of the conqueror, was still accounted as royal and, as ruler of the minimum number of countries required, presumably could be ranked as a *chakravartin* independently of the conqueror's status. In this case four kinds of beast weight tail might be required (one for the conquering *chakravartin* and three for the subordinate *chakravartins*). Four kinds might exist but too few specimens have been seen to be sure. However, whether there are three or four kinds, the symbolism is unaltered.

[67] Cooper, p. 183. Christie, p. 130, 131. South, pp. 12, 16. Lessing, p. 114.

[68] Jobes, p. 501.

[69] Zimmer, p. 103. Williams, p. 170.

[70] Ball, p. 79. Cooper, p. 60.

[71] Ball, p. 84.

[72] Liebert, p. 87. Cooper, p. 61. Ball, pp. 82, 83. Jobes, p. 502.

[73] Eberhard, p. 94. Williams, p. 170. Ball, pp. 73, 74, 79.

[74] Zimmer, p. 105.

[75] Irwin, III, p. 643; IV, pp. 750, 753.

[76] Zimmer, pp. 13, 14.

77 Christie, pp. 57, 67.

78 Cooper, pp. 134, 158. Eberhard, p. 276, refers to the magic square made up of small squares. Anderson, pp. 76, 130, refers to the *ts'ung*, the Chinese cube bored through by a circular tube which symbolised the earth and the earth mother. Mackenzie, p. 87, describes the pyramidal base as typical of south China about the 5th century AD.

79 Irwin, III, pp. 641, 643; IV, pp. 745, 749.

80 Bussagli, 'Farther India', pp. 922, 923.

81 Irwin, I, p. 781.

82 Taw Sein Ko, p. 178. Gutman, p. 281, refers to the connection of the Asokan pillars with the *chakravartin* and the adoption of the Asokan concepts of the *chakravartin* by the Buddhist kings of south-east Asia. On the Asokan pillar at Sarnath in India the abacus is square in cross-section like the square base on the Group 2 beast weights. Where the cross-section of the pillar or the weight base is octagonal, the eight sides symbolise the eight directions of his rule (Gutman, p. 284, quoting Irwin).

83 Cooper, pp. 134, 160. Taw Sein Ko, p. 178.

84 Htin Aung, p. 20.

85 Herrmann, p. 838. The *chakravartin* was symbolised by the wheel. It was the *chakravartin's* function to 'turn the wheel of the law' and so regulate both the atmospheric and earthly forces, the heavenly and worldly forces, which would ensure the continuing prosperity of the kingdom (Gutman, p. 281). Moreover, the extent of his power was visualised as a circle of power, i.e. the mandala (Kulke, p. 3) to which the ts'ung is similar.

86 Anonymous, *Mandalay Palace*, p. 25.

87 Yule (Cordier, ed.), p. 143, quoting Ibn Batuta.

88 Rubruck, pp. 12, 25, 296.

89 Coomaraswamy and Rowland, p. 99. Jacobs, p. 337.

90 Basham, p. 84.

91 Beale, p. 63 note 4.

92 Jairozbhoy, p. 92.

93 Swearer, p. 369. Lehman, p. 575. Mahoney, pp. 6, 7.

94 Aung Thwin, pp. 334, 335.

95 Jacobi, p. 337.

96 Mahoney, p. 6.

97 Lehman, p. 576.

98 Liebert, pp. 53, 259.

99 Williams, p. 215. Eberhard, p. 89. Another word for earth is *di* or *ti.*

100 Williams, p. 368. Jobes, p. 1591. Shulman, p. 49. Eberhard, p. 106.

101 Mackenzie, p. 89. There was an earth god named Tu-ti (Lessing, p. 106). At first a male, then a female, it finally appeared as an elderly couple (Wilhelm, p. 643). According to Zimmer, 1972, pp. 127, 128, the archetypal parents, Father Heaven and Mother Earth, are known in China as T'ien and Ti.

102 Williams, pp. 279, 456, 462. A *kuei* is an inhabitant of the subterranean (and other) prisons called *ti-yu*, i.e. hell. *Kuei* is translated into English as spirit, goblin, demon or devil. Another meaning of the word is sceptre of supremancy or authority, of which the Chinese ideogram is that of 'earth', thrice repeated (Williams, p. 239).

103 Fontein, p. 513. *Lokapalas* were represented as armoured soldiers of fierce appearance who lived on Mount Meru and guarded the regions of the cardinal directions (Larousse, pp. 369, 422).

104 Watson, 1973, p. 144.

105 Williams, p. 368. Giles, p. 27. Jobes, p. 1591, states that the constellation Sagittarius is represented by the griffin in the Chinese zodiac. However, there is no griffin in the Chinese solar zodiac and Sagittarius is represented by the monkey (Jobes, p. 1725). There is a griffin in the Chinese lunar zodiac (Shulman, p. 49) but it is in the northen palace while Sagittarius is the most southerly of all the zodiacal constellations.

106 Savill, p. 240.

107 South, p. 12.

108 Backus, p. 128.

109 Backus, pp. 159, 160.

110 Gutman, 1987, pp. 19, 20. Refers to the god-king symbolism on the ancient coins of South-east Asia.

13
The Symbolism of the Anserine-Feline Combination

The lion, together with its horned varieties, and the anserine birds are linked through their common symbolic association with the sun, as are other creatures both real and mythical. They are also linked through the opposition of their attributes. The lion is the symbol of destructive earthly material power, the anserine of creative heavenly spiritual power. All the birds and beasts mentioned in the sections dealing with the identities of the creatures are associated with the sun or the heavens. Some also are associated with the earth and soil. Most of them have been and are royal symbols. Some of them, like the dragon with the phoenix,[1] also form combinations like the lion with the duck. So why was the latter combination so widely adopted across Eurasia? The short answer appears to be that in Neolithic times the predatory territorial feline represented strength and death upon earth, while the preyed-upon celestial migratory anserine represented flight from the strong, from dying and the recreation of life upon its return in the spring.[2] Later they came to represent fertility symbols, the lion representing the sun and the gander the rain, both together being essential for the growth of crops. In China the beast (dragon) and the bird (*feng huang*) have a contrary symbolism, the dragon symbolising rain and the *feng huang* the sun. Other animal combinations have a similar symbolism, e.g. the earthly snake and the celestial predatory eagle, which also is used widely. But the large feline could be destructive to man himself, unlike the eagle, and the lion, of all large felines, looked sun-like when seen full-face. The harmless waterbirds, the easiest of all birds to observe, were obviously heavenly and migratory. This polarity of death and life after death, of barrenness and fruitfulness, became transmuted into different forms with time and culture. Examples are the material and the spiritual, the monarchy and the priesthood, Caesar and God, ferocity and gentleness. It seems that the combination of feline and anserine, tiger and bird, lion and goose, *to* and *hamsa*, was unique in its polaric contrasts between the lion's symbolism of the monarch's fearsome power of earthly destruction and the anserine's symbolism of the hope of re-creation.

Another reason for the adoption of the feline-anserine combination was that the ancient Egyptians were among the first of the urbanised peoples to display and record this symbolism and so their representations tended to be copied. One of their portrayals of the symbolism which developed in Heliopolis is given here. The Egyptian god, Geb or Seb, often is shown with his head surmounted by a goose.He represents the Chaos gander whose female, according to one group of legends, created from Chaos the first egg[3] which then became the sun god Ra. When Ra grew old, men rebelled against him so he sent the lioness-headed goddess Sekhmet to destroy most of them, leaving only a few. This she did.[4] Here apparently is the first record of the association of the anserine bird with creation and the predatory feline with destruction. Similar myths existed in Sumeria and China and all three regions may have derived it from an earlier widespread Neolithic source.[5] The legends may have arisen because the sedentary agricultural peoples of the Nile, the Tigris. the Euphrates and elsewhere just emerging from their hunter-gatherer period of wholly animal gods would have feared the lion as no other animal was feared, not only because of its occasional destruction of their livestock and themselves but also because of its probably former role as a god. Moreover, early man worldwide had to live near water. He observed the anserine's migratory habits, arriving with the spring, the period of new life, and departing to unknown regions in the autumn as the year commenced to die. In an animist world where everything has or was a spirit, it was obvious that these freely wandering anserine birds of water, land and air were associated with birth and death,[6] that they carried away the souls of the dead to regions unknown and created new life on their return in the spring. Depictions of souls (man-shapes) being carried to Paradise in the bellies of flying birds exist[7] in the art of many ancient cultures.

In China of the early part of the 1st millenium BC it was believed that the union of the rain-producing dragon and the sun-producing *feng huang*, both of the heavens, was essential for good harvests on earth. In spring ceremonies were held in the croplands to celebrate both this union of heaven and earth and the hope of abundant crops.[8] These beliefs were incorporated in Confucianism and Taoism. In west central China of the 6th century BC a tiger-bird (*feng huang?*) combination existed which might have served a role as a soul-protector.[9] When Buddhism reached China many of the previous beliefs were adopted by it or persisted alongside it.[10] Unions similar to the above were common apparently among the nomadic Indo-Aryans of the steppes.[11]

The information located by the writers specifically concerning the pre-Buddhist symbolism of the lion and gander combination in India is negligible. However, it may be worth observing here that where the pre-Asokan (3rd century BC) pillars bear representations of the gander (shown pecking the earth - and symbolising water?), they are associated with a lion capital. At Sarnath the one pillar of probably Asokan age with a lion capital bears no ganders on its abacus but instead carries reliefs of the four animals which are combined on the weight beast.[12] Whether these pillars represent pre-Buddhist or Buddhist symbolism is not clear though the motifs themselves are pre-Buddhist and show Persian or west Asian influences. More is known of the relevant Hindu and Buddhist symbolism. Brahma was transported on a gander and from this stemmed the particular expressions of spirituality with which the *hamsa* was associated. Gautama Buddha was symbolised by a lion which, later, on the weights, was to be modified by the concept of the *chakravartin.* It is much the same symbolism as existed in Egypt, the power of earthly destruction combined with that of re-creation through the heavenly spirit.

The choice of the lion and duck combination specifically for the weights may have been intended not only to convey the sun/rain, death/rebirth symbolism known to the animist, the non-Buddhist and the specifically Buddhist symbolism relating to the material and the spiritual but also to convey the message of giving what is rightly due, i.e. to god, the ruler, the buyer and the seller. The same idea is evident in Christ's comment "Render therefore unto Caesar the things which are Caesar's and unto God the things which are God's". Chinese balances with the dove on the arm convey a similar injunction, one which was derived from the *Jatakas.* Somewhat similar is the association of the idea of measuring and the ever watchful third eye of Tou Mu, the earth goddess.

One of the reasons for the association of the Buddhist monasteries with trade was that the monks could read and write, their calling implied their trustworthiness and they were concerned with the moral behaviour of people. They could draw up contracts under conditions of spiritually-guaranteed trusts. Similar conditions to these existed in ancient Egypt where animal weights were in use and, to only a slightly lesser degree of assurance, it can be said that they existed in Mesopotamia. From the foregoing it seems that similar sets of circumstances may have existed in Egypt, Mesopotamia and Burma at times when similar animal symbolisms existed, so leading in each case to the adoption of similar animal effigies even for their weights.

Notes

1 Lanciotti, 'Myth and Fable', p. 491.

2 Campbell and Lack, p. 200.

3 In Egypt and Assyria duck weights may occur in egg-like form (Ridgway, p. 245. Petrie, 1926, plate 9) while representations of the goose coiled into an egg-like outline occur in China (Eberhard, p. 87). In Siberia a swan was the mother creator of the Buriat tribe.

4 Larousse, pp. 14, 36.

5 Armstrong, p. 46. Wilhelm, p. 643.

6 Walker, 'Hindu World', p. 155.

7 Vajda, pp. 792, 793.

8 Williams, p. 2.

9 Mackenzie, pp. 90-92.

10 Ball, p. 30.

11 Raevskii, p. 146.

12 Irwin, I, pp. 709-710.

14
The Sign and Style System Symbolism

The purposes of the signs and style systems:

The Burmese term *shwe-arlay* (gold weights) may be a mistaken rendering of *sri arlay* (king's weights) especially since gold was weighed far less often than the other metals. (Gold was mainly used in the form of leaf which was obtained in large part from Yunnan and used almost entirely for decorating religious structures). Thus the main signs may have been intended to signify that the weights were those used by the king's agents or brokers. The deduced connection with the king may be given some additional support by the use for the weights, in the 1700s, of the expression *sandaw-hmi* (up to the royal standard). Additional support for the view that the main signs indicated that the weights were the king's weights in use by the king's brokers comes from the 1972 survey results. It was apparent that of the Group 2, periods C and D, bird weights selected as being official, only about 10% – 20% bore signs. Had the rejected bird weights (almost entirely of these periods) been included, these proportions might have been halved. Of all the other periods with sign-bearing weights, between 50% and 100% bore recognisable signs. This happening suggests that before the introduction of the Group 2 weights, large numbers of weights were not required, their use being confined to certain kinds of persons, e.g. the king's agents (weights with a sign) and local (?) merchants (weights without a sign). Foreign merchants may have preferred to use their own weights, at least for verification. After the introduction of the Group 2 weights, the signed king's weights became a much smaller proportion of the total number of weights being made.

The auxiliary signs of Groups 4 and 5 (and 2(r) also?)may indicate a place, e.g. of weight issue, usage or re-authentication. This view receives support from the fact that the simplest style of all, the Group 4 period H beast weight, is that which is most frequently associated with auxiliary signs. That is, variations in this style did not exist and so could not be used to indicate place. Thus, variations in shape (and number?) of auxiliary signs appear to have taken their place. Further support for this view is given in Appendix 3 where it is shown that the signs on the *kakim* ingots and elephant-shaped weights make evident the assocociation of the 4-rayed star sign with Chiengmai suggesting that the 4-rayed star sign on the weights, both main and auxiliary, was similarly associated with Chiengmai. This last interpretation may be supported in turn, by the occurrence of the 4-rayed star sign for the last time on what is inferred to be the earliest of the Group 2 beast weights. The cessation of this sign may be associated with the Burmese loss of Chiengmai to the Siamese in 1785 AD thus tending to confirm the former association of the 4-rayed star sign with Chiengmai.

The style systems may have indicated place, just as the auxiliary signs appear to have done. The three tail shapes of the beast weights on Group 5 have been inferred to be associated with the three sub-divisions of Ramannadesa, as possibly were the one , two or three birds on the bases of the Group 6 bird weights. It seems that in the case of the Group 5 bird weight styles and the Group 4 auxiliary signs, the systems may have become too complicated, or out-dated by events, and were abandoned in favour of simpler systems with fewer classes.

The symbolism of the sign locations:

The main signs on both beast and bird weights always are located on the base, either to the front of the weight-animal or the right front. It was an Eastern requirement that the monarch on his throne at the centre of the world should face a particular direction. In Burma this was towards the east whence rose the sun, which symbolised the king's ancestry.[3] This was the case among the minority peoples of China also, though among the Han Chinese and the Mongols of the 13th century AD,[4] it was towards the south. The beast on the weight, centrally placed, so symbolising the monarch at the centre of the universe, is aligned exactly at right angles to the front face. Presumably the beast is supposed to be facing east and the right or right front side is facing south or south-east.

The location of the signs may be related to the representations of the Buddha and his disciples, to Burmese court procedures or to the location of the lord of the 37 *nats* (see 'Religions'). The Lord Buddha and two of his chief disciples are frequently shown together in the same relative locations. On the rights is the wise Saraputta, who was charged especially with duties relating to matters of Buddhist doctrine, the Law. On the left is Moggallana, who supervised the monkhood and the supernatural.[5] In the Burmese court the monarch's left or north side was the side of honour, or perhaps better, honourable behaviour, because it also carried the idea of femininity and non-aggressiveness.[6] Put another way, this was the side of weakness. The Mongols regarded the left as the side of honour in peacetime. In wartime the side of honour was changed to the right.[7]

It may be that the frontal location represented the position at court occupied by the most important of those people carrying out the king's bidding. These were the heir-apparent and the chief ministers. It could also represent all creatures everywhere listening to the Law, the Buddhist doctrine. The first may be the more likely since the mainly beast weight signs (circle, four-rayed star and beast) appear to be most closely associated with the symbolisms of the earthly universal monarch.

The main sign on the right front of the weights could symbolise the location at the

Burmese court of the king's bodyguard, that is, those who safeguarded the interests of the monarch and his people. This, being the right or south, was the side of physical power. It could also indicate the Buddhist law, suggested by the location of Sariputta. The right front main sign was first brought into use on the bird weights of Group 4 during a period when the evidence suggests that no or few beast weights were being made. Since the bird is most closely associated with a religious symbolism it seems likely that the right front indicated the Buddhist law. This implies that the wholly or mainly bird weight signs (*gha*, six-rayed star and bird) also have a religious significance. Perhaps an even more likely explanation of the choice of the right front location for the main sign is the following. The immediate head or lord of the 37 *nats* dwells on the top of Mount Popa, which lies about 50 kilometres south-east of Pagan. He guards not only the whole kingdom but also the palace and indeed every house within the kingdom. Thus the south-east is the sacred direction and the south-east corner of a house is the most proper place for the household shrine.[8]

One Group 4 bird weight class has the auxiliary sign on the underside of the base and some Group 6 bird weight classes, large size only, have auxiliary signs on three sides. On the Group 4 beast weights period H, auxiliary signs occur mostly on the right side but sometimes also on the left. The Group 2(r) beast weight class frequently has an auxiliary sign on the rear side of the base. The location of some of the auxiliary signs on the left front may have been due to lack of space on the right front or to indicate weakness. The meaning of the emplacement of the sign on the back of the weight is not known.

The nine-rayed star:

In Central Asia the cosmological schema has three regions: the sky (heaven), the earth and the underworld. The sky and the underworld are connected by a central axis (pillar, tree) through the earth which terminates at the Pole Star.[9] The auspicious number nine is derived from the combination of 3 x 3. Thus in this region nine is frequently employed, e.g. nine celestial levels in the sky, nine gods, nine branches of the cosmic tree.[10] In Hinduism there are nine underworlds.[11] In China the Taoist/ Buddhist goddess Tou-Mu has nine sons, each of whom is a star.[12] In Buddhism nine represents the supreme spiritual power, the level of meditation at which one attains Nirvana.[13] In Burma the word *ko* means 'nine' and 'to seek protection by worshipping'. In the worshipping of the *nats* (nature spirits), both meanings are combined by the offering of nine candles and nine different kinds of food. There is a ceremony of the nine gods,[14] usually held when there is sickness in the house. This brings to mind the use in healing of the Bodawpaya weight, bearing its nine-rayed star sign. The Shans refer to the nine countries and the 999 chiefs (*sawbwas*).[15]

The nine-rayed star occurs only on the Group 2 beast weights and the Group 3 bird weights. These commenced in 1752 AD with King Alaungpaya only after the Shans came firmly under Burmese control. The founder of the Konbaung dynasty, Alaungpaya was born in 1714 AD at Shwebo. His three sons, each of whom ruled Burma, were also born there. It was made the capital of Burma between 1752 and 1765 AD. Near Shwebo is the ruined Pyu city of Halin where Alaungpaya gained his first victory over the invading Mons. Halin is associated with its locally well-known legend in which a noted mango tree had nine arms bearing nine branches which each had nine twigs, etc.[16], with which he and his sons would have been well acquainted. Moreover, though he was a chief of but a small village, he claimed that by going back nine generations he could prove he was of royal blood.[17] This was of importance since it showed that he was descended from the Pagan king who claimed descent from Asoka. Therefore he was a universal monarch and a rightful ruler in the line of continuous descent.

The eight-rayed star:
The eight-rayed star of the elephant weights of Chiengmai must be connected with the elephant's symbolism of the number 8[18] and the role of the eight elephants in supporting the earth. The symbolism of the defeated mighty enemy, the elephant, and its association with the number 8 may explain why the eight-rayed star never appears on the weights of the victors, the Burmese, except very rarely as a mis-strike. Other relevant symbolisms are referred to in the accounts of the octagonal base plans.

The circle, the circular depression, the circle and dot:
The symbolism of the circle is considered under the base plans in connection with the Wheel of the Law (the *chakra*), the *chakravartin*, the heavens and the sun. The sign of the circle surrounding a central dot occurs on the back of the base of about 20% of the Group 2(r) beast weights. It may have marked the 1802 AD revenue inquest.

The six and five-rayed stars:
The *chakra* is divided into five or usually six sections, in each case by the same number of spokes, so suggesting the five and six-rayed stars. The spaces represent the various realms of existence, i.e. the gods, humans, animals, spirits, infernal beings and semi-divine being.[19] Other meanings of the six-rayed star mostly correspond with those of the hexagon described under the base plan symbolisms. Another possibility, however, has been referred to in the account of the symbolism of the Chinese word *tou. Tou* is the name of a group of six stars in the constellation of Sagittarius in which the earth goddess, Tou-Mu, resides and which is represented by the griffin.[20] It is not known if the Burmese symbolism is similar but the connection of this explanation with two similar ones for the four-rayed and nine-rayed stars is striking. Moreover, the connection becomes stronger when it is recalled that both Chinese and Hindu elements of astronomy/astrology were incorporated in that which was practised by the

ordinary Burman.[21]

Turning to the five-rayed star, it could also represent, for example, the five Buddhas of a short Buddhist era (*Buddhagaba*) or the five disciples of Buddha present at the first sermon or the five *bodhisattvas* (potential Buddhas), each of whom has five supreme virtues.[22] The location of the five-rayed star on the underside of the base may have been intended to symbolise the lesser role of a *bodhisattva* or the monarch in his role as a Bodhisattva, the overlying anserine bird symbolising the Buddha. This last interpretation would agree with the symbolism suggested for the chest decoration on the beast weights. Burmese Buddhism developed partly from that of the Pala-Sena dynasties of Bengal which attached much importance to the number 5.[23] China too had a cult of five. In particular the *feng huang* is associated with the number 5 in a variety of symbolisms. The symbol of Mount Meru surrounded by its four continents, that is, five constituents in total, symbolises universality both in Hinduism and Buddhism.[24] Among the Indo-Aryans, the king was required to make a step in each of the five cardinal directions (including the zenith) upon his consecration in order to assume control of the cosmic order and human behaviour within his universe, i.e. his kingdom.[25] Similar practices exist among the shamans of Central Asia.

The four-rayed star and the square depression:

The idea of totality and other symbolisms of the four-sided geometric shapes, the square and rectangle, have been referred to under base plans. Other requirements in Burma connected with the monarch's role as lord of the four-square world were that he had to have four principal queens, four chief ministers and four privy councillors.[26] In the Buddhist paradise four streams flow from the tree of life, each bearing one of the four major virtues.[27] In China a group of fours star in the form of a cross (like the sign on the weights), which is located on the skirts of the constellation Virgo, is represented, significantly, by the earth dragon. Another group of four stars in the shape of a measure is named Ti or 'earth'. Relevant here are the comments above under the six-rayed star.[28]

The birds:

The bird sign on the bird weights, Groups 1 and 2, could be an anserine bird or a peacock. The peacock could have been chosen as an emblem of royalty after the beast weights had been discontinued in the first half of the 19th century.[29] King Mindon (1853-1878 AD), for example, used the peacock on his coins. The small size of the bird sign and its sparrow-like shape may have given rise to the name *karawaik* of the Group 2 v-tailed bird weights which, along with the similar Group 1 bird weights, mostly bear this sign.[30] The bird signs on the bird weights of Groups 5 and 6 look like anserine birds and presumably have the same symbolisms as the *hamsa*.

The *gha* sign:

This sign is one of the letters in the Burmese alphabet but is rarely used in words. It is an auspicious character and can signify 'great'. Pali is the sacred Buddhist language of south-east Asia and Sri Lanka. In Burma it is written in the Burmese script. In Pali Buddhist texts written in Burmese, letters of the alphabet are use instead of numerals, as in Latin numbers. Number four was ဃ, the *gha* sign.[31] The *gha* sign never occurs on the beast weights. It occurs only on the bird weights and these have the spiritual connotation. Thus, in this case, there is the symbol for four (totality) in the Buddhist sacred language associated with a symbol of the divine spirit. The conclusion is inescapable that the meaning is the spread of the Buddhist doctrine to all creatures everywhere.

The arrowhead or umbrella:

The ⋀ sign could be described as a claw, an arrowhead or the profile of an umbrella (*hti*). The claw appears to have no relevant symbolism. Among the ancient Indo-Aryans, royal power was symbolised by a vertical arrow.[32] Among the Scythians and Siberian tribes it was a shamanic symbol of magical flight.[33] Many Scythian arrows had heads shaped like the trefoils[34] which appear as mouth appendages on several classes of Burmese weight. In China the breaking of an arrow in half symbolised the finalisation of a proposed agreement, for example, between a buyer and a seller.[35] The umbrella was the most important symbol of social status,[36] the higher the rank the more umbrellas, with the Burmese monarchy certainly having seven and perhaps eleven. It is also possible that the sign is an archaic form of a Mon letter and indicates a different symbolism.

There may have been a regional significance in the arrowhead since there is a report of the 'broad arrow' being carved on the breast of the king's secretary in south-west Java in 1588 AD. Similarly, in the Siam of 1688 AD the king's officers were branded on the outside of the wrist with an "anchor"[37]

The Group 6 script:

The script on two of the four large panels comprising the signs on the Group 6 weights appears to be Yuan Thai (north Thai) or a form of old Mon, the meaning of part possibly being 8000 or 10 000. A sinuous snake-like sign is present, along with the script, which presumably is the Sanskrit *Na*, a sign that is widespread on the old currencies of Siam and elsewhere. The Mons of Lower Burma were linked with the ancient Theravada Buddhist Mon state of Haripunjaya (Lamphun) for about 1000 years. Haripunjaya was close to Chiengmai, the capital of Lan-Na. Lower Burma also had alliances with Chiengmai. At the end of the 14th century AD the Buddhist sect established in this region was a leading cultural force. Possibly too the monasteries may have been engaged in ensuring fair practices in trade as they had been in China and Central Asia. If the weights were Mon weights acceptable in both Siam and Burma, then the script on the weights may have been intended to make them acceptable to the northern Thai (though perhaps only 10% or less of the population would have been literate) while the *hamsas* may have been intended to make them acceptable to the peoples of Burma. Such a combination may have been possible under the considerable Chinese pressure which was being exerted on the region at the time through 'comforters' (Chinese political advisors). China had been forced to use south-east Asia as a route instead of Central Asia for part of its trade, especially the rivers and their valleys, their outlets in the Mon region of Lower Burma being of particular importance. Moreover, it would have been natural to use a widely known scale similar to that of the Group 6 weights (see 'Mass Scales').

The beasts:

Beast shapes within rectangles occur only on the Group 5 beast weights. Presumably the symbolism is that of the weight-beast.

Notes

1 Htin Aung, 1959, p. 121. The use of the word *shwe* (gold) brings to mind the gold cult of the Sakas (Litvinskii, pp. 520, 522) when taken in conjunction with the use of gold on religious and royal artifacts in Burma. This was an association that was common among the Sakas, the Parthians, the Kazakhs and in India. The Sakan gold cult probably arose from the abundance of alluvial gold in the vicinity of the Altai mountains, the name meaning 'golden mountains' (Hudson, p. 43).

2 Anonymous, *Weights of Historical Importance.* Judson, p. 361. *Sandaw* = a royal standard. *Hmi* = to reach.

3 Anonymous, *Mandalay Palace*, pp. 8, 22, 25.

4 Rubruck, p. 75.

5 Lowry, plate 3. Lessing, p. 71.

6 Eberhard, p. 162. Cooper, pp. 96, 138. In Siam the right side was that of honour, de la Loubere, p. 56.

7 Rubruck, p. 75. Cooper, p. 96. Williams, p. 162.

8 Lehman, p. 577.

9 Eliade, 1972, pp. 259, 266.

10 Eliade, 1972, p. 274.

11 Cooper, p. 118.

12 Savill, p. 240. The Mons, the Pyu, the Shans and the Burmese must have brought with them the symbolism of the regions from which they came. The widespread distributions of the snake cult and the bronze drums throughout south-east Asia and south China are but two pieces of relevant evidence for this statement. Hence the relevance of the symbolism derived from the region which is now China.

13 Eliade, pp. 406, 407.

14 Htin Aung, 1959, p. 7 ff. This is a comprehensive account of the symbolism of nine as the Burman sees it. There are several local legends where this symbolism is employed.

15 Harvey, p. 323.

16 Williamson, pp. 249, 250.

17 Harvey, pp. 219, 221, 367.

[18] Liebert, p. 87.

[19] Herrmann, pp. 830, 831.

[20] Williams, p. 368.

[21] Temple, 1910, p. 29.

[22] Ball, p. 87. Herrmann, p. 832.

[23] Goetz, pp. 24, 25.

[24] Christie, pp. 67, 87, 93, 112. Lessing, pp. 17, 36, 56, 70, 73.

[25] Irwin, I, p. 750.

[26] Anonymous, *Mandalay Palace*, p. 19.

[27] Cooper, p. 114.

[28] Williams, p. 367.

[29] Iyer, p. 82. Ball, p. 84. Jobes, p. 1246.

[30] Williams, p. 442. In old Chinese writing the pictogram for the word 'bird' much resembles the small bird sign on the Groups 1 and 2 bird weights. If *kalavinka* means 'sparrow' (Thomas, 1980, p. 92), then the dealers may have called the whole weight by this name, i.e. a sparrow weight.

[31] Prinsep, vol. 2, plate XL opposite p. 72.

[32] Litvinskii, p. 521.

[33] Eliade, pp. 152, 388.

[34] Rice, 1969, p. 117.

[35] Eberhard, p. 22.

[36] Praz, p. 719. Htin Aung, 1975, p. 91.

[37] Hakluyt, p. 292. de la Loubere p.79.

15
The Chronological Sequence of the Weights

The dating principles:

The dating of the weights depends on the following principles:

1. The weights conform to a number of recognisable styles.
2. Of a number of similar styles, some characteristics are identical and others differ in systematic ways.
3. The styles of the bird weights can be arranged in a continuous sequence. Those of the beast weights can be arranged in a few discontinuous sequences. (See table 'Style Characteristics by Group')
4. The sequences are chronological in nature for the following reasons: firstly, weights are known to have been used in Burma from before the 12th century AD until the 19th century AD; secondly, the combination of the many kinds of style characteristics into their systematic order for use in different places during one brief period or even a few periods would have been difficult to impossible (see table 'Style Systems'); thirdly, the weights of Group 1 look young while, those of Group 7 look old; fourthly, the average masses of the style groups increase in a way that matches the changes in the masses recorded by European traders over the centuries (see table 'Variations in the Burmese *Kyat* Mass with Time'); fifthly, by arranging the style groups accordingly it can be seen that the different locations and shapes of the main sign on the sides of the base follow a similar order (see table 'Signs on the Weights').
5. The changes in the styles were related to events occurring in Burma's history. (See tables 'Appropriate Dates for Increases in the *Kyat* Mass; Approximate Dates of Events Accompanying New Beast Weight Issues: The Periods of Groups 1, 2, 3, 4, 5, 6 and 7'). The relative frequencies of occurrence of the weights also were so related.
6. The symbolisms of the weights, especially that of the *chakravartin*, to a considerable degree were related to the styles and hence to Burmese historic events.

Other methods of investigation, in addition to those of style and symbolism, were employed. Analysis of the histories of the reign of the Burmese kings from 1531 until 1885 AD gave information firstly on their need for concern with the weights, e.g. wars and increased revenue, religious structures, irrigation works and the like; secondly on

TABLE 19: Style Characteristics by Group

Group		Animal Appearance	Animal Proportions	Base Height	Animal Style	Tail Shapes				Base Plan Shape	Mouth Appendage	
						Birds			Beasts			
Birds	Beasts					Underside	Spread	Rear V U			Absent/ Present	Shape
1		Fowl, standing	Elongated	High	Elongated knobs, Elaborate	Hollow	Lateral	V		Hexagonal	A	
2		Standing fowl or duck	Normal	High Medium	Knobby, elaborate Knobby, simple	Hollow	Lateral	V U		Hexagonal/ Octagonal	A	
	2	Crouching' standing	Normal	High to medium	Knobby				Rod	Square/ Octagonal	P	Rod Vee Trefoil
3		Standing fowl	Normal	Low	Separated head knobs	Hollow	Lateral	V		Octagonal	P	Rod Bird
4		Standing duck	Squat	Low	Short knobs	In-filled	Lateral			Octagonal/ Circular	A P	Trefoil Bird Spatula Knobbed spatula
	4	Crouching/ Standing? lion	Squat	Low	Simple styles except on large weights				Rod	Octagonal/ Circular	A P	Rod Vee Trefoil
5		Standing duck	Normal to tall	Low	Overlapping head knobs	Hollow or in-filled	Lateral	V		Octagonal Circular Polygonal	P	Rod Bird
	5	Crouching/ standing? lion	Normal to squat	High to low	Weakly knobbed				Rod (c) (d) (e) (f)	Octagonal Circular Lozenge Rectangular	A P	Rod(s)
6		Brooding hen	Normal	High	Weakly knobbed comb		Vertical Longitudinal	(a)		Octagonal Circular	A F	Bird
7		Standing duck	Normal	High	Weakly knobbed comb		Vertical Longitudinal	(b)		Octagonal Circular Polygonal	A P	Rod Bird

Rear = As viewed from rear. (a) = Scrolled triangle (side view). (b) = As (a) or wide scrolled strip. (c) = Short, wide, pendent (horse). (d) = Short, erect, cone (bud). (e) = Short, wide, single loop. (f) = Short, wide, double loop. (A) = Absent. (P) = Present.

the possibility of their being able to have concern with the weights, e.g. duration of their reigns, the state of the realm; and thirdly on the actuality of their concern, e.g. recorded activities, weight standardisations and revenue inquests. Seven kings were identified as being likely to be connected with the weights.

The date when the capital changed its location in relation to styles, signs and *kyat* mass changes, also yielded some information. On six occasions, perhaps seven, there was an association between the date of the capital change and a style change. Only in one instance of a change in the location of the capital was there no identifiable correlation between the two kinds of change. Style changes also took place which were unassociated with a capital change. After the styles had been identified and the sequences recognised, then the 1972 count and weighing of 1078 weights became worthwhile and enabled comparisons to be made and inferences to be drawn.

Study of the differences between the relative proportions of the groups, of the bird and beast weights in each group (and, where possible, class), and of classes yielded information which could be correlated with historic events.

The shapes and locations of the main and auxiliary signs yielded information of some value, particularly in regard to the locations of intended usage of the weights.

The frequencies of unsigned weights and unofficial copies yielded indicative information. Those of the Group 2 bird weights, periods C and D, and beast weights 2(r), period C, appear to coincide with the loss of the last Burmese possessions in north Thailand in the early part of the 19th century. Annandale illustrates a Group 2(u), period C weight (the most copied class) which was sent to India in 1868 AD. Similar coincidences exist in classes of Group 4.

The styles have been arranged firstly, by groups in which all the weights have certain characteristics in common and secondly, by classes within a group, the members of which have other characteristics in common.

With the establishment of the chronological sequence, it became possible to define the duration of a style group or class or system as a period. A period can be of any number of years. The periods have been lettered beginning with A, the most recent. These periods have been used as units for the purpose of dating the weights. The dating of the periods has been done with reasonable confidence for the beginning and end of a group or territorial expansion. For periods of lesser duration there is more doubt.

Group 1, period A (Plates 24 and 53, Table 20):

These weights would have ceased with the fall of the Burmese monarchy in 1885 AD and the introduction of British weights. They could have ceased in 1864 AD when payment of taxes had to be made in coin and not in kind as always previously. They could have commenced firstly in 1852 AD when Burma lost the second third of its original territory to the British; secondly in 1864 AD; thirdly in 1875 AD when Burma's last king ascended the throne. Stylistically, these were the last of the weights to be made. They are the least worn and patinated in appearance and because of the brassy colour, are recent. The long head knobs, so easily broken off, found on most of these rare weights also help to make it evident that these weights could not have been in existence long. They were seen in use at Inle Lake in 1917 AD.[1] The writers have seen them bearing the date 1923 and with the name of a person who had ordered them. Such markings would not have been permitted before the end of the monarchy. The possible dates of official fabrication are 1852/1875 AD to 1875/1885 AD.

Groups 2 and 3, periods B to E (Plates 18 to 23, 41 to 43, 53, Table 20):
The style of the bird weights of Group 2 must have ceased when that of Group 1 commenced, at some time between 1852 and 1875 AD. The beast weights would have been expected to have ceased after about 1796 to 1802 AD when north Siam had been lost. In fact they may have continued in use in Arakan (annexed in 1785 AD) where this type of weight is the only one to have been seen by the writers. A specimen was obtained thence in 1837 AD.[2]

In 1826 AD and again in 1845 AD[3] the beast weights were reported as being used in Burma. However, the latter report may have been copying the former. Partly because Burma lost Arakan, Tavoy and Tenasserim to the British in 1825 AD, partly because of Burma's incessant military activities, including the annexation of Assam and Manipur in 1819 AD, partly because this is the most common of the beast weights, the year of cessation of beast weight manufacture has been put at 1825 AD.

The bird and beast weights each fall into a number of styles which can be correlated to form four periods. Support is given to the dating of the square-based beast weight of periods B and C because even in 1970 AD it still was called a Bodawpaya weight. Moreover, the 4-kyat weight, characteristic of part of this class, was still in use in 1825 AD[4].King Bodawpaya ruled Burma from 1782 to 1819 AD. The dating of period D receives probable confirmation from an illustration of this beast weight published in 1776 AD.[5]

The period E, U-tailed bird weights and beast weights are similar to the Group 4 weights in their trefoil-mouth appendages and the use of six and four-rayed star signs. They are similar to the Group 2 weights in their *kyat* mass. However, while the beast weight is clearly Group 2 in style, one of the bird weights is equally clearly Group 4 in style, so creating some doubt about its dating. The Group 3 bird weight is clearly similar to Group 2 in style and *kyat* mass. The preferred solution to the problem of correlating these weights has been influenced by the possibility that the six-rayed star sign may be connected with the Shans (who submitted to King Alaungpaya in 1755 AD) and that the predatory appearance of the Group 3 bird weights and the later Group 2 V-tailed bird weights may have been made as a reminder to the Mons of the massacre at Pegu in 1757 AD.

Group 4, periods E to J (Plates 15 to 17, 37 to 40, 54, Table 21):
The Group 4 weights are quite uniform in appearance except for the characteristics shown on the table 'The Periods of Group 4'. Minor confirmatory evidence is available. The bird weights are known as Mon ducks, suggesting the period 1531 to 1635 AD when the former Mon capital Pegu was the capital of Burma. One Group 4, post-period J weight has been seen bearing the date 1739 AD in Burmese script and era. The Group 4 weights are inferred to have commenced between the time of the fall of Pegu in 1531 AD to the beginning year of the weight standardisation in 1563 AD and ceased to be made at the time of its fall in 1757 AD.

Groups 5 and 6, bird weights, periods K and L (Plates 13, 14, 34, 35, 36, 55, Table 22):
The partial concurrency of Groups 5 and 6 is inferred from the following evidence apparent on the weights of both groups:

1. Three, two or one bird(s) on the base: No Group 5, two-bird weight has been seen. It is assumed that it did exist.
2. Convex wings: Apart from Groups 5, 6 and 7 bird weights, the wings of all other bird weights have a concave shape.
3. Embossed bird signs: No other weights have them.
5. Feminine characteristics: Apparent on most subsequent weights.

The evidence for the durations of and changes in the Groups 5 and 6 weights is the following:

1. In 1452 AD Shinsawbu became queen, the reign beginning a period of spiritual and material revival, culminating in the reign of King Dammazedi from 1472-1492 AD. In Group 5 the use of the mouth emanation on a *hamsa* would symbolise, appropriately, material prosperity and spirituality.
2. The feminine characteristics associated with both Group 5 and Group 6 weights would be appropriate for their introduction at the time of Queen Shinsawbu's coronation. The use of the *hamsa* sign on the weights would symbolise the spiritual revival of the times.
3. The three-bird weights of Groups 5 and 6 could correspond with the period when the queen had two monk advisors, i.e. from 1452 to 1460 AD. The two-bird weights could correspond with the period when she retired and Dammazedi acted as regent, between 1460 and 1472 AD, while the one-bird weight could have been manufactured after she had died. However, it seems more likely that the symbolism refers to the three countries of Ramannadesa.
4. A double chevron ornament occasionally appears on the mouth appendages of some Group 5, period K and Group 4, period J weights and no others. A four-rayed

star has been seen once on a Group 5 weight, it being frequent on Group 4 and later weights. Circular base plans are found on weights of Group 5, periods K and L, and Group 4, periods I and J, after which they are no longer produced. So the Group 5 weights probably terminated between 1531 AD (fall of Pegu) and 1563 AD (the first year of King Bayinnaung's standardisation), when they were replaced by the Group 4 weights.

The Group 5 weights of more than about 5 *kyats* which do not have signs on the base (period L) and the Group 6 weights which have circular plan bases may be somewhat older than the other weights of their groups. The evidence for the Pegu location of the group 5 weights lies in:

1. the legendary association of the *hamsa* with Pegu, though the legend may 'have arisen during or after the queen's reign;
2. the presence of the three overlapping forehead knobs, symbolising the union of the three provinces;
3. the replacement of the earthly beast weights of King Razadarit by weights symbolising the spiritual associations of the *hamsa* during this period of spiritual purification.

The evidence for the Ramannadesa (Martaban?) or Siam location of the Group 6 weights is as follows:

1. The presence of a possibly north Siam or Old Mon Script on two base sides.
2. The manufacture of bases with thin walls and stone cores, a technique for which south Siam (Ayudhya) was noted in the mid-15th century AD.
3. The mass scale which, from 300 to 35 grams, appears to be similar to the elephant weight, the *tok* (and *kakim*?) silver ingot scale of Lan-Na (14th-16th century AD) and the Burmese scale.
4. The mass of the 100 unit member is about that of the Siamese (Ayudhyan) 'catty'. The catty or *kati* was a term used internationally throughout the Far East for the Chinese *kin* or *chin* weight of nominally 625 grams, but, in fact, variable. The Siamese catty was unique in that it was about twice the mass of the nominal mass, i.e. about 1250 grams.
5. The Group 6 mass unit of about 11,2 grams is not greatly different from the 12 gram unit mass of Lan-Na; hence there may be a connection between the two. Compare the religious connection between Pegu and Haripunjaya (Lamphun) and the trade connection between Yunnan and Martaban. Also In support of this statement, Lan-Na may have used base metal *kud* as weights.[6] In 1554 AD possibly similar 'round' weights of *fruseleyra* (called *caturna*) were in use in southern Burma by "Siamese" merchants only, as was customary. Apparently there were two *kyat* masses, one of about 12 grams and one of 14 grams.[7] This

usage could be expected since in 1514 AD the silver *pod-duang* were the main source of Pegu's silver supply.[8]

Group 5 beast weights, periods K, L and M (Plates 32, 33, 52, 53)

The three Group 5 beast weights are considered to have been made in Ramannadesa during the reign of King Razadarit (1385-1423 AD). They agree firstly, with the separation of the Mon kingdom into three provinces, each warranting a *chakravartin* representation; secondly, with the king's appellation the 'Lion King', a common title but one for which he is noted; and thirdly, with the visit of a Chinese mission during his reign, a visit which may have resulted in an agreement regarding the use of the Chinese decimal scale and other weight-related matters. It was about this time that the Siamese and Cambodians 'received' Chinese weight standards (presumably the Far East commercial scale which is still in use).

King Razadarit reigned at a time of considerable activity by the Chinese to develop their trade routes to the south and their mining activities in Burma and elsewhere. It would be unlikely that the Rammanadesa kingdom did not receive the Chinese standards. It may be that at this time there was an adjustment in the *kyat* mass from about 11 or12 grams to about 14,5 grams, so that the *bol*of 5 *kyats* (about 73,5 grams) became equal to 2 *liang* (about 75,0 grams), the mass of the most common *sycee.*

The former monk, King Dammazedi, some 60 or so years later, also exchanged missions with Yunnan and was concerned with the Chinese trade. The dominance of the spiritual motive at this time and the peaceable reigns suggest that he did not cause beast weights to be made. He became a *chakravartin* but not until his death.

In 1548 AD there were three different *chakravartins* in the region. These were the rulers of Burma and, in Siam, the combined kingdoms of Lan-Na/Lan Chang and Ayudhya. Each of these could have issued beast weights, each with a different tail shape. However, the mass scales in Siam would have been binary, not decimal, which is the scale of all the Group 5 beast weights, hence the beast weights were not made in Siam.

There is a small group of seemingly genuine beast weights with a pendent tail which appears not to belong to Group 4 or Group 5. They could have originated after the death of King Dammazedi (1492 AD) and before King Bayinnaung's weight standardisation.

Group 7 bird weights, period M (Plates 31, 55, Table 22):

The Group 7 bird weights have been inferred, provisionally, to be the oldest of all the weights. This is because of the following.

1. Their workmanship is crude compared with that of all weights except possibly some Group 5 beast weights.
2. The tail shape suggests that they are older than the Group 5 bird weights and of the same age or older than the Group 6 weights; in short, older than either.

3. These weights with a dome-shaped base suggest an early Chinese influence unless the shape represents one of the early Indian *stupas*.
4. The fact that they do not carry a sign indicates that they may be older than the Group 5 beast weights.

Dating a particular weight:
The procedure for dating a weight is to identify the group first, then the style characteristics of the period. The table 'Style Characteristics by Group' should suffice for the first. The sequence of three tables entitled "Characteristics of the Weights ofthe Periods of Group(s) . . ." usually will suffice for the second, provided the signs are present. Doubts mostly can be removed by referring to the line diagrams and photographs of the weights and their components, the table entitled 'Decorative Motifs' and that entitled 'Signs on the Weights'.

Notes

1 Annandale, plate XLIII.
2 Annandale, p. 197.
3 Temple, 'Indian Antiquary', June 1898, p. 141.
4 Wilson, Appendix, p. LXI.
5 Mollat, p. 440, quoting from *Alphabetum Barmanum.* Rome, 1776.
6 Le May, pp. 17, 18, 22, 23 Kneedler, p.12.
7 de Campos, p. 122, and the table 'Variations in the Burmese *Kyat* Mass with Time'.
8 Pires, v.1, p.100.

Following on as from page 208

Groups 1, 2 and 3. Plate 53 illustrating Table 20

Group 4. Plate 54 illustrating Table 21

Groups 5, 6 and 7. Plate 55 illustrating Table 22

Group	Period	Birds Laterally-spread tails ∧ – tails	∩ – tails	Beasts
1	A	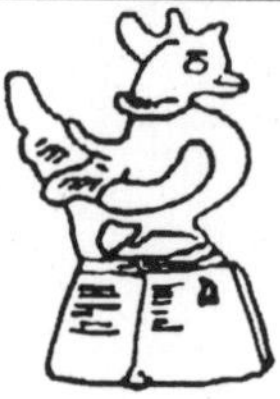		
2	B	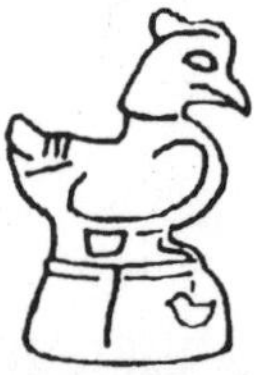	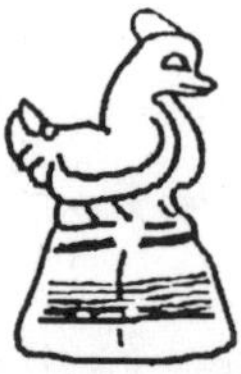	
	C	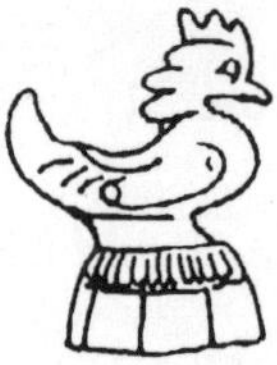	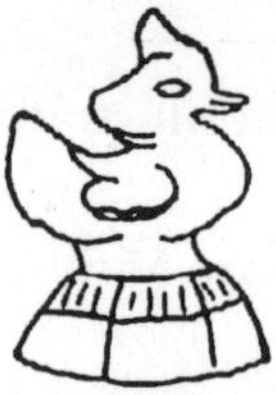	As for Period D?
	D	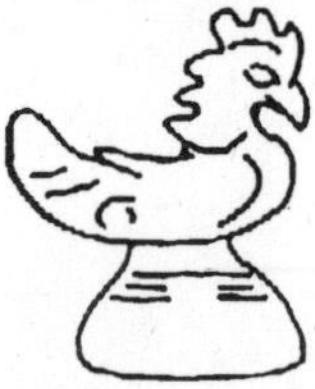	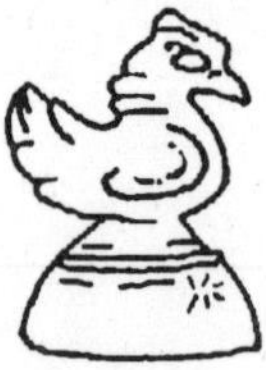	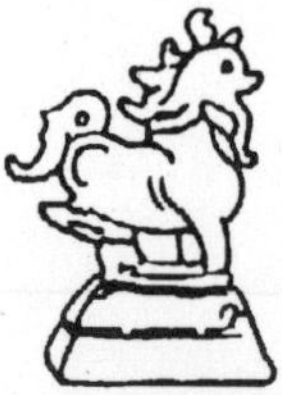
	E			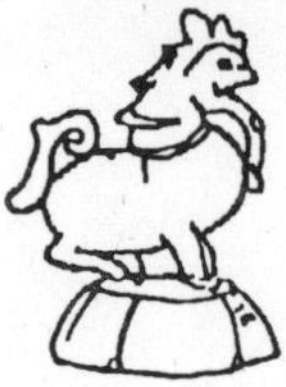 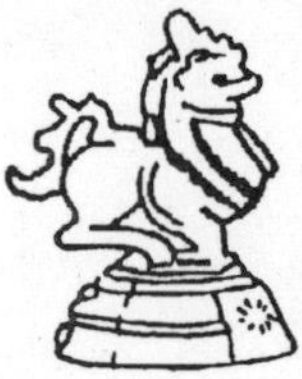
3	E	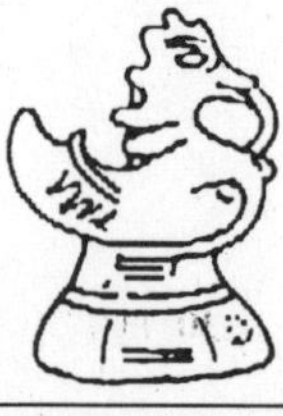		

Note: The sketches are representations of one class of a style system from each period. See Style System plates 41 to 43 for other classes

Plate 53: Weight Shapes of Groups 1, 2 and 3 by Periods A to E

Table 20: Characteristics of the Weights of the Periods of Groups 1, 2 and 3: United Burma

Period Year	Period Letter	Characteristics	Birds: U	Birds: V	Beasts	Territorial Gains	Territorial Losses	King	Dates
1875/85									1885
							Upper Burma		
		Base Plan	------	Hex.	------			Thibaw	
		Frequency/1000	------	2	------			------	1878
	A	Mouth append.	------	—	------				
		Sign	------	Bird	------			Mindon	
		Base horiz. lines	------	6	------				
		Kyat mass		16,3			Lower Burma	------	1852
1852/75									
		Base Plan	Hex.	Hex.	------				
		Frequency/1000	Incl. in Bb	Incl. in Bb	------				
	Ba	Mouth append.	—	—	------			------	1837
		Sign	6	Bird	------				
		Base horiz. lines	4	4	------			Bagyidaw	1825
		Kyat mass	15,9	15,9			Arakan, Tavoy etc.	------	1819
1819/24						Manipur Assam			
		Base Plan	Hex.	Hex.	Square				
		Frequency/1000	49	45	Incl. in C				
	Bb	Mouth append.	—	—	R				
		Sign	6	Bird	9				
		Base horiz. lines	4	4	4				
		Kyat mass	15,9	15,9	15,8		N. Siam	Bodawpaya	1802
796/1802									
		Base Plan	Hex.	Oct.	Square				
		Frequency/1000	278	26	140				
		Mouth append.	—	—	R				
	C	Sign	6	Bird	9				
		Base horiz. lines	SV	SV	SV				
		Kyat mass	15,9	15,9	15,8	N. Siam Arakan		------	1785 1782
1785									
		Base Plan	Hex.	Oct.	Oct.			Singu	
		Frequency/1000	84	26	45		N. Siam	------	1775
	D	Mouth append.	—	—	V				
		Sign	6	Bird	9				
		Base horiz. lines	3	3	3			Hsinbyushin	
		Kyat mass	15,7	15,7	15,7	N. Siam		------	1767 1763
1867									
			Group 4	Group 3					
		Base Plan	Oct.	Oct.	Oct.	Tavoy,			
		Frequency/1000	19	13	36	Tenasserim			
	Ea	Mouth append.	L	R	L	Manipur		------	1760
		Sign	6	9	4				
		Base horiz. lines	2	3	3				
		Kyat mass	15,9	15,3	15,4				
1757			------		------	Lower Burma Shan States		Alaungpaya	1757 1755
	Eb		?	------ ------	?				
		KONBAUNG Dynasty commences							
1752									1752

1. U, V indicate the tail shapes.
2. The dashed lines in the column headed "King" indicated the beginnings and ends of reigns.
3. The dates in the column headed "Dates" indicate those of the reigns (opposite ends of dashed lines) or those of the territorial changes.
4. The frequencies/1000 of the U-tailed bird weights of periods C and D are probably too high because these appear to have been much copied before 1868 AD.
5. SV = Short Verticals
6. Sign, b, 9, 4. See Table 21 Note 5.
7. Mouth appendage L = Trefoil, V = Vee, R = Rod

Group	Period	Birds: Laterally-spread tails ∧ – tails	Birds: Laterally-spread tails ∩ – tails	Beasts
4	F G			
	H I			
	I			
	J			

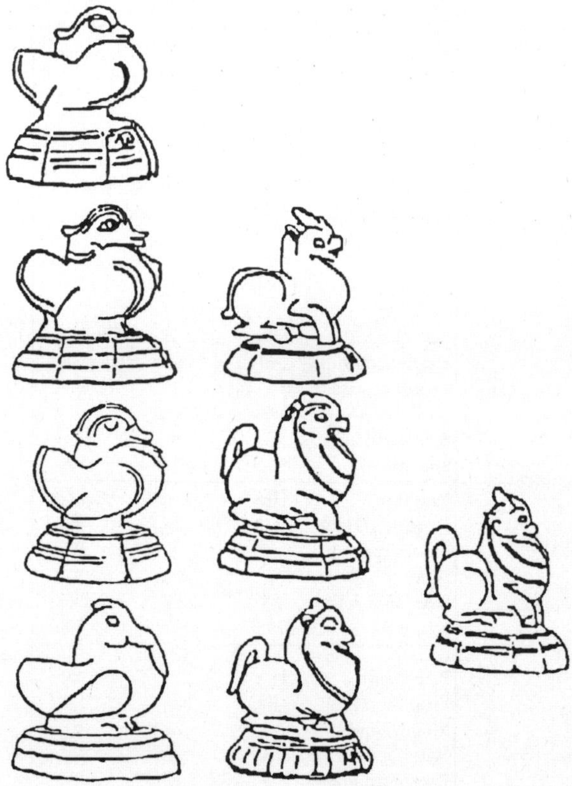

Weight sketches are representations of one class of a style system from each period. See Style System plates 37 to 39 for other classes.

Plate 54: Weight Shapes of Group 4 by Periods F to J.

Table 21: Characteristics of the Weights of the Periods of Group 4: United Burma

Period Year	Period Letter	Characteristics	Birds	Beasts	Beasts Gains	Beasts Losses	King	Dates
1752								1752
		Frequency/1000	1	—		Lower Burma		1740
		Base plan	0	—				
		Mouth append.	L , A	—				
	F,G	Main sign location	R F	—				
		Auxiliary sign shape	—	—				
		Main sign shape	W	—				
1717						N. Siam		1717
		Frequency/1000	1 5	75				
		Base plan	O	O				
		Mouth append.	L , A	A			------	1648
	H	Main sign location	F	F			Thalun	
		Auxiliary sign shape	—	6, –			------	1629
		Main sign shape	W	C,S				
1626					N. Siam			1626
		Frequency/1000	(inc. in H)	66				
		Base plan	O	O				
	Ia	Mouth append.	L,A	L,V,A				
		Main sign location	F	F	Burma re-united		Anaukpetlun	1615
		Auxiliary sign shape	5,6	6,4				1610
							------	1605
		Main sign shape	W,Cd	Cd, 5d		Burma disunited		–1595
1595						N.Siam		1586
		Frequency/1000	12	(inc. in Ia)			------	1581
		Base plan	O,C	O,C	Shan States		Bayinnaung	
		Mouth append.	Sk,A	L,V,A	N. Burma			
	Ib & Ic	Main sign location	F	F	Manipur			
		Auxiliary sign shape	6,7	C,T	N. Siam			
		Main sign shape	Cd	Cd,Sd				
1551							------	1551
		Frequency/1000	8	10				
		Base plan	O,C	O,C				
		Mouth append.	S,A	R,A			Tabinshwehti	
	J	Main sign location	F	F				1544
		Auxiliary sign shape	—	—	Central Burma			
		Main sign shape	Cd	Cd	Ramannadesa			
1539								1539
1531		NYAUNGYAN Dynasty commences						1531

1. Kyat mass of all periods is about 14,5 grams
2. Base shape: O = Octagonal plan. C = Circular plan (rare)
3. Mouth appendage: V = Vee, ogive. A = Absent. L = Trefoil. Sk = Spatula knobbed. S = Spatula. R = Rod.
4. Main sign location: RF = Right front. F = front.
5. Auxillary sign shape: 6 = 6-rayed star. 5 = 5-rayed star. 4 = 4-rayed star. C = Circle. T = Hti (arrow head).
6. Main sign shape: W = Gha sign ဃ. Cd = Circular depression. Sd = Square depression.
7. Bird wing shape: Period Ib and Ic to FG, concave ear shapes. Period J, convex comma shapes.
8. Beast tail base attitude: Period H, horizontal. Period Ia to J, vertical.
9. The dashed lines in the column headed "Kings" indicate the beginning and end of reigns.
10. The dates in the column headed "Dates" indicate those of the reigns (opposite ends of dashed lines) or those of a territorial change.

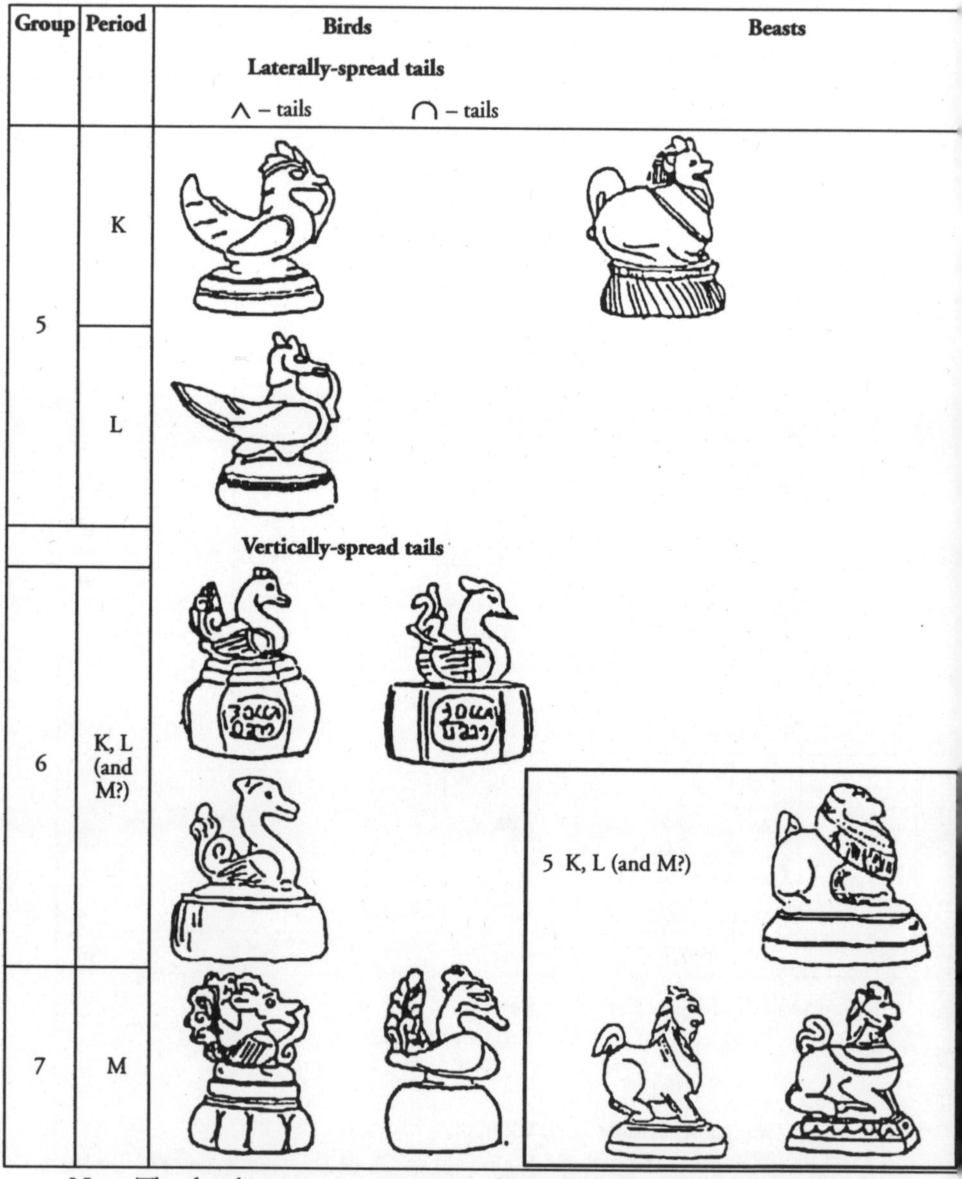

Group	Period	Birds		Beasts
		Laterally-spread tails		
		∧ – tails	∩ – tails	
5	K			
	L			
		Vertically-spread tails		
6	K, L (and M?)			5 K, L (and M?)
7	M			

Note. The sketches are representations of one class of a style system from each period. See Style System plates 31 and 35 for other classes

Plate 55: Weight Shapes of Groups 5, 6 and 7 by Periods K to M.

Table 22: Characteristics of the Weights of the Periods of Groups 5, 6 and 7: Lower Burma (Ramannadesa).

Period		Characteristics	Birds		Beasts	
Years	Letter		Group 5	Group 6		Group ? post-1492?
1531/1563						
		Frequency/1000	12	6	The Group 5 beast weights could be here and not earlier as shown	O
		Tail attitude	L	V		P
		No. creatures/base	1, 3	1, 2, 3		1
		Manes	1 – 7	N		1 – 3
	KL	Base plan	O, C	O		O, C
		Older?	C, P	C		
		Mouth appendage	R	N		R
		Main sign shape	B	N, S, B		T, C
		Older?	N			
1460			Group 7	Group 6	Group 5	
1460						
		Frequency/1000	O		O	
		Tail attitude	V	Here also?	V	
		No. creatures/base	1		1	
		Manes	N		N 4	
	M	Base s plan	C, P		C, R, L	
			N R		V, R, L	
		Mouth appendage			B, T, 2	
		Main sign shape	N		T	
1385/1423			Group 7			
1385/1423						
?			Some older?			

Queen Shinsawbu reigned between 1453 and 1460 AD and then delegated her rule to Dammazedi who became king in 1472 AD reigning peacefully until 1492 AD.

King Razadarit reigned between 1385 and 1423 AD. He united Ramannadesa.

1. Not present: N
2. Main sign shape: B = Bird in depressed panel. S = script in depressed panel. T = Beast in depressed panel. C = Circular depression. On the Group 6 weights signs appear only on the 80 kyat weights with octagonal bases.
3. Main sign location: All signs on front except in Group 5 where they occur on front, back and two sides but only on octagonal 80 kyat weights.
4. Mouth appendage: V = Vee, ogive. L = Trefoil. R = Rod. B = Bifurcated rod. T = Trifurcated rod. 2 = Twin strips or rods. The rare small bird on the chest of the large one is not indicated on the table.
5. Base plan shape: O = Octagonal. C = Circular. P = Polygonal. R = Rectangular. L = Lozenge (elongated octagon).
6. Tail attitude: V = Vertical. L = Lateral. P = Vertical tail base, pendent tail.
7. The kyat masses of each of the Group 5 bird and beast weights average 14,4 grams but in each case there might be two groups of about 14,6 and 14,0 grams. The Group 6 bird weights have an average kyat mass of 11,2 grams perhaps in two groups of 11,4 and 10,2 grams. The Group 7 bird weights have an average kyat mass of 11,2 grams.

16
The Transference of the Anserine-Feline Motifs to Burma

Introduction:

The mythologies which are associated with the symbolism of the animal weights probably go back to Neolithic times and, under similar environmental conditions, probably were widespread even then. Similarities between eastern and western Eurasian mythologies have been recorded by many authors. Many others have recorded similarities in materials from Central Asia to Europe and China. A few others have studied the similarities of weight shapes throughout the Near and Middle East. Some have followed the similarities in weight names from Babylon to Mongolia and Thailand. A smaller number of people have followed the courses of mass units and mass scales from Assyria to India and China. There are other examples in science, technology and astrology. So in recognising the possibility of the ultimate origin of the Burmese weights as being in Assyria, no original idea has been produced. However, in investigating this possibility modern evidence has been assembled which supports this already long-established hypothesis.

Though the lion-duck motif may have been first used on weights in ancient Mesopotamia, the feline-bird combination of motifs may have been adopted in east and south Asia from nomadic invaders from the steppes. Since no other lion-duck weights are known to exist between Mesopotamia, Persia and Burma and both occurrences are at the southern extremities of nomad invasion routes, routes along which the feline-bird combination of motifs is common, this suggestion is not pure speculation. To support the suggestion there are many pieces of evidence to connect the peoples of Burma with the steppes east of the Altai, some of which have been mentioned previously while others are included in the following accounts.

The information contained in the previous chapters has drawn attention to the similarities between the characteristics of the Burmese weights and those of artifacts from other regions. These similarities sometimes may have had a common origin. Sometimes they may have arisen independently. There is much evidence to support the view that most of them were transferred from other cultures to the peoples of Burma who modified them to accord with their own culture.

Transference could have been by trade or gift (more important in ancient times than now), by religious proselytising, by copying of examples and so on. For transference to have come about, artifacts and the motifs appearing in their shapes, styles and decorations must have moved from person to person, so practicable routes must have existed between places where people were relatively concentrated. The following account describes some important influences on the designers of the weights and the more important ways in which the motifs and symbols embodied in the weight could have been, and probably were, transferred either from ancient Egypt and Mesopotamia to Burma or from the steppes to both Mesopotamia and Burma or through some combination of the two routes, first to Mesopotamia, then to Burma.

Population Centres

Urban centres, trade and Mahayana Buddhism were closely associated.[1] In India the cities of the Ganges basin began to increase in numbers and populations from about 500 BC,[2] in north and central China from about 200 BC and in Central Asia a little later. Urban centres in all of these regions reached their peaks of prosperity and size about the 2nd or 3rd century AD. Afterwards they declined, many cities in north-west India, along the silk route and in north China being deserted or reduced in population.[3] About the 5th century AD the cities of north-east India began to flourish again. Those of north China were recovering by the 7th century AD where not entirely abandoned. Probably the cities of Central Asia were recovering also but it was not long before Islam began to adversely affect the already declining Buddhist trade of the region. Urban centres in Yunnan may have begun to develop slowly after about the 1st century AD and in Burma about the 4th century AD[4] or perhaps earlier. However, then, as now, more than 95% of the population of two to three million were rural.

Transport and Travel

Along the Eurasian routes transport was by slave, ass, mule, horse and camel. Judging from recorded historical practices on the Central Asian section, the leader of the caravan usually would have been the nomad chief from whom the animals had been hired. The size of the caravan would have varied but with many pilgrims, traders and guards, sometimes may have reached 5000 beasts, e.g. camels, horses, etc. After the conquest of Chinese Turkestan in 751 AD it is likely that there could also have been caravans of perhaps thousands of people making the yearly pilgrimage to Mecca. The timing of the caravan's departure usually would have been determined by the availability of pasture and water.[5] On the Central Asian route carts may only rarely have been used in the caravans even though they had existed in Mesopotamia since about the 35th century BC and in China since the 13th century BC. Nevertheless the nomadic Saka of Pazyryk in the Altai mountains in the 5th – 3rd centuries BC used them to transport their families and buried both horses and carts in the tombs of the rulers.[6] In 7th century BC India materials and trade articles were carried by caravans of 500 or more ox wagons. Such were used by the Aryan invaders of about 1500 BC. The carts had spoked wheels and rawhide tyres to facilitate travel over the soft soil.[7]

Long before 580 AD and for hundreds of years afterwards Ta-li in Yunnan was sending caravans of 1500 horses each year to China as tribute. The horses had a reputation for general excellence and especially endurance.[8] The proto-Burmese were already skilled horsemen when they reached Burma. Subsequently such was their fame in this respect that the Manipuri word for horse became 'the Burmese animal'.[9] It is relevant to mention here that the Scythians were among the earliest peoples to master horse riding and that in ancient Inner Asia the horse was the basis of the peoples' economy and military power. Evidently horses and horse riding were especially associated with the ancient nomads on the north-west Chinese borders, the nomads of the Yunnanese grasslands and the Shans or Burmese.

In most of eastern India, Burma and Siam, goods mainly were transported by oxen, either in packs or carts. In pre-Christian era India, the caravans might reach 10 000 animals, in ancient China, 1 000 or more and in Burma and Siam where the Shans were the chief itinerant traders, several hundred oxen or elephants would be employed.[10] In Yunnan, cattle had little value neither for transport, milk nor meat. Because of the hilly country, carts had little value. Pack ponies, mules and even asses were the only animals used, caravans of 2 000 animals or more being employed in the 15th and 16th centuries AD.[11] On the Yunnan-Burma route the Yunnanese operated the caravans and controlled the trade. Formerly, slaves were often used as pack animals, while in the 20th century pedlars often carried their goods for hundreds of kilometres on their own shoulders.[12]

The horse was domesticated about the 35th century BC and the Bactrian camel about the 4th century BC. (The one-humped camel was tamed about the 12th century BC). A camel bearing a load of less than 180 kg could travel three days without water in hot weather on soft sand along the silk route at an average rate of between 30 and 45 km/day. It could do this for prolonged periods.

Caravan travelling times (without rest intervals or delays) were approximately as follows.

From	*To*	*Distance (km)*	*Time(days)*
Caspian Sea	Kashgar	2300-3000	70-100
Kashgar	Chiu-chuan	2000-2700	60-85
Chiu-chuan	Lanchou	700-900	15-25
Caspian Sea	Lanchou	5000-6600	145-210

Kashgar was at the west end of the Tarim basin, Chiu-chuan was near the east end and at the end of the Great Wall. Lanchou was at the junction of the silk route and the route to Yunnan and Burma.[13] Lanchou is about 2500 km from Kunming (Yunnan-fu), i.e. about 60 travelling days by foot and animal transport.

A mule and pony caravan journeying from Yunnan-fu (Kunming) to Bhamo in 1930 AD took 31 travelling days to cover the approximately 1200 km (800 miles).[14] This represents a rate of movement of 39 km/day. Postal carriers in ancient China often travelled at the rate of about 240 km/day and sometimes over 300 km/day.[15]

In 1934-5 the Communist army in China fought a continuous rearguard action against the Kuomintang over a distance of 10 000 km in the often mountainous country of west and north-west China during a period of 368 days, of which 268 were spent in marching.[16] Overall the marching rate averaged 27 km/day. Omitting static periods the rate of movement averaged 37 km/day.

It is evident from elsewhere in the text that the ancient, partially nomadic peoples of Yunnan may have had contact with the steppe nomads of the north-west China regions now known as Gansu (Kansu), Xinjiang (Sinkiang) and Qing Hai (Ch'ing Hai). One of the groups of nomads who dwelt in the regions of north-west China in the early part of this century, the Kazakhs, regularly travelled many hundreds of miles between their summer and winter grazing grounds.[17] Thus it is apparent from this assemblage of relevant information that the distance itself between Lanchou and Bhamo of about 3700 km (about 90 to 100 travelling days) would have presented less of a problem to migrating nomads than the 'Long March' did to the Chinese communists.

Trade

General:

Among the conditions required for regular and extensive trade by land were urbanisation, adequate security and communications between towns which could be used by pack animals or wheeled transport for at least part of the year, even though unpaved. In turn these depended upon a country being large enough to produce an agricultural surplus so as to afford a social and legal organisation and a road system.[18] Where distances were great and pack animals were used, as in the great caravans, only small, highly-valued, non-perishable goods could be transported.[19] The long distance trade then was in small quantities of luxury and exotic items intended for wealthy, privileged minorities.[20]

Concerning sea trade, ships were small and in the east only one trip a year could be made usually, the directions of the journeys being dependent on the monsoons. In the 11th and 12th centuries AD, China's sea trade was at its most extensive, its ships sailing to the Persian Gulf, Arabia, India and Indonesia.[21] By the 15th century AD it had restricted its direct trade to Malaya, Indonesia and south-east Asia. In particular it traded with Siam and Malacca,[22] partly because it had closed its own ports to foreigners and dealt with them only in foreign ports.[23] Most of the merchants in Siam and Cambodia were Chinese,[24] as they may have been in Malacca. In activities such as tin mining the Chinese bought the rights from the Siamese king.[25] The political influence of China in the region was so powerful at this time that in the early 15th century AD it was able to force Malacca once again to re-submit to Siam.[26]

It seems likely that the 'Far East Commercial Scale' based on the Chinese. mainly decimal, scale but using mainly modified Malayan names came into existence throughout China's main trading regions before the 14th century AD. The trading dominance of the Chinese in Siam, Cambodia and Malaya explains why this scale was used in these countries for commerce while the smaller Chinese influence on sea trade in Burma, especially after the early 16th century, resulted in the exclusion of the scale and the continuation in use of the Burmese weights both for commerce and currency, though modified in the 15th century AD to suit the Chinese.

Early in the 16th century AD the Portuguese captured Malacca in Malaya and this led to another great increase of sea trade throughout the south-east Asian region and between East and West. This was a trade in which Pegu participated, its wealth helping to pay for the creation of the Burmese empire in the mid-16th century AD.

Western Asia to India and Central Asia:

There were trade links between the Turkmenistan, Mesopotamian and Indus valley civilisations between 3000 and 2000 BC.[27] Lapis lazuli was being exported from Afghanistan to Sumeria and China before 2000 BC[28] and jade went from east Turkestan to Switzerland and China before 1500 BC.[29] Though it is unlikely that King Solomon of Israel was importing peacocks from India around 950 BC,[30] nevertheless there was trade with Israel not long afterwards. Indian spices were reaching Egypt before 700 BC.[31] The Indian Vedas, after accounts of the prolonged conflict between the Aryans and the Scythians, report that the two antagonists formed an alliance and together began sea-borne commerce to distant lands.[32] This, presumably, would have been after the pre-1700 BC Aryan conquest of northern India and before the 6th century BC uniting of Persia and north-west India, perhaps about 700 BC. These traders may have been the first of those later to be known as Persians or Parsees. During this period the lion and duck weights were in use in Persia and adjacent regions. It is likely that these weights may have been carried to the 'distant lands' if only for the use of the Aryan/Scythian/Persian traders themselves. There was active trade between the Scythian nomads of south Russia and the relatives they had left behind in the Altai mountains from about the 8th century BC onwards, a trade which was extended to north China.[33] Indian gemstones were reaching the west by sea in the 6th century BC.[34] There was a regular salt trade between Syria, Persia and northern India before the 4th century BC.[35] Before 250 BC there was regular maritime trade between India, Egypt and Europe.[36] By about 240 BC an expedition of Indian merchants was discussing trade matters in Babylon.[37]

With the rise of the Roman, Seleucid, Kushan and Han empires, varying between about the 3rd century BC and the 3rd century AD, there came a considerable expansion of Eurasian trade which must have helped to pay for the concomitant expansion of the Indian stone architectural art, including its animal art. The Romans were trading with India by the 1st century BC along with the Greeks, Arabs and Abyssinians, who had long been established.[38] Roman maritime trade was considerable between the 1st and 3rd century AD,[39] the overland route then being blocked by the conflict between Rome and Parthia. This blockage led to a considerable increase in Indian and Persian shipping and trade.[40] It probably helped in the subsequent Indian expansion into south-east Asia and Indonesia. A similar increase in sea trade occurred from the 14th century AD when the Chinese lost control of the Central Asian route and the Europeans found their way to eastern Asia around Africa.

Central Asia to China:

From about the 2nd century BC the Buddhist building programmes from India to Persia to Central Asia to north and south China were important channels for the introduction of Western styles and designs.[41] Mahayanist material culture and the need to make donations to the monasteries in order to acquire merit[42] were the dominant determinants of material requirements. In particular, donations were made to decorate the *stupas* and temples, such decorations being the then current fashion. These seven treasures, symbolising the sapta-ratna of the Buddhists, were gold, silver, lapis lazuli, crystal (quartz), coral or ruby, carnelian or agate and pearl or amber. Also traded were horses, dancing girls, glassware, incense and perfume.[43] They were paid for in silk, this being a currency in China as well as an exportable product.[44] Though such goods as the forementioned were also required as status symbols, religious values tended to override material values and so dominate the trade of this time. Monks and pilgrims travelled with the traders. Buddhist monasteries provided accommodation for the travellers, monk and trader alike. The monasteries acquired capital and acted as banking and lending organisations.[45] Their sanctity gave them reliability in this respect so that transactions were often made in the monasteries between peoples from different countries in order to make their agreements inviolable. In view of the foregoing it would be expected that the honest weighing of the small, highly-valued articles listed above also would be a concern of the monasteries. Such an expectation is supported by the Buddhist injunction to use fair weights and is illustrated by the *Sibi-jatakas* (some of the Buddhist birth tales) which illustrate the praiseworthiness of accurate weighing of what is properly due.[46]

However, it is inferred that trade along the Central Asian silk route may not have been responsible for the transmission of the lion-duck motif to Burma because very little animal art has been found in the oasis cities of the Tarim basin. Nor, apparently, is there much in Tibet. The Kushans who controlled the western approaches to the

Tarim basin from about the 2nd century BC appear to have lost the animal art they had had as the Yueh-chieh. Nor are animal-shaped weights known from India, part of which was under their control. The Sogdians, based on Samarkand, were among the main intermediaries in the silk trade.[47] As near neighbours of the Mesopotamians and Persians, perhaps the Sogdians could have been aware of their stone lion and duck weights and perhaps even of the leaden bull-shaped weights used in ancient Bactria. There is but little knowledge of the weights used along the silk route but one may speculate that some may have been on a binary scale and in geometric shapes following Indian practices. This view receives some support from the facts that first, the three undated presumed weights (no masses given) recovered in Central Asia are either bronze octagonal slabs or bronze-coated rectangular slabs[48] and second, that many Indian and other Buddhist missionaries travelled along the silk route and established monasteries which were concerned with trade. They, of course, would have been familiar with Indian practices and like all traders until recently, would have carried their own weights with them.

India to South-East Asia:

Indian and Burmese records do not mention commerce as do those of the Chinese[49] but there was trade between India and south-east Asia certainly during the last three centuries BC[50] and probably before. The trade along the pre-6th century BC 'Great North Road', running from near the Ganges delta to Persia and farther west, was probably at its peak in the 5th-6th centuries BC when the prosperity of Achaemenid Persia also was at its peak. Trade must have been an important influence in spreading Buddhism because it provided much of the wealth for the merchants who, in turn, were the main patrons of Buddhist holy places and monasteries. That these entrepreneurs did not direct their attentions to Burma, China and south-east Asia also is difficult to believe. Maritime trade was flourishing between the 1st and 6th century AD.[51] By the 2nd century AD there was trade between Rome, India and south-east Asia.[52] The Persians appear to have dominated this trade until about 800 AD when the Arabs (*sic*) may have replaced them.[53] Nevertheless, Persian trade influence remained strong in Siam and south Burma for several centuries after. In Cambodia and Vietnam however, trade in the ports was largely in the hands of the Chinese, presumably Mahayanist Buddhists.[54] Bengal and Madras were important regions trading with Burma.[55]

Exports from south India to south-east Asia, perhaps from before the 3rd century BC until the 18th century AD, included a variety of cotton textiles and silks.[56] Buddhist images and sculptures were sent from north-east India.[57] In Pagan, for example, there are several reliefs and panels which appear to have been imported from this region. Tobacco, indigo, jute-gunny, sandalwood, rosewater and mercury were also being exported before the 17th century AD.[58] Involuntary exports were Bengalis captured by the Arakanese for sale as slaves.[59]

Burmese exports:

There is relatively abundant archaeological evidence of ancient trade in Burma and other parts of south-east Asia.[60] It is likely that in the early centuries AD exports from Burma would have included beeswax, honey, ivory, buffalo horns, rhinoceros horns, yak hair fly whisks, gold panned from the rivers and silver.[61] By the 12th century AD elephants were being sent from Burma to Ceylon (Sri Lanka) on Mon ocean-going junks.[62] By the 13th century AD they were being exported to China along the Burma-Yunnan road. There was already some trade in gems.[63] Burma, like Sri Lanka, was especially noted for its rubies, the king of Pegu displaying specimens of remarkable size and colour. Much later jade became important. By the 15th and 16th centuries AD the export/import trade had grown considerably.[64] The usual products were still being exported but to them were added musk, lac, benzoin, amber,wood oil, sandalwood and camphor.[65] Large ships could carry exports of tin and lead.[66] The Burmese ports also had for sale products from Arabia, Persia, Europe, China and Indonesia.[67]

Slaves too were for sale.[68] In the 15th and 16th centuries AD Pegu was said to be dependent on the import of Siamese silver, much of which it exported to Bengal.[69] It is not surprising then that in 1541 AD the merchants at Martaban included Siamese, Portuguese, Greeks, Venetians, Moors, Jews, Armenians, Persians, Abyssinians, Malabaris and Sumatrans. From the 16th century AD onwards trade was variable in amount. It could be profitable but was frequently interrupted by internal disturbances and often was small in amount.[70] In the 17th century AD the European trade with Arakan amounted to about six ships each year. Proportionately the same trade with Burma as a whole might have amounted to 40 ships a year. By the 1650s trade with Burma had greatly declined. Before the 18th century AD trade had increased, rice, *ganza* (illegal), raw cotton, cutch, pepper, timber and planks[71] had been added to the list of exports. At about this time exports to Yunnan were mainly bales of cotton but edible birds' nests, salt and many of the usual items also went north. Though by 1769 AD trade had greatly declined it had recovered shortly after.

China to the West:

There are numerous examples of early Chinese trade with the West, small and indirect though it probably was. Examples are the import of lapis lazuli from Afghanistan before 2000 BC and of jade from east Turkestan before 1500 BC. Wheat was grown in China after seed had been imported from the Near East between 1000 and 500 BC.[72] Chinese silk was reaching Europe before the 2nd century BC.[73] Chinese junks were trading with India not later than the 1st century BC.[74] From hereon until about the 14th century AD most of the relevant history of China's trade has been mentioned in the previous accounts.

South-West China to Burma:

Chinese silk was being exported to India in the 4th century BC, possibly by the Yunnan-Burma route.[75] There was trade between Szechwan and Central Asia in the 2nd century BC which must have included north Burma and north India.[76] The independent Tien people of south and west Yunnan controlled the export of goods southwards until 109 BC, when the Chinese decided to control it and occupied Yunnan.[77] Even so far back as 140 AD censuses showed that the key economic regions of China had shifted southwards, resulting in rapid, marked increases of the population.[78] From 700 AD until after 1130 AD records show that Yunnan (Nanchao and Ta-li) was a well-developed pastoral and agricultural state with outstanding irrigation and terracing works producing staples, fruits, medicines, perfumes and its famous horses. It had an important mining industry which supplied salt, gold, silver, tin, mercury, cinnabar, and other minerals while its manufactures also achieved fame, e.g. various coloured textiles, including felts and silks, and its swords and armour.[79] The Pyu, the Mons, the Burmese and the Shans would have been exposed to the practices and products of the Yunnan region during their transit of it and perhaps copied them as their ancient irrigation works suggest, e.g. Lake Meiktila. No doubt some of the Yunannese products went to Burma in exchange for its products. Salt, for example, was used as a currency in north Burma in the 9th century AD.[80] Cinnabar was used in 12th century AD Pagan.

Lack of written records, of interest in trade by Burmese chroniclers, devastation caused by earthquakes and by Mahometan uprisings in Yunnan after the 14th century AD present problems in assessing the trade along this route during the 1st millenium AD. It was important because the Chinese built a town near the border in 69 AD to protect the trade.[81] It was internationally known because the Romans were using it in the 2nd century AD. The growth of Buddhism in China must have resulted in an increase in trade of a similar type to that along the Central Asian route. There would have been perhaps an even greater trade in Buddhist artifacts since the distance to the increasing population of southern China from Bodhgaya in India, the Buddhist *sanctum sanctorum*, was so much less than on the Central Asian route while travelling was easier along this route than through Tibet.

Though Nanchao appears to have stopped the India-Burma-Yunnan-China communications between 342 AD and about 750 AD, for which the evidence of the diminution in the numbers of Chinese pilgrims appears to provide support in part, the loss possibly could have been compensated for by a growth in Yunnan's own demand for Buddhist-related services. In support of this inference are the records showing that with the increasing extent of Buddhism, by the 9th century AD Buddhist temples were scattered all over Yunnan.[82] There is also evidence that there was much economic development in northern Burma at this time. This may have been due to a surplus of rice resulting from the increase in the area under irrigation in the regions close to Pagan. Possibly it would have been sold in south Burma and elsewhere. For example, in 849 AD the chief of Pagan at last was able to afford a wall around the town. By about 200 years later sufficient wealth had been accumulated to begin the Pagan temple building programme of the 11th to 13th century AD and to finance Burma's first empire, commencing in the 11th century AD. There appear to be no more likely sources of income to meet these costs than the Ramannadesa and the India-Burma-Yunnan trade. Many donations to the temples of Pagan were made by the wealthy and not-so wealthy of the times and were inscribed there.[83] A similar practice existed in ancient India. Inscriptions show that it was customary for Indian merchants and mercantile guilds to make gifts to religious causes of all kinds.[84] That the Yunnan-Burma-India trade was a contributor to Pagan's wealth is evident from the sale of Yunnanese horses and Burmese or Shan oxen to India in the 13th century AD.[85] Moreover, cinnabar and mercury were on sale in Pagan and these too must have come from Yunnan, while Burma's amber was reaching the Tien people of Yunnan before 2 c. BC.[86]

The wars between the Yunnan states and Annam (Vietnam) clearly did not affect the Buddhist building programmes elsewhere because in Cambodia the Angkor Wat temple complex (like Pagan, on a route junction) also was built about this time. Both places must have been important entrepots for trade goods. Trade with Yunnan presumably was important also about 1400 AD when the Chinese were interested in the Irrawaddy and Salween rivers to transport their goods to the Bay of Bengal. Pegu and Yunnan exchanged missions in 1472 AD when the former's prosperity was relatively great. By the mid-18th century AD Yunnan was exporting to Burma silks and raw silk, velvet, China tea, gold-leaf, white-insect wax, horses, hams, copper, steel, liquor, cinnabar, needles and thread.[87] (See Plate 56)

It has been stated that the China-Burma trade was small in the 17th and 18th centuries AD. Support for this view is given by the loss of Chinese influence over Burma after about 1530 AD and because, at about this time and afterwards, Chinese mining activities were often officially prohibited and the government deliberately opposed overseas trading activities. However, the considerable amount of *ganza* (melted cash?) reaching Burma, presumably from Yunnan, is evident from Burma's usage of the metal internally as currency, weights, gongs, bells, cannons, figurines, images and other artifacts. Also the amounts exported by Europeans (up to 36 tonnes in one vessel) suggests that there may have been more trade along the Yunnan-Burma route than usually believed. In fact, duties on overland trade with China yielded considerable revenue[88] while in the 1780s villagers were migrating to Bhamo (on the Yunnan-Burma border and route) in search of quick riches.[89] The effect of the prolonged Burmese occupation of north Siam and Laos may have been a reason for

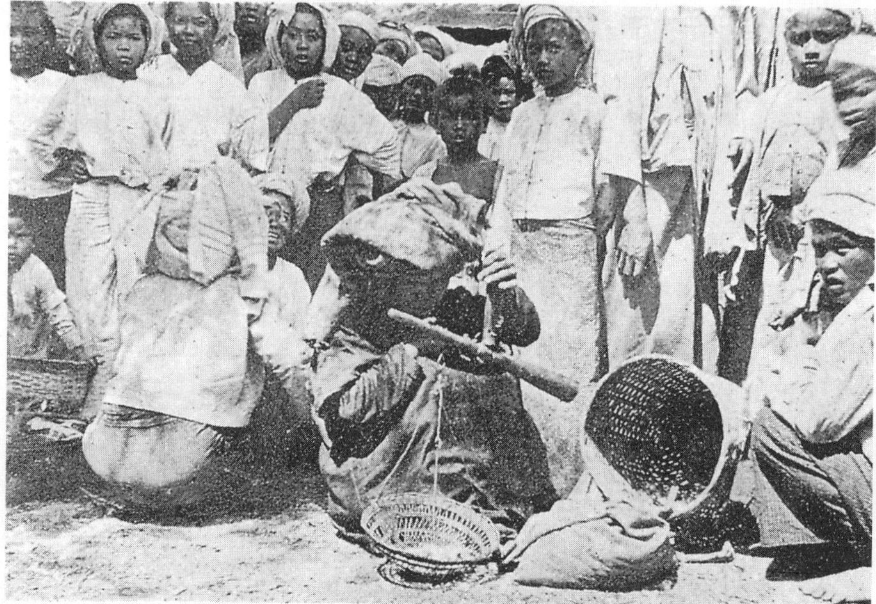

Plate 56: Burmese Weighing Beams in Use
Fort Stedman Bazaar, 1917

this since Burma appears not to have suffered the shortages of Chinese copper so acutely as Siam. However, Burma may have had access to Indian copper, at least in Pagan times.

The Movements of Peoples

General:

In ancient times the acquisition of knowledge which was not obtained from personal experience depended upon talking with other people or seeing their behaviour and artifacts. This came about by population movements such as migrations, invasions, flights, colonisations, deportations, slavery, trade, travel and the like. The forms of knowledge of concern here are those connected with animal art, particularly portable bronze animal art. The movements of most interest are those of the foot-travelling tribes, mostly before the 1st millenium BC, the horsed nomads best recorded from the 1st millenium BC and after, and the trader and missionary, who are best known from the middle of the 1st millenium BC and thereafter.

Mainly west to east movements:

Between about 3000 and 2000 BC much of the Mesopotamian-Persian region was affected by eastward-moving peoples and armies, especially those of Assyria which introduced its animal art to the region. The movement of the Aryans, apparently from south Russia or perhaps Uzbekistan, into Iran and India,[90] taking with them their Vedic religion, is believed to have taken place between about 2000 and 1200 BC. There may have been a similar eastward movement of Indo-Europeans about 1000 to 700 BC, usually termed the Pontic migration.[91] Perhaps as a result of these migrations the Yueh-chieh people settled at the east of the Tarim basin and the bronze animal art workers of Ordos settled in the great bend of the Yellow River (Huang Ho) in China between 1500 and 800 BC.[92] The Pontic migration affected the bronze art of the west of China, perhaps on the main route southwards of the proto-Burmese peoples. The 4th century BC saw the Greeks moving from Macedonia to the Indus but the effects of their settlement in the Bactrian region appear not to be directly relevant to this work.

In India the Indo-Aryan invasion of about 1800 BC was probably followed by waves of similar invaders from the steppes pushing ever eastwards,[93] each driving the previous waves before them,[94] increasingly adopting local ways[95] and, so far as caste allowed, intermingling with the indigenous population. As they went they established aristocratic republics along the nothern borders of India,[96] from one of which came Gautama Buddha, the Sakyamuni.[97] They may have reached the India-Burma border and become part of the present day Nagas.[98] About the 2nd century BC the returned Yueh-chieh (See Note 1, Identities and Origins), now the Kushans, invaded India and commenced to drive the Greeks and Sakas before them. It seems likely that, in time, the Kushans and Sakas may have intermingled or, at least, both become known by the same name, i.e. Saka (Europeanised to Scythian). Some of these people appear to have reached Assam before the 1st century AD, e.g. the Indo-Aryan Kalita people, and perhaps some *kshatriya* adventurers would have carved out states for themselves in Assam. There even appear to have been battles in the east of Assam between the invading Sakas, the already present Mongols and the indigenous peoples.[99] But apparently these invaders may not have practised Buddhism (or Hinduism?) since no evidence of previous or current Buddhism was evident to the Chinese traveller, Hsuen Tsiang, in the 7th century AD. According to tradition, Aryans of the *kshatriya* caste migrated into Burma from India via Manipur and Assam, beginning perhaps during the second half of the first millenium BC. Here they also may have made themselves chiefs of small states, e.g. Tagaung, which today forms the centre of the Kadu people who speak a Sak-type language.[100] Indian chiefs appear to have been present in Upper Burma in the 1st century AD and by not later than the 5th century AD motifs derived from those employed on the coins of the Saka kings of northern India had appeared on the coins of South-east Asia including those of the Pyu.[101] Pagan inscriptions of the 11th and 12th centuries mention the Sak people at least six times.

The most frequently mentioned people are the Kulas i.e. Indians. Kula in Sanskrit means "caste people", i.e. of Indo-Aryan antecedents. A large proportion of these actually bore Indo-Aryan names and some were referred to as being of a white colour.

It may be that the Kulas, both light and dark-skinned, so dominated the northern part of Burma that by the 13th century AD, Marco Polo considered it to be part of Bengal.[102]

Even today there is a tribe in Arakan with the name of Sak and another nearby in Bengal, called Chak. In Siam there are traditions, or evidence, of ancient Indian or Saka influence such as Kshatriya-led campaigns and rulers,[103] the use of a blood-brother ceremony like that of the Sakas[104] and the occurrence of a principality within Lan Na with the name of Sak.[105]

The last major west to east movements were those of the Moslems, beginning in 637 AD and continuing for centuries thereafter. They assisted in the termination of Buddhism and animal art.

Mainly east to west movements:

The movements of the pastoral steppe nomads long before 2000 BC brought about a cultural continuity from China to Ireland,[106] possibly connected with the domestication of the horse about 3500 BC. There were repeated migrations from east to west, the last major movement being the Mongol occupation of most of Eurasia in the 13th century AD. A movement important to this work was that of the Scythians of the Saka group of tribes, part of whom, beginning about the 8th century BC,[107] moved westwards from the Altai mountain region of Central Asia with their animal art and ultimately created a new home for themselves in south Russia, where they survived until about the 2nd century AD. Temporarily they occupied parts of Iran, reached Egypt in the 7th century BC, moving into northern India before the 6th century BC. The Scythians were followed from Central Asia first by the Sarmatians and then by the Huns. With neither of these was animal art the dominant form it was with the Scythians.

Another important east-west migration was that of the Yueh-chieh peoples (later the Kushans) who, about the 2nd century BC, under pressure from the Huns (Hsiung-Nu), migrated from the then western border of China,[108] i.e. the eastern end of the Tarim basin, to its western end where they more or less occupied Bactria, other states of the time, including part of Persia, and, together with the Sakas, India up to Bengal. By this time the Kushans appear to have lost their animal art but they played a most important role in the spread of commerce and Mahayanist Buddhism through Central Asia to China.

Mainly north to south movements:

Beginning in Siberia between 3000 and 2000 BC there was a movement of Mongoloid peoples southward into China.[109] The earliest (about 3000 BC) location thought to be the home of one of the proto-Burmese peoples is between the Gobi Desert and north-east Tibet.[110] It includes the Ordos (Shaanxi), Honan (Henan)[111] and Kansu (Gansu) regions. The Mons, another of the Burmese peoples, may have originated in the Ordos region.

The Chiang, a nomadic, sheep-herding, Tibeto-Burman people,[112] in the 5th century BC appear to have been located in north-east Tibet and adjacent regions until they moved south, possibly to avoid Chinese attacks. The Hsiung-Nu,[113] another nomadic people, also dwelt in these regions and extended up to the Altai mountains.[114] Neighbours of the Chiang and Hsiung-Nu were the Yueh-chieh people who, like the Scythians of the Saka group of tribes, were of Indo-Aryan origin. When they migrated, one part of the Yueh-chieh went to the western end of the Tarim basin while the other part moved south into oblivion,[115] so that its effect on the animal art of the region can only be surmised.

There is little evidence of large scale migration by Thai peoples either from north to south China or from south-west China into modern Thailand at any time in history.[116] This may have been the case also with the Tibeto-Burmans.[117] One can conceive of the movement of small bodies of people southwards along the west side of China for purposes of trade or exploration under the pressure of expanding populations, depletion of land and timber availability, attack from the nomads and Chinese,[118] flight from the Tibetan and Chinese oppression of the Bons after 751 AD and flight of the Buddhists about 840-845 AD. By the last centuries BC the Mon and Khmer peoples extended from Lower (southern) Burma to the Mekong River delta,[119] while the Pyu who followed the Mons were in northern Burma, central Burma and perhaps south-west Yunnan,[120] which may have been the only truly non-Thai part of Yunnan. Tibeto-Burman language speakers would have arrived in Bengal long before this time.[121] The Burmese themselves however, did not arrive in Burma until between the 7th and 9th century AD. The Shans of Thai stock were widely spread in northern Burma, perhaps from the 1st century AD onwards, which they populated and mostly dominated until about 1530 AD.[122]

The effects of the west to east Pontic migration appear to have been transmitted southwards to affect the style of the animal art of the pre-2nd century BC bronze-working culture of the Tien people who dwelt around the lake of that name in east Yunnan. They also spread to the Dong S'onian people of north Vietnam, who, about the 6th century BC,[123] had established the first bronze-working culture in south-east Asia.

Deportations and voluntary travel:

Another kind of movement of peoples was by deportation, often for slavery but also to obtain craftsmen and to occupy unpopulated land for defense reasons. As examples, in 832 AD the ruler of Nanchao took 3000 of the Pyu, including many craftsmen, from north Burma to populate part of his homeland.[124] Slavery was a fundamental part of Nanchao's economy. Nanchao was important for its crafts, e.g. textiles. Its sumptuary laws may have provided the model for those of Burma.[125] In 1057 AD many craftsmen from Thaton, now in south Burma, were deported to Pagan where

Mon culture became dominant for a time. Soldiers were settled in the areas they were sent to control. Prisoners were sent to places according to the need for their skills and settled there. Wars in south-east Asia were often glorified slave raids.[126] All these practices were common throughout Eurasia whenever settled civilisations existed. Moreover, as in Africa, slave raids were extended to great distances. In Eurasia such raids brought to the markets not only the artifacts and motifs of distant cultures but also their craftsmen, thus spreading knowledge.

Lastly, there were the voluntary travels of individuals to distant countries for various reasons. Just as today, there were the traders, the missionaries, the pilgrims, the students, the ambassadors, the entertainers, the artisans, the refugees, the seekers after new places to live, the fortune-seekers, the caravanners, the sailors and many others. These grew in numbers as opportunities appeared, security permitted, routes became available and the means of transport developed. For example, from the 1st century AD knowledge of the periodic alternation of the monsoons became widespread and so greatly increased the trade and travel between India and the Red Sea ports, hence between Europe and the Far East.

Routes

General:

Three maps accompany this account of the routes. These are 'Ancient Routes of Possible Transference of the Lion-Duck Motifs'; 'Main Sea, Caravan and Buddhist Pilgrim Routes, Pre-1 c. BC to 15 c. AD'; and 'Eurasian Steppes and Savannas'. They are intended to illustrate how the lion-duck motif combination could have been, and probably was, distributed before and after the growth of urban centres, trade and anthropomorphic religions. The first map also indicates in part the limits of forest and desert between which most movements of people and livestock would have been confined. The considerable distance between the two presently known regions of lion-duck weights, i.e. Egypt-Mesopotamia and Burma, also is evident from this map. Both regions are at southern ends of nomad migration routes, one on the east and one on the west.

The word 'route' is used here in the sense of a known way between locations. Paved roads did not exist in India, for example, until the 15th century AD, though in China they were extensively constructed from the 3rd century BC or earlier.[127] Drought, lack of pasture, insecurity and way-blockages could cause deviations from the usual route. Moreover, while today one thinks of a route or road as a narrow strip bearing wheeled vehicles deriving energy from petrol stations, at that time a route had to be a zone able to support frequent caravans of pack animals needing vegetation and water. When route networks are mentioned, these, except for main roads in China, usually refer to unpaved zones between towns, military posts, caravanserais and the like which could

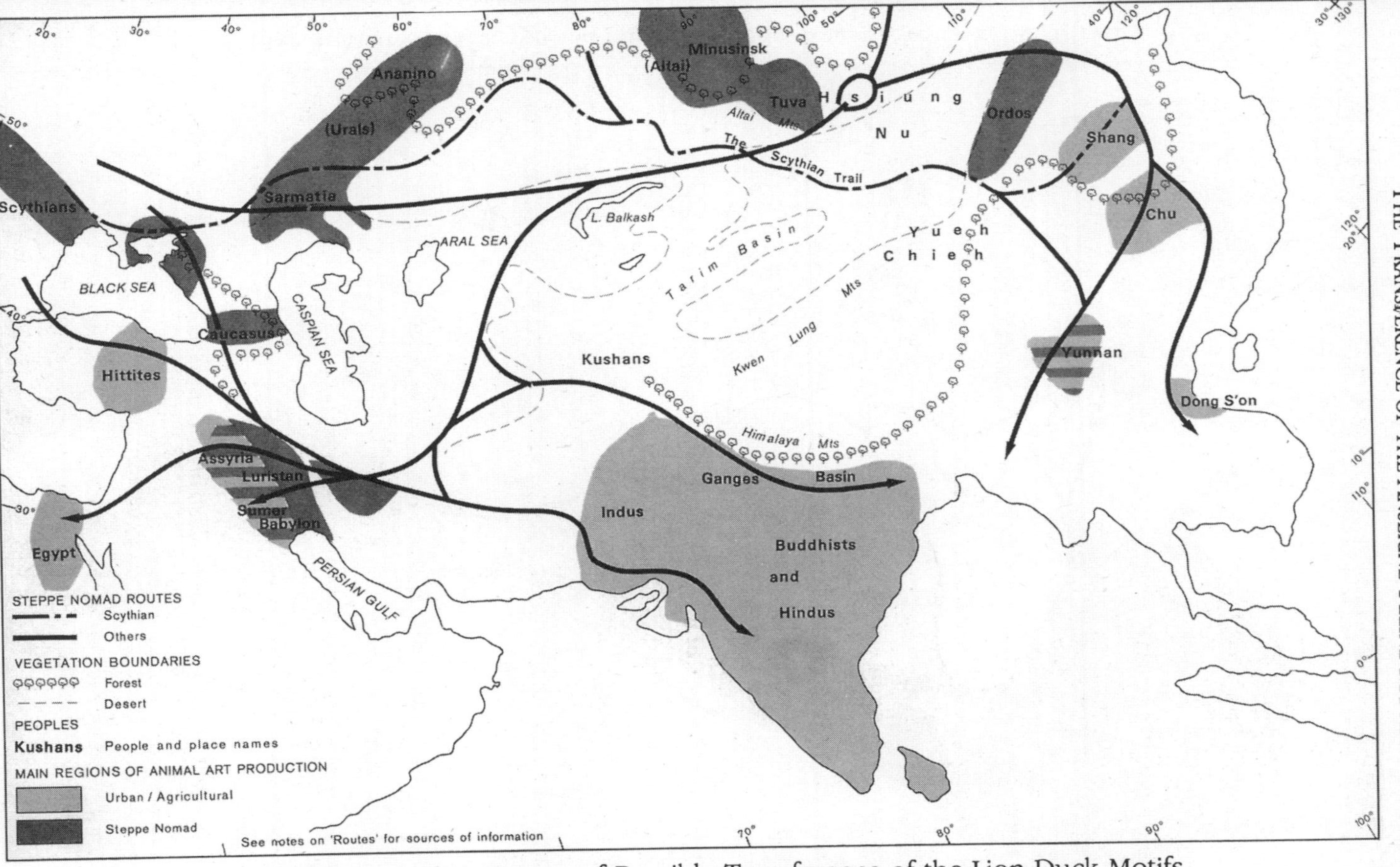

Map 2: Ancient Routes of Possible Transference of the Lion-Duck Motifs

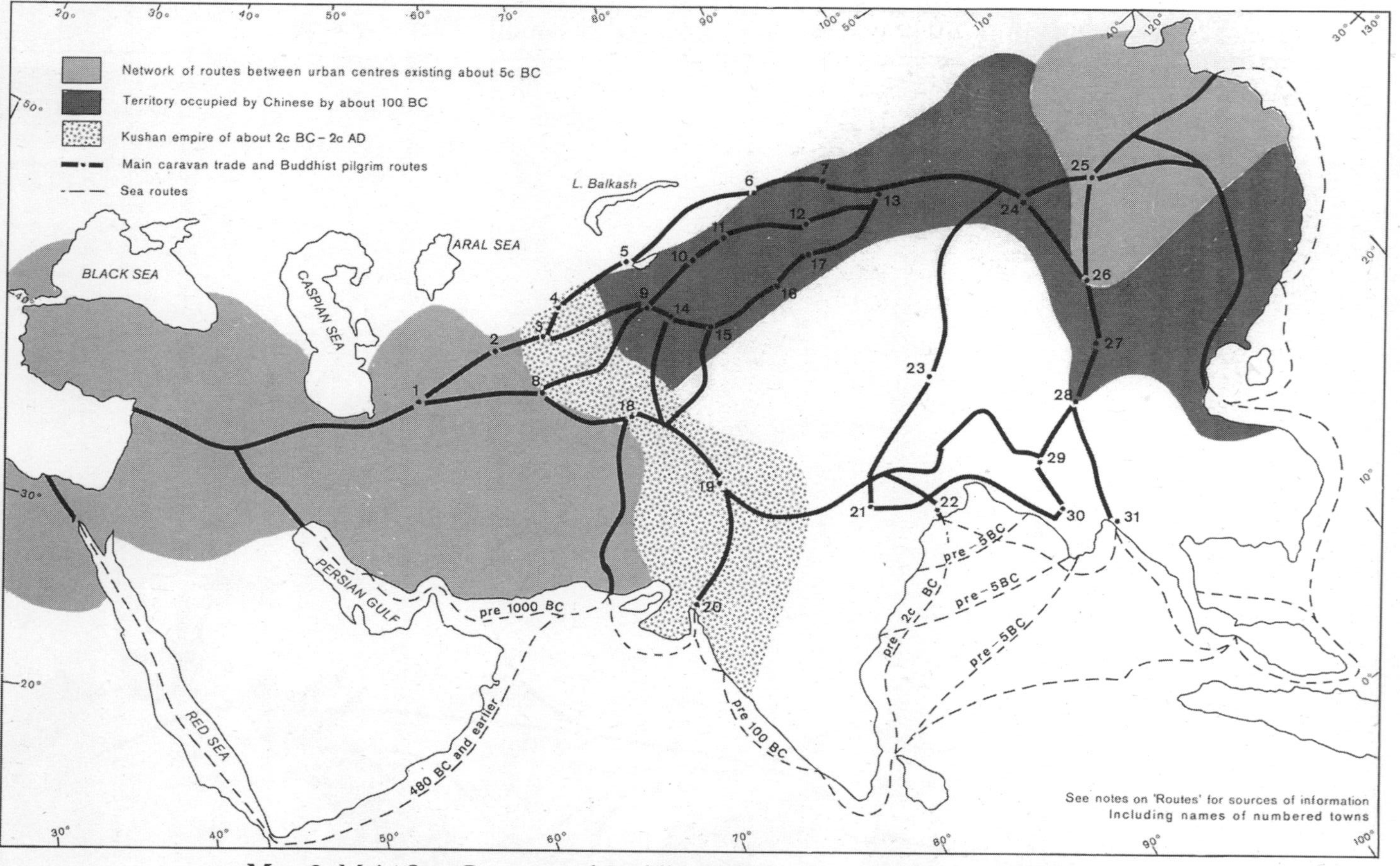

Map 3: Main Sea, Caravan and Buddhist Pilgrim Routes, Pre-1st c. BC to 15th c. AD

be unusable during the rainy season. They could also refer to narrow mountain tracks alongside steep valleys.

The frankincense, myrrh and other routes to Egypt, Israel, Rome and Greece from Assyria, Babylon, Arabia and the Persian Gulf which were in use before 3000 BC, and long after, are well known. However, it is relevant to mention here that by about 500 BC the Persian Achaemenid empire had organised a route network which linked its far-flung provinces extending from Egypt to the western border of the Central Asian Tarim basin.[128] By about the 1st century BC the Romans had extended their roads to meet those of the Parthians,[129] thus making possible journeys along well-established routes from Rome almost to China.

The Central Asian routes:
The north-west India, Central Asia, northern China route which ran along the northern edge of the Himalayas was in regular use by the 1st century BC and probably long before.[130] There were two branches. One passed to the north of the Gobi Desert, through Mongolia and western Siberia. The second passed to the south of the Gobi Desert. The first branch was subject to Scythian and Sarmatian influences while the second mainly transferred Hellenistic influences. Although there were temporary stoppages, they continued and increased in use because of the common interest of all concerned in maintaining the trade along these routes. When there were blockages a shorter, but more difficult, route from Kashmir to the Tarim basin could be used.[131] Between about 900 and 1200 AD there were many trade centres in the Gansu-Xinghai (Kansu-Ch'inghai) region then occupied by the still mainly Buddhist Hsi-hsia (Tangut) state. Not only was this state on the old silk route through the Tarim basin but also at least three of the great trade routes to and from China passed through it, including that to Tibet and India. It is from this region that a part of the Burmese people may have originated at an earlier time.[132] After the Mongols conquered the greater part of Eurasia in the 13th century AD, they organised a route system from China to Russia and elsewhere.[133] By the 15th century the spread of Islam, dessication of the Central Asian route and greater use of sea routes had initiated a continuing decline in trade along it. At the Chinese end of the route good road and river transport systems had been developed in north and central China by between 500 and 300 BC, reaching parts of south China perhaps four or five centuries later.[134] At the other end of the route a trunk road system was constructed in the north and centre of India between about 250 and 200 BC, extending north-west to join the existing caravan trails which linked it to the Central Asian and Mediterranean routes.[135] However, the 'Great North Road', which was an economic and cultural lifeline between India and West Asia, had been in existence since long before the 6th century BC.[136] It passed through Patna near Bodhgaya, along the Himalayan foothills to avoid the forested plain, past the country of the Sakyas from which Gautama Buddha originated, and so to Punjab, north-west India and Central Asia.

The Burma-India route:

The earliest evidence for a land route from China to Yunnan to Burma to north-east India can be dated to about 2 500 BC to 1 500 BC. Best dated from a particular kind of stone axe, it was associated with megaliths and mounds.[137] The border between India and Burma is marked by the Chin mountains. An easy crossing is afforded by the Pangsau Pass in the extreme north-west of Burma.[138] This was the route taken by the Burmese armies which invaded Assam in 1816 AD and 1819 AD. Since the route was well watered, even during the dry season, it was probably part of the main Yunnan-India trade route since it led to the Brahmaputra river by means of a tributary about 40 kilometres distant. It was probably the route most used by the Shans when they occupied Assam mainly about the 12th century AD and would certainly have been used for the Yunnan-India trade in horses of about the same time. It was used by Chinese Buddhist pilgrims in the 4th century AD. As a trade route it must have been in use at the latest from about 500-300 BC.

Buddhism and Hinduism were in northern Burma before the 7th century AD and possibly before the 3rd century AD,[139] implying a pilgrim traffic along this route to and from Burma and the main Buddhist centres in India. From the capitals of the kingdoms in Burma a route passed across the river Irrawaddy near Prome, thence to Arakan and so to India.[140] There was another route to India also, along the Chindwin valley to Manipur.[141] Although the 12th century AD may have seen a stoppage of traffic due to the Mohometan invasion of north-east India, these were usual routes both for pilgrims and for traders bringing Buddhist artifacts from India to Burma. 11th and 12th century Pagan shows many features as evidence of Bengal/Bihar and Nepalese influence, e.g. stupas, reliefs and frescoes. Before the 14th century AD the Yunnan-Mecca land route had been established by the Mahometans, the journey taking about a year. For these and other reasons, this route is believed to have had a history about as long as that of the north Burma-Yunnan route.

The Burma-Yunnan route:

In Neolithic times people travelled from Kansu to Yunnan and northern Burma bearing tanged axes. It has been suggested that Indians may have entered China before 1000 BC.[142] By about 500 BC the Chinese had already incorporated into their agricultural economy particular species of domestic animals and crops obtained from the Bengal and the south-east Asian regions.[143] In 109 BC the Chinese removed the control of the route between south-west Yunnan and Burma from the Tien people who occupied west Yunnan.[144] In 69 AD in order to protect the route,[145] they established the town of Yung Chang (modern Pao-shan) 90 km east of the north-eastern Burmese entrepot town of Bhamo. The main Bhamo-Yunnan route ran through Bhamo, across the Salween River to Yung Chang, across the Mekong River to near Tali-fu, then east-south-east to Yunnan fu (modern Kunming), then north to

Cheng Tu, the capital of the Shu kingdom in the 3rd century BC. In ancient times the road was well made, where necessary being excavated and paved with granite blocks up to 30 cm deep and two meters and a half wide. Ferries and boat bridges were present on the Salween River. Until after 1930 AD the Mekong River was crossed by a chain bridge which was built a thousand years earlier. Elsewhere on the route were several excellently built stone bridges which by 1930 AD were over 1000 years old. Some may have been built to acquire Buddhist merit or Taoist absolution from sin.[146]

The made road and paving extended from near the Yunnan-Burma border to Cheng Tu and beyond. From Cheng Tu an alternative unmade route continued along one of the tributaries of the Yang-tze River on to the ancient famous horse pastures of Kansu and the steppes, to Lanchou and the silk route. It seems likely that the Bhamo-Yunnan-fu road may have been constructed to the standard design about the time that Yuan Chang was built. Along the Burmese and Laotian borders the Chinese also maintained watch towers.[147] Bandits, steep high slopes, sometimes narrow tracks just one pack mule wide, washaways, earthquakes, floods and malaria in the lower valleys have always caused problems[148] and during the 10th and 11th centuries AD the road systems in China deteriorated. The subsequent Mongol and Ming dynasties repaired and improved them but from the 18th century AD there was a gradual decline in the quality of the roads.[149]

The Burmese fought a losing battle with the Mongols at Yung Chang in 1277 AD from which time they exported along this road elephants for the armies of the Great Khan. The old pavilion at Yunnan-fu, formerly used as stables for the elephants, was used as the British Consulate from 1899 AD onwards.[150] The route came to be used by Buddhist pilgrims and missionaries, craftsmen, traders, soldiers, migrants, government officials and foreign embassies of many nations, including Romans, in 97 and 120 AD.[151] There were frequent missions from India and the Shan chiefs to China until about 760 AD, some of which must have used this route.[152] From about this time northern Burma and, on occasion, most of Burma was under the sovereignty of Nanchao, a successor state to the Tien of south and west Yunnan, so that the routes between Nanchao and Burma were in use by the Nanchao armies. In the 11th century AD the Burmese armies were using the route. In the 12th and 13th centuries AD the two main routes to China, even from western India, were by sea and overland to Bengal, Burma and Yunnan.[153] The growth of Pagan at this time must have been due in large part to its position on the intersection of the India-Yunnan overland route and the Bay of Bengal/Irrawaddy river route. In 1287 AD the Mongol armies, with their Shan allies, entered Burma from Yunnan and thereafter until about 1530 AD. The considerable degree of control China exerted on the region required frequent use of the routes by officials and missions both of China and of the countries it ruled. Nor had Buddhist pilgrim and other religious traffic ceased because the route was still being used for these purposes in the mid-14th century AD.[154] Even though Yunnan was then partly converted to Islam it was still mainly Buddhist, especially in the west. That this route was flourishing in the 1780s is evident in that villagers were flocking to Bhamo in the hope of making their fortunes quickly. The route has continued in use until the present time, with occasional interruptions. This is shown by Burmese, European and Chinese records.[155]

Burma's internal routes:

The south Burma-north Burma route, especially by the river Irrawaddy, was also an alternative route to India and Indonesia which was utilised by travellers along the Yunnan-Burma route. The Pyu of the 1st century AD and before must have used the Irrawaddy River route since they appear to have been resident in western Yunnan, north Burma, Laos and south Burma. From the building of Yung Chang in 69 AD, not far from the Irrawaddy, it may be inferred that there was knowledge of the trade routes in the region. The people of Yunnan are known from written records to have been aware of the Irrawaddy route before 766 AD and journeys were being made to south Burma along it before 832 AD. Before the 11th century AD there was trade along the Irrawaddy and coastal route between Pagan and Bengal.

The rivers had long been in use for travel and transport but acquired greater importance after 1368 AD when the Chinese lost their control of the Central Asian silk route. This revived their interest in outlets to the south-west, one of the outlets being the Irrawaddy River, their goods being taken to Cosmin near Bassein, which now replaced the former port of Prome. Another may have been the Salween or, more probably, the route alongside this rocky river which was also subject to a bore. The port here was Martaban. A road led east into Siam, enabling access to other routes into China. During the period 1438 to 1465 AD the Chinese subdued the Maw Shans who were blocking the way and thereafter ensured that the Yunnan-Irrawaddy route was kept open.[156] Further south were roads across the peninsula. One leading from the port of Tenasserim was well known because of its use by the Romans in 131 and 166 AD.[157]

The sea routes to Burma:

The sea route from the Red Sea to the west coast of India was in use by the Arabs and Persians during the 1st millenium BC. the Macedonians used it in the 4th century BC and Greek traders from Alexandria in Egypt were using it before the 1st century BC, being joined by the Romans a little later.[158] The sea routes from east and south India to Burma and south-east Asia were probably in use in the 1st century BC or even much earlier since the coastal route would seem to have presented few major problems. By 500 BC Indian sailors probably made their first contacts with Burma.[159] By the 1st and 2nd centuries AD Roman envoys and Roman entertainers were passing through

Burma to China while Roman traders were visiting Burmese ports.[160] The Indians made much use of these routes, especially during the 4th and 5th centuries AD.[161] Thereafter Arab and Persian sailors dominated the routes, increasingly being joined by the Chinese.

Notes

1 Liu, pp. 36, 107.

2 Liu, p. 6.

3 Liu, pp. 27, 37, 48.

4 Harvey, p. 309.

5 Yapp, p. 861.

6 Rice, 1969, p. 117. Litvinskii, p. 521.

7 Kosambi, p. 125.

8 Backus, pp. 9, 30, 119, 163.

9 Htin Aung, p. 55. See also Davies, "Yunnan . . ." p.317

10 For India, see Basham, pp. 39, 219-225, 275, Allchin and Allchin, plate 18c. Ball (Tavernier), p.39. For Burma, see Harvey, p. 183. Crawfurd, Ava, pp. 190 – 192. For Siam, see Bock, pp. 189 – 192, 268, 289. Mouhot, v.II, pp. 39 – 40, 125 – 126. de la Loubere, pp. 38 – 39. For China, see Weins, pp. 30, 31.

11 For Yunnan, see Harvey, p. 258. Davies, p. 312. Le May, Arcady, p. 249. Bock, pp. 189 – 192. 230 – 232. Gill, pp. 355 – 356.

12 Colquhoun, p. 377, Weins, pp. 31, 32.

13 Hakluyt, pp. 80, 89. Other similar data concerning caravan travel times in Central Asia are given in Yule (Cordier, ed.), "Cathay and the Way Thither", vol 1, pp. 134, 140, 163, 174, 225, 285, 292, 293, 297; vol. 3, p. 48. Manriques, v. II, pp. 265, 340, 351, gives similar information for the Kandahar-Baghdad route.

14 Bradley, p. 138.

15 Boxer, p. 265, quoting da Rada. Polo, p. 226.

16 McWhirter, p. 216.

17 Bacon, p. 559.

18 Goodrich, pp. 50-53. Duche, p. 16, states that in Mesopotamia and elsewhere by the 18th century BC commercial practices were well developed. Banks, cheques, letters of exhange and limited companies were all in use. It seems likely that such practices would have been known to the Kushans and Sogdians (north-east of Persia) who established trading colonies all along the northern silk route up to Lo-yang in China and who were largely reponsible for the introduction of Buddhism to China (Emmerick, p. 402).

19 Liu, pp. 178, 180. Backus, pp. 18, 19.

20 Duche, pp. 14, 30, 32.

21 Barbosa, v. II, pp. 214-215.

22 Pires, v. I, p. 103. van Linschoten, v. I, p. 148.

23 Floris, p. xxiv.

24 Pires, v. I, p. 103.

[25] Bowrey, p. 240, quoting Hamilton, *East Indies.*

[26] Barbosa, v. II, p. 170.

[27] Maity, pp. 16, 17. Petrie, 1934, p. vi.

[28] Anderson, p. 126.

[29] Ridgway, pp. 105, 109, 111.

[30] Philips and Vinacke, p. 600.

[31] Davies, 1959, p. 16.

[32] Ball, p. 166.

[33] Rice, 1969, p. 116. Gimbutas, p. 269.

[34] Anderson, p. 67.

[35] Watson and Vinacke, p. 598.

[36] Rapson, p. 162. Davies, 1959, p. 16.

[37] Campbell, pp. 71, 72. Rapson, pp. 104, 107.

[38] Hall, 1968, pp. 7, 14. Stamp, p. 556.

[39] Philips and Vinacke, p. 600.

[40] Harvey and Than Htun, p. 440. Lieberman, p. 22.

[41] Rawson, 1984, p. 39.

[42] Liu, p. 99 ff.

[43] Liu, pp. 84, 92.

[44] Liu, pp. 66-70.

[45] Liu, pp. 120, 121. Yet in the 6th century AD the government had to control the commercial activites of the Buddhist monasteries because of corruption (Liu, pp. 151-155).

[46] Cowell, v. 4, p. 250.

[47] Yule (Cordier, ed.), v. 1, p. 23.

[48] Stein, pp. 121, 316.

[49] Lyons, p. 354. Taw Sein Ko, p. 330.

[50] Harrison, pp. 8, 10, 11.

[51] Dobby and Wyatt, p. 680.

[52] Smith and Watson, p. 258.

[53] Goodrich, pp. 125, 129, 150.

[54] Groslier, p. 28.

[55] Bagchi, p. 33.

[56] Luce, *The Tan (AD 97-132) and the Ngai-lao*, p. 233.

[57] Harvey, p. 10. Lieberman, pp. 27, 118, 119.

[58] Hall, 'The Dagh Register of Batavia', pp. 112, 113.

[59] Hall, 'Studies in Dutch Relations with the Arakan', pp. 80-83.

[60] Bronson, p. 326. Pires, v. I, pp. 92-108, gives details of the numbers of voyages made each year in the small ships of the time between ports of the region in the early 16th century, e.g. 6 or 7 junks from Siam to China; 15 or 16 junks and 20 to 30 *praus* from Pegu to Martaban to Malacca, Pase and Pedir; 1 ship from Gujerat to Martaban and Dagon, etc. Bowrey, p. 24, mentions 4 or 5 ships from India to Queda (Kedah). From Pegu to Martaban the monsoon allowed ships to travel south to Malacca in February, reaching there in March or

April. They left Malacca in July and reached Martaban in August. Thus ships made but one return journey each year.

61 Hall, 'The Dagh Register of Batavia', p. 102.

62 Luce, 'Economic Life...', p. 336.

63 Htin Aung, p. 57.

64 Htin Aung, pp. 101, 102.

65 Harvey, 1925, p. 157.

66 Lieberman, pp. 18, 27, 119, 120.

67 Harvey, 1925, p. 122.

68 Hall, 'Studies in Dutch Relations with the Arakan', pp. 80-83.

69 Pires, v. I, p. 100.

70 Harvey, 1925, pp. 10, 191, 357.

71 Harvey, 1925, pp. 258, 357.

72 Goodrich, p. 27.

73 Rice, p. 90.

74 Goodrich, 1969, p. 579. Grun, p. 22.

75 Yule (Cordier, ed.), v. 1, p. 6.

76 Goodrich, 1957, p. 43. Goodrich, 1969, p. 580. As one example tea was introduced to China from Burma possibly about the 3rd century BC.

77 Yule (Cordier, ed.), v. 1, pp. 39, 51, 65. Weins, pp. 112 – 116, is especially good on the Yunnan-Burma-India trade from 2nd century BC and later.

78 Cotterell, p. 114.

79 Backus, pp. 18, 19.

80 Luce, "Economic Life. . .", p. 337.

81 Backus, p. 22.

82 Backus, pp. 128, 129, 159, 160.

83 Luce, Economic Life etc. pp. 370 – 375.

84 Basham, p. 218.

85 Frampton, pp. 288, 294. See also "The Origins of the Mass Scales" and the accompanying notes.

86 Weins, pp. 110 – 115.

87 Harvey, 1925, p. 258. Sangermano, p.173. Davies, "Yunnan . . ." p. 317. Bock, pp. 209, 230 – 232.

88 Lieberman, p. 125.

89 Htin Aung, 1967, p. 190, 191.

90 Rapson, p. 30. Litvinskii, p. 517.

91 Hallade and Heine-Geldern, p. 45. Jettmar, p. 216, considers that the concept of Pontic migration is too simple. He thinks of the movements as resulting from a zone of unrest in the south and west, partly due to the transition from static steppe agriculture to mobile steppe nomadism arising from the employment of the horse for transport.

93 Allchin and Allchin, pp. 152, 323.

94 Basham, p. 329.

95 Jairazbhoy, pp. 66, 67.

96 Burrow, p. 137. Kosambi, pp. 108 – 114.

97 Hamilton, p. 36.

98 Ball (Tavernier), p.4 Allchin, pp. 790 – 791. Hutton, p. 1147.

99 Gait, pp. 2 – 21.

100 Phayre, (History etc.) pp. 2 – 7. Bird, p. 369. Narain, p.928. Luce, 1959, pp. 56 – 58.

101 Gutman, 1978, p.12.

102 Frampton, pp. 289 – 293. Luce, 1959, pp. 70 – 73.

103 Le May, (Arcady), pp. 12, 13. Phayre, (History), pp. 3, 4, 15. Gait, p. 14. Mouhot, v. I, pp. 172, 173.

104 de la Loubere, p. 76.

105 Cresswell, p. 16.

106 Bunker, 1970, pp. 13-15. Hudson, p. 61.

107 Rice, 1969, p. 116 ff.

108 Burrow, p. 138.

109 Campbell, 1970, pp. 374-375.

110 Luce, 'Economic Life ....', p. 323.

111 de Lacouperie, 1885, pp. xliii, xlv.

112 Elisseeff, p. 33. Hambis, p. 821.

113 Liu, pp. 13, 14.

114 Denwood, p. 225.

115 Liu, p. 2, 3.

116 Harrison, p. 5.

117 Backus, p. 50.

118 Harrison, p. 5.

119 Wyatt, pp. 5, 6.

120 Backus, pp. 50-52, concludes that there is no evidence to support the usually held view that the Nanchao people were Thais (e.g. see Rawson, 1983, p. 218). They were Tibeto-Burman in speech except in the south and south-east of Yunnan. Sainson, p. 30, makes the point that certainly in south-west Yunnan the people were truly non-Thai. It seems quite possible that the Burmese people may be offshoots from the ancient people of Yunnan, sharing much the same origin.

121 Goetz, p. 22. von Furer-Haimendorf, 1969, p. 129. Keyes, p. 513.

122 Temple, 1926, p. 18.

123 Dobby and Fall, p. 2. Coedes, 1968, pp. 7, 263. Bussagli, *Farther India*, p. 917.

124 Harvey, p. 15.

125 Backus, p. 119.

126 Harvey, pp. 347-349.

127 Bradley, p. 10. Carlson, p. 174. Polo, pp. 325, 326. van Linschoten, v. I, p. 138.

128 Goodrich, 1957, p. 35. Toynbee and Myers, p. 120.

129 Hall, pp. 14, 17.

130 Hudson, p. 59. The nomadic trade between China and the Ural Mountains of before and after the 8th century BC was probably along what came to be known as the northern silk route (Gimbutas, p. 269). By 110 AD there were maps of the Central Asian caravan routes and tables of the travel stages and distances (Goodrich, 1957, p. 68). Other information concerning town locations and names along the Central Asian routes is given in Yule (Cordier, ed.) "Cathay and the Way Thither", v. 1, pp. 43, 51, 58, 72, 183.

131 Liu, pp. 32, 67, 82. There were several routes from China to India via Tibet, all difficult but used (Yule, Cordier, ed., v. 4, p. 176). Long before the 8th century AD there were also difficult routes through Tibet from west China (Szechwan) to India (Assam) and from Nanchao (Yunnan) to Assam (Liebenthal, pp. 12-15).

132 Krader, p. 800. Shabad, pp. 293, 294. Goodrich, 1969, p. 581.

133 Goodrich, 1957, pp. 174, 175.

134 Janse, p. 26.

135 Liu, p. 4.

136 Irwin, I, pp. 717, 719.

137 Basham, p. 324. Gait, p. 4.

138 Peale, pp. 69 – 82.

139 Scott, 1910, pp. 37-39.

140 Taw Sein Ko, p. 330. The way by land from Prome (Sri Ksetra) to India appears to have been in use during Pyu times, i.e. the 1st millenium AD. During the same period three shorter routes across Burma to India existed from Burma's eastern border with Yunnan to north-east Assam. By the 3rd century AD Chinese Buddhist pilgrims were travelling along them to India and Indian Buddhist monks from Nalanda were journeying to Yunnan (Hall, p. 23. Liebenthal, pp. 7, 8, 10-12. Crawfurd, p. 196).

141 Bagchi, p. 17. Griswold, *Siamese Art*, p. 17.

142 Yule (Cordier, ed.), v. i, pp. 2, 3.

143 Goodrich, 1957, p. 27. Goodrich, 1969, p. 580.

144 Harrison, p. 10. Goodrich, 1957, p. 62, refers to the India-Burma-Yunnan route as the "great channel along which trade and ideas flowed from India to China".

145 Hall, p. 23. Backus, p. 22.

146 Bradley, pp. 3, 6, 9, 85, 90. Goodrich, p. 68. For more information, including illustrations, on the Chinese and Yunnanese roads see Davies, e.g. pp. 51, 52, 312. Rock e.g. plate 3. Creel, p. 316. Fugle-Meyer, pp. 3, 13, 19.

147 Boxer, p. 106, quoting da Cruz.

148 Bradley, pp. 22, 74. See also Liebenthal", pp. 9, 10, quoting Hui Lin's *Glossary* of 817 AD and I. Tsing's *Memoires* of 687 AD. Probably most trade and travel was done during the first few months of the years since climatically these were the best months.

149 Goodrich, pp. 174, 197.

150 Bradley, p. 10.

151 Taw Sein Ko, p. 329. Bagchi, p. 17. R.B. Smith, p. 448. Parker, p. 30. Harvey, pp. 77, 85, 87, 117. Hall, pp. 23, 142. Backus, p. 18. and others.

152 Yule (Cordier, ed.), v. 1, pp. 66, 69, 70, 72.

153 Yule (Cordier, ed.), v. 1, pp. 131, 132.

154 Yule (Cordier, ed.), v. 1, pp. 4, 17, 75.

155 Annandale, p. 195, in 1917 refers to observing Chinese merchants from Yunnan and Kansu at Inle Lake in the Shan states. Pires, v. I, p. 111, writing of the early 15th century AD, remarks that traders went overland from Pegu and Siam to take pepper and sandalwood, etc. to China. Wright, p. 272, quoting from Symes, *Mission to Ava*, comments that in the mid-19th century AD traders went to Bhamo, a Yunnan-Burma border village, and traded their wares to Chinese merchants who took them into China.

156 Hall, pp. 163, 164.

157 Htin Aung, pp. 8, 9.

158 Liu, p. 121. Yule (Cordier, ed.), v. 1, p. 52.

159 Basham, p. 226..

160 Hall, p. 23. Lyons, p. 355. Janse, p. 21.

161 Coedes, p. 19. Yule (Cordier, ed.), v. 1, pp. 66, 87, 88; v. 4, p. 19.

162 MAP ANNOTATIONS:

Map2: Ancient Routes of Possible Transference of the Lion-Duck Motif

Information sources for this map were Bussagli, Steppe Cultures, p. 375. Carter, pp,14,15. Blawatsky, col. 214. Griaznov and Goloshnikov, pp. 240, 241. Overmyer, p. 259. Toynbee and Myers, plates 12B, 19, 21, 26. O'Neill, p. 393. Kosambi, pp. 136, 137.

The approximate durations of the animal style art regions shown on the map are as listed below.

East Asia

Shang	18c. BC	—	12c. BC
Ch'u	5c. BC	—	3c. BC
Dong S'on	6c.BC	————	1c. AD
Yunnan	5c. BC	————	2c. AD
China	As above	————	12c. – 15c. AD

Steppes

Ordos	8c. BC	————	6c. AD
Altai	8c. BC	————	1c. AD
Scythia	8c. BC	————	2c. AD
Urals	8c. BC	————	1c. AD
Caucasus	7c. BC	—	4c. BC
Black Sea	7c. BC	—	3c. BC

India

Indus	30c. BC	—	15c. BC
Indian Hindu and Buddhist region	3c. BC	————	5c. AD to about13c. AD

West Asia/North Africa

Egypt	pre-30c. BC	—	4c. BC
Sumeria/Babylonia/Assyria	40c. BC	—	4c. BC
Luristan	26c. BC	————	8c. AD
Persia	pre-6c. BC	————	12c. – 15c. AD

Animal themes continued as decorative art motifs in China, Persia and India but, with the replacement of animism by anthropomorphic religions and anthropocentrism during the first millenium AD, they gradually lost both their symbolism and importance.

The forest boundaries on this map are an attempt (Griaznov) to represent their locations as they might have been perhaps near the end of the second millenium BC before changes due to humans and climate affected them appreciably. The boundaries of the northern forests probably have changed little but those of the temporate and tropical forests probably have changed greatly, so resulting in an extension of the grasslands and cultivated lands. Both diminishing rainfall and mountain glaciers have resulted in an extension of the deserts.

MAP 3: Main Sea, Caravan and Buddhist Pilgrim Routes, pre-1 c. BC to 15 c. AD.

The chief sources of information for this map were Toynbee and Myers, plates 19, 20A, 23, 24, 28. Hambis, cols. 819, 820. Liu, maps 1, 6. O'Neill, pp. 385, 393. Backus, p. 111. Elisseeff, pp. 5, 6.

The place names corresponding to the numbered dots on the map are listed below.

Central Asia: 1 = Merv. 2 = Bukhara. 3 = Samarkand. 4 = Tashkent. 5 = Kyz-art. 6 = Urunchi. 7 = Qomul. 8 = Balkh. 9 = Kashgar. 10 = Aksu. 11 = Kucha. 12 = Qara Shahr. 13 = Tun huang. 14 = Yarkand. 15 = Khotan. 16 = Endere. 17 = Miran.

India: 18 = Taxila. 19 = Mathura. 20 = Barygaza. 21 = Bodhgaya. 22 = Tamralipti.

China 23 = Lhasa. 24 = Lan-chou. 25 = Hsian. 26 = Ch'eng-tu. 27 = Ta li.

Burma: 28 = Bhamo. 29 = Pagan. 30 = Prome. 31 = Thaton.

MAP4: Eurasian Steppes and Savannas

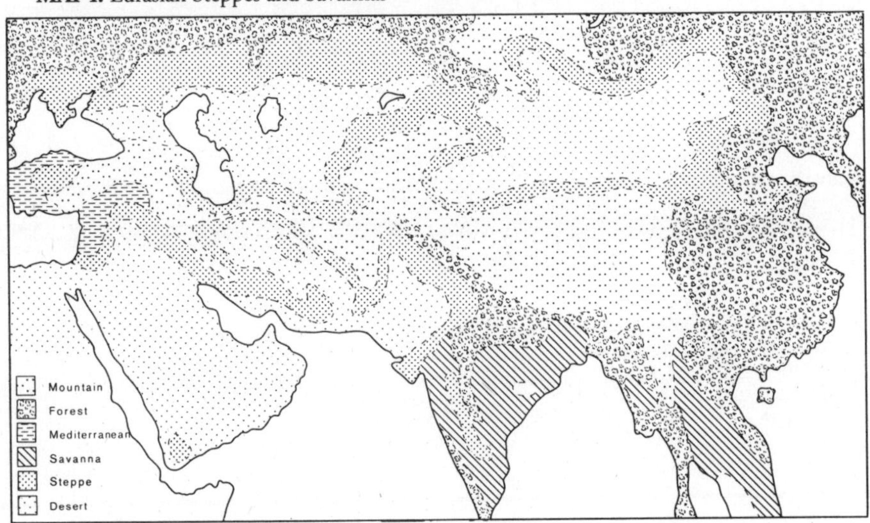

Referring to maps of the present day Eurasian distribution of steppes and savanna and, with justification, assuming that approximately they were similar in the first millenium BC it is quite obvious which routes the steppe nomads, and perhaps later Buddhist pilgrims, may have travelled. There are almost uninterrupted belts of grasslands around the deserts and mountains, from the Altai mountains to the Black Sea, from the Altai to the south of the Caspian Sea and Mesopotamia, from the Altai to the Pamir mountains, north, north-west and west India and from the Altai to Ordos, Yunnan and Dong S'on. See "Encyclopedia of the Environment" editor Colin Tudge, London, 1988; "The Desert" J. Cloudesley-Thompson, New York, 1977. "The International Book of the Forest", Mitchell Beazley, London, 1981. The map below illustrates the foregoing comments. Much of the eastern forest had been cleared by the 5th century BC.

17
Summary and Conclusions

Characteristics:

THE MASS UNIT of the weights, i.e. the *kyat*, has varied from about 12 grams or less before the 15th century AD to about 16,3 grams at the end of the 19th century AD. The accuracy of the unit mass obtained for any one style group varies from 90 to 110% of the mean value. In addition to the observed mass units of the weights, records of mass determination made by European travellers in the past enable a chronological sequence of the changes in unit mass to be established.

THE MASS SCALES of southern Burma before about 900 AD were probably binary. Between the 12th and 13th century AD or even earlier there may have been a mixed binary/trinary/decimal scale in the north. About the 15th century AD a mainly binary/decimal scale existed in southern Burma. From the 16th century AD onwards the mass scale was mainly a decimal one throughout Burma. This scale in kyats is ⅛, ¼, ½, 1, 2, 5, 10, 20, 50, 100 and 250.

THE MOST FREQUENTLY FOUND WEIGHTS lie within the range of 2 and 50 *kyats*, the commonest being the 20 *kyat* weight which constitutes about one half of the genuine weights.

THE WEIGHT MATERIAL is of a copper-tin alloy admixed with variable amounts of lead and zinc. Though all other weights are solid, some of the Group 6 weights have a base which has a stone core. There are also marble weights shaped like mushroom heads.

In DIAMETER AND HEIGHT respectively the most common animal weights, i.e. those between 2 and 50 *kyats*, vary from about 20 x 30 mm to 60 x 100 mm.

THE WEIGHT ANIMAL SHAPES are of three main forms, one a composite, horned, open-mouthed, feline-like beast, another is an anserine-like bird with a knobbed crest (the Group 6 weight looks hen-like rather than duck-like) and the third is an elephant in various styles (Appendix 3, not summarised here). The individual parts of the weights have characteristic shapes and attitudes. Both animal representations are on plinths, the plans of which may be circular, octagonal; square to rectangular (beast weights only); polygonal, hexagonal (bird weights only). The overall shapes of the bases commonly are in the forms of truncated pyramids and cones, rarely of truncated cylinders or domes. The shapes of the weights vary in style. The styles often fall into assemblages of classes in which systematic variations of shape are evident.

SIGNS are found frequently on one side, rarely more, of the weights. They are often similar in shape to the base plan, to the weight creature or to a pattern of rays. System is apparent in that certain signs are found only on the beast weights or on the bird weights and on particular sides of these weights, the result being that they can be placed in a sequence.

A CLASSIFICATION of the weights resulted from their separation into seven groups characterised by major similarities of shape while classes within each group were characterised by minor shape similarities. The bird weight groups fell into a continuous style and sign sequences but the discontinuous sequence of the beast weight groups was established firstly by comparison of these groups with those of the bird weights and secondly confirmed later by their symbolism.

PARTS OF DIFFERENT ANIMALS WERE USED AS MODELS to create the weight shapes. The beast representation is a combination of parts of four animals. The lion of west and south Asia or the tiger of central, east and south-east Asia supplied the model for the head and torso. Horns were modelled either on those of the bull or on those of the antlered muntjac of eastern and southern Asia. The long mouth appendage came from south China. The upstanding tail base was taken from the Yunnanese horse while the feet (and legs?) were modelled on those of the elephants of Yunnan and Burma. The Burmese name *to* was derived from China.

The bird weights of Groups 1 to 5 are in the form of a mandarin duck or drake (*aix galericulata*) which is known only from China and Japan. It appears to show the influence of the Chinese mythical bird, the *feng huang*. The Group 7 bird weight is also anserine in form but is not modelled on a mandarin duck but probably on the Indian goose (*anser indicus*). The Group 6 weight model is not identifiable with certainty. The bird is commonly known as a *hamsa* (Sanskrit) or *hintha* (Pali) though *karawaik* is used for a particular style.

THE AGES OF THE WEIGHTS were established mainly, but not only, in the following ways:

(a) recognition and establishment of a style and sign sequence apparently in chronological order.

(b) comparison of the mean unit masses of the groups as measured by the authors, with those unit masses obtained from measurements recorded by European travellers overt the centuries.

(c) relating the occurrences of the beast weight groups to periods of territorial expansion.

(d) relating the Group 6 bird weights to the historical events of the 15th century AD.

The approximate ages of the groups are:

Group 1 (birds)	1852 - 1885 AD
Group 2 (birds, beasts)	1752 - 1852 AD
Group 3 (birds)	1752 - 1763? AD
Group 4 (birds, beasts)	1531/1563 - 1752 AD
Groups 5 and 6 (birds)	1460 - 1531/1563 AD
Groups 5 (beasts) and 7 (birds)	1385/1423 - 1460 AD

History

THE EARLIEST MASS UNITS of 12 grams or less are of Indian origin. That of about 14 grams could be derived from a north Indian *karsha* (copper unit) mass or, speculating, from a Yunnanese unit. From this time onwards increases in the *kyat* mass unit appear to be of Burmese origin. The unit mass of any one style appears to have been constant throughout the region. The weights were issued under the authority of the monarch whose duties included the ensurance of their satisfactory accuracy. Since an important use of the weights must have been to weigh silver alloy ingots of which the silver content could not be assessed to much better than 10%, there was little need for more accurate weights though standardisations were made occasionally.

THE EARLIEST MASS SCALES of southern Burma probably came from India. In turn these were probably derived from the Indus valley civilisations of about 3000 to 1700 BC. In northern Burma in the early part of the Christian era, the first mass scales may have been any that were used by itinerant traders from India, south-east Asia or southern China. It seems likely that about this time the region of north-east India, north Burma and Yunnan comprised a kind of 'free trade' area which shared many characteristics in common. These may have included a widely accepted mass unit and mass scale and the utilisation of silver ingots and cowries as currencies. In Rammanadesa of the 15th century AD the Group 6 scale may have been intended for use, or at least was used, both there and in Siam. The wholly decimal (except, perhaps, below one *kyat*) scale introduced about the 15th century AD throughout Burma was derived from China, probably to facilitate Chinese trade.

THE WEIGHING OF THOSE METAL INGOTS USED AS CURRENCIES, i.e. silver, *ganza* (usually a copper alloy) and lead was an important use of the weights. Other than in Arakan and the ancient countries of the Pyu and Dvaravati, coins were introduced for but two brief occasions before the 1860s. Since most of the usual silver alloy ingots formerly used in the region weighed less than 50 *kyats*, it is reasonable to suppose that most sets of weights, i.e. those used by householders in the 19th century, ranged only up to this mass. Because of this the observed frequency distribution of the weight magnitudes probably reflects that in which they were made originally, allowing for losses of the smaller weights. Lead was relatively low-valued, the value ratio of silver:lead being about 1:1500. Very heavy masses would be weighed using either a balance and the marble domed weights or a steelyard or *bismar*. The usage of ingots as currency probably originated from Chinese practices while the early Pyu, Dvaravati and Arakanese silver coins probably were devised by Indians who long had been accustomed to a coinage.

BRONZE METAL WORKING was familiar to the peoples of the Indus valley and the steppes of 2000 BC but it was not until between 1500 and 800 BC that portable metal animal representations were commonly produced and then on the steppes and in China but not India. About 800 BC and later there were several foci of production in the steppe regions from west Persia to China and later, Vietnam. One of the reasons for this was the easy access of the semi-sedentary steppe nomads, including the Sakas, to relatively considerable and readily worked deposits of gold, copper and tin, which led to the early development of metal-working skills. It is probable that the Shang Chinese learned their bronze-working skills from the nomads while, still later, these were transmitted to south-west China, apparently both by the Chinese and the steppe nomads. In the 18th century AD and probably earlier the Burmese weights were made using the lost wax technique by specialised weight makers located near the capital from which the weights were distributed. The technique may have been learned from the Indians or Chinese or both during the second half of the 1st millenium BC. The technique for coating the stone-cored weights of Group 6 probably came from Yunnan. While it is likely that the lead and tin were mined locally, the copper probably came mainly from Yunnan, though before the 13th century AD Indian and Japanese copper may also have been used.

ANIMAL REPRESENTATIONS including those of the felines and the anserine birds were made over much of Eurasia from Stone Age times. Stone effigies of the horse, lion and bull, components of which symbolise the *chakravartin* on the weights, are found in 3000 BC Sumeria and Babylonia. Among other animals they appear as clay figurines made by the people of the Indus valley about 2000 BC. No such effigies made by the Indo-Aryan invaders of India about 1700 BC have been found, perhaps because these peoples were too few, scattered and nomadic. Stone lions and ducks are known as weights from Egypt, Assyria and western Persia between about the 17th and 6th centuries BC. Lion or tiger and bird motifs, including the anserine birds, appear among the products of the metal-working centres of the steppes and China almost as early as those on the Mesopotamian stone weights and increase in use thereafter. Animal motifs in India in permanent materials, became evident with the Asokan pillars of the 3rd century BC, one of which bears representations in relief of the lion, bull, horse and elephant. However, animal representations did not become important until the arrival from the steppes, in about the 2nd century BC, of the Kushans and Sakans and the concomitant increase of wealth and trade, interest in Buddhism, commerce and stone-working. The first three Asokan motifs are similar to those associated with the Saka group of tribes from Central Asia but which are in the forms of small metal tigers, stags and horses. However, in India, nomadism having been abandoned in favour of sedentary life, the effigies made were in stone and for the first time the stag became a motif while the horse once again became popular. In Burma portable animal lion and duck figurines appear before the 4th century AD. In India by about the 5th century AD animal art was in decline and by 1200 AD was confined to a few scattered temples in north-east India. Stone reliefs of animals are known in Burma from Pagan (11th to 13th century AD) but it is not until the introduction of the animal weights about the 15th century AD that lion and duck animal figurines in bronze reappear. The two earliest groups of Burmese bird weights appear to be Indian in model and style. The subsequent groups are Chinese in model but Burmese in style, the latter becoming accentuated on the more recent weights.

THE MIGRATION OF MOTIFS AND PEOPLES are closely connected. From before 2000 BC there were migrations from west Eurasia to the east by Indo-Aryans and later there were return movements. The migrations of the nomads were due to their early mastery of horse riding and their access to the vast strips of steppe grassland bordering the deserts, the mountains and forests and radiating from the Altai mountains of Central Asia. These strips terminated in south Russia, Anatolia, west Persia and Mesopotamia, west India and north-west China. When the forests were sufficiently cleared in China there was easier access to the savannas of south-east Asia. In all of these steppe regions, especially their terminations, the lion-duck motifs are found. Only in west India are they unimportant and do not occur in bronze. Nor were cowries employed there.

THE TRANSFER OF MOTIFS BY EURASIAN OVERLAND TRADE was taking place long before 2000 BC but to a small extent. It grew apace from about the 2nd century BC. This came about because of increasing commodity surpluses, urban developments and connecting routes, the taming of the Bactrian camel and the spread of Buddhism with its trade connections. However, the writers, as yet, have found no record or evidence to suggest that the lion-duck motif on weights was carried along the branches of the old trading (silk) road through the Tarim basin of Central Asia. Only three weights in geometric shapes, have been found to date. It seems that the motifs of the bronze feline and avian were carried to Burma either along the nomad routes and then south along the west side of China or from India, but not, in either case, on weights. It seems probably that there was an increased demand for weights in the late 14th century AD because of a growth in the China-Burma-India trade caused by the Chinese need to develop their trading routes to the south, in turn caused by the blockage of the Central Asian overland route. Such weights would need motifs. Because the China-Burma-India route was already a trade and Buddhist pilgrim route it seems likely that the Buddhist monkhood would have been involved in the choice of motifs for the weights since for centuries it had been concerned with, primarily, the moral aspects of trade, though well aware of its practices.

Symbolism:

THE SYMBOLISMS of the weights and their parts appear to be derived, first from the Stone Age Eurasian animistic beliefs of the partly forest-dwelling hunter-gatherers, second from the sky gods of their crop-farming successors, third from the magico-religious beliefs and practices of the later Turko-Tartar and Indo-Aryan steppe nomads, especially the Saka group, forth from urbanised Indian Buddhism and its reaction against the earlier beliefs and practices and fifth, from Burmese political motives.

The contributing sources utilised animal shapes to symbolise meanings. The symbolisms, the sources and the paths travelled to reach Burma, mostly can be identified with a fair degree of assurance though there has been a melding of both shapes and symbolisms during the journeys. The symbolisms of the feline and anserine motifs served specific commercial, religious and political purposes intended for subjects who might be animists, shamanist nomads, Buddhists, Confucianists or Taoists. There was also a need to symbolise the fair exchange of what was properly due from one trader to another, from a devotee to his spiritual guide and from a subject to his ruler.

THE SYMBOLISM OF THE BEAST WEIGHTS combines those of a potential Buddha (*bodhisattva*) and a universal monarch (*chakravartin*) together symbolising an earthly god-king (*devaraja*) characteristic of Burma and south-east Asia. The introduction of a new beast weight style symbolised either the creation of a new Burmese empire or an expansion of it and usually, a new dynasty.

THE BIRD WEIGHTS SYMBOLISE the heavenly perfection and purity of the Buddhist belief. There may also have been a symbolic connection with Ramannadesa's (and Burma's) only sovereign queen, Shinsawbu (15th century AD). That kind of weight bird with a small bird on the chest probably portrays one of the Buddha's birth stories which enjoins fair weighing. Possibly it was used for standardising.

THE ASSOCIATION OF BEAST AND BIRD representation conveys, among other meanings, that of earthly and heavenly power, material and spiritual power. The god-king symbolism and the dominant usage of the weights before the 19th century AD by the king's officers justifies the common name of the weights "sri arlay" (the lord's weights) usually rendered as "shwe arlay" (gold weights). If conditions were similar in both cases the foregoing explanations may suffice to explain why the Mons of Lower Burma should have chosen to use the combined feline-anserine motifs on their weights almost 2000 years after the Mesopotamians and Persians are thought to have abandoned them. Nevertheless, one wonders whether the Mons accepted these motifs upon the advice of a Persian Buddhist from Yunnan. Under Mongol rule from 1258 AD presumably there must have been some of them in Yunnan. Perhaps there were even some who were acquainted with the former use of lion and duck weights in

Persia, weights which may have continued in use for centuries after their official abandonment, just as the motifs themselves did and as the Burmese weights do today.

APPENDIX 1
The Domed Marble Weights of Burma

The white, evenly crystallised weights were made at the Zingyin (Sagyin) quarries about 25 miles from Mandalay. These quarries were in use at least as early as the 17th century,[1] but probably much earlier, for the making of Buddhist images. The weights were certainly in use in 1878 AD and probably before.[2] In amplification, the mass scale shown below is very similar to that of the Group 6 mass scale based on 100, suggesting that both scales may have existed at about the same time. Their production ceased about the beginning of the 20th century.[3] Apparently the Myowun's officials had to check these weights to see that they were made to the same standards as the bronze weights.[4] They were made up to 10 *viss*. They were considerably cheaper than the metal weights. Their main use may have been by produce brokers when weights in excess of 100 *kyats* were required. Made in the shape of a flat-based dome, the smooth upper surface shows a radiating pattern of 4, 8 or 16 lines, as shown on the illustration. Large specimens can occasionally be seen in village market places. They were referred to as 'turtle weights'. The shape presumably indicates the Chinese view of the heavens, the lines indicating the geographical directions from the central world axis.

Table 23: The Masses and Dimensions of the Domed Marble Weights

Mass		Diameter	Height
Kyats	**Grams**	**mm**	**mm**
100	1639	130	70
50	806	100	55
25	427	83	45
10	158, 165	57, 59	33, 32
5	86	50	30
3, 2½, (?)	43, 55	40, 40	24, 22
1, 1½, (?)	22	32	18
1	16	27	15
½	7	22	14
¼	4, 5	20	10

The only set of these weights seen had the masses and dimensions shown on the following table. Plate 57 illustrates a partial set of the marble weights.

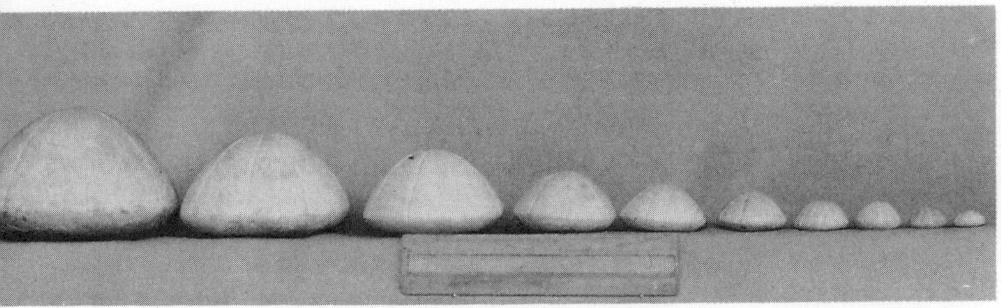

19th? century **100 – ¼ kyat**

Plate 57: Marble weights. Partial set

Notes

1 Lowry, plate 8.

2 Forbes, p. 132.

3 U Myaing of Kyauksittaung, the stone-working village at Mandalay, was 75 years old when the writers met him in 1972. He could remember the making of the weights.

4 Kyaw Tun.

APPENDIX 2

Unofficial Weights and Weight-Like Figurines

Unofficial animal weights were, and are, made to fulfil a need, such as during occasions when official weights were not available, to fulfil a special order by a trader or a monk, to act as a donation, e.g. to a monastery, to bury in a *paya*, to sell to tourists, for use by astrologers, sorcerers and so on. Weights made with the intention of deceiving are considered to be unusual. Bronze figurines in the form of solar zodiacal (Chinese year) animals including bulls, elephants, monkeys, etc. are rare in Burma, though occasionally they are used as weights in Thailand. They are employed also to stand in front of the household shrine.

Unofficial animal weights (about 50% of all those seen) were made as copies of commonly used official weights and mostly belong to a few particular style classes, especially those of the bird weight classes 2(u), period C, and 2(v), period D. There appears to be a more varied range of unofficial and abnormal animal weights in Thailand than in Burma. Unofficial weights cannot be distinguished from official weights unless some abnormal feature is present. Such features are summarized below:

(a) unusual style or combination of unrelated parts,

(b) mis-shaping of parts, poorly styled and asymmetric weights,

(c) occurrence of sets, especially of an abnormal mass scale,

(d) a mass which does not conform to a customary one,

(e) unduly large numbers of a class without a customary sign,

(f) having an unusual sign such as a name,

(g) unknocked, unpitted or unworn surfaces,

(h) poor casting (air-holes, etc.),

(i) abnormal colours and patina,

(j) elaborate and unusual ornamentation,

Plates 58, 59 and 60 show examples of weight- like figurines and copies of the royal weights.

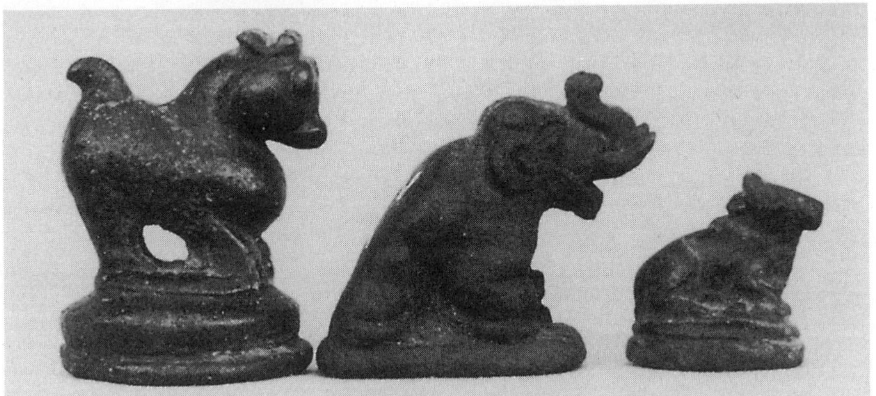

Standing goat or deer, sitting elephant, reclining bull.

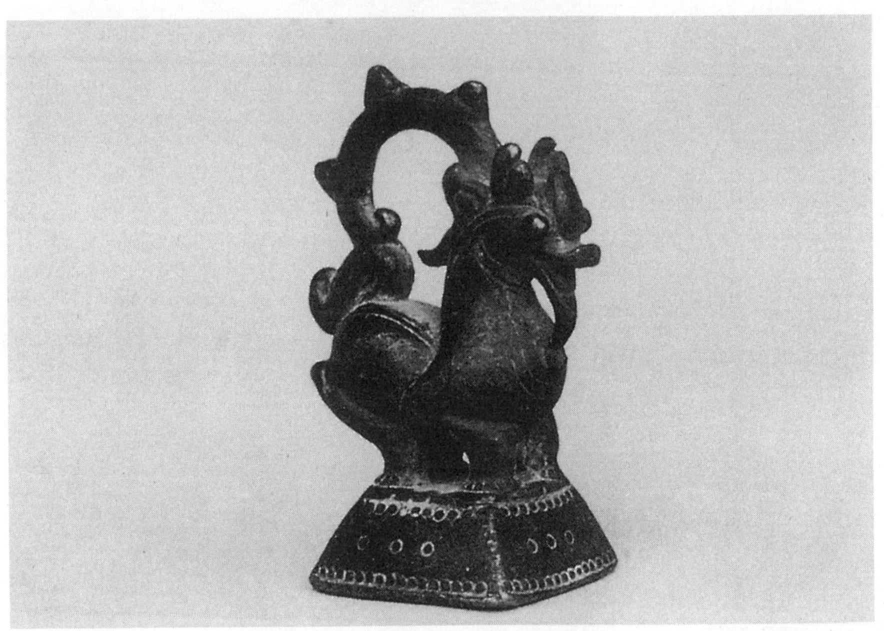

Copy of Group 2B beast weight

Plate 58: Figurines and Copies of Royal Weights

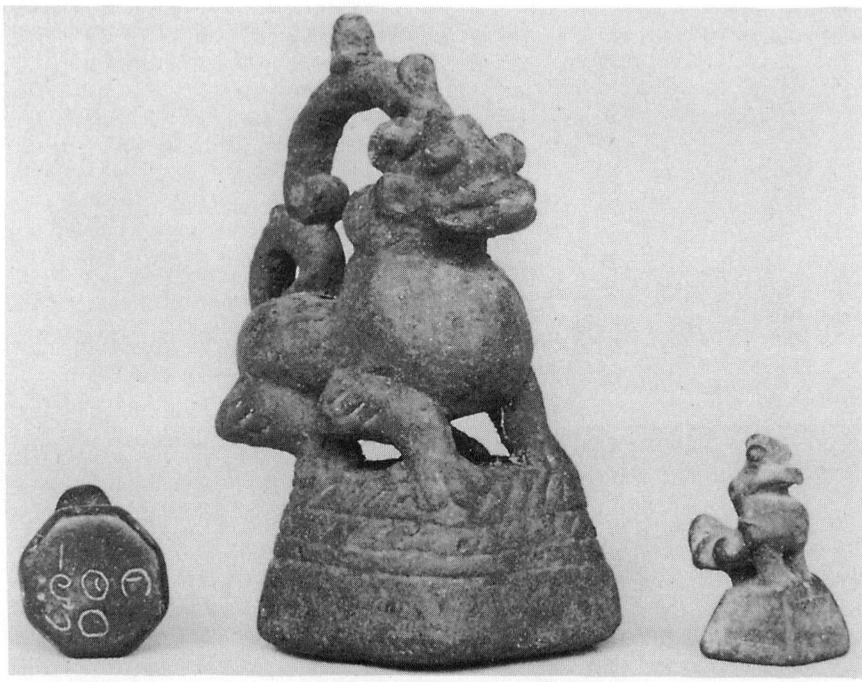

Copies of (left) Group 2 bird weight bearing trader's name: (centre) Group 2 beast weight in iron: (right) Group 2 bird weight on a square base

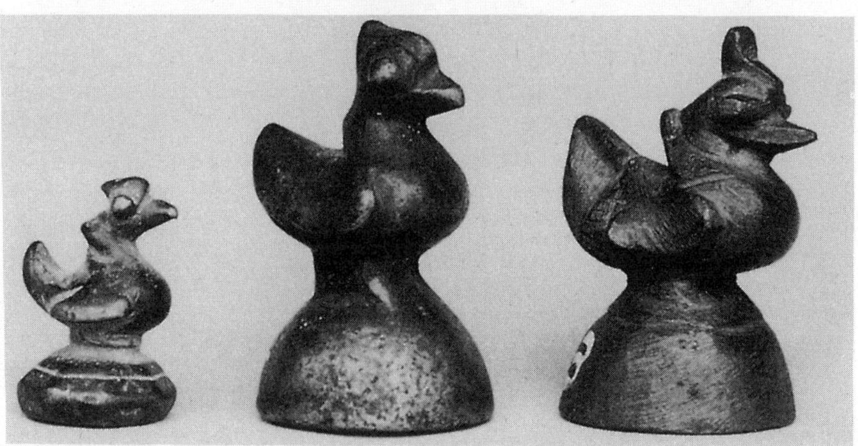

Copies of (left) Group 2 bird weights bearing trader's name: (centre and right) Group 2 bird weights made in 1930 and 1937.

Plate 59: Figurines and Copies of Royal Weights

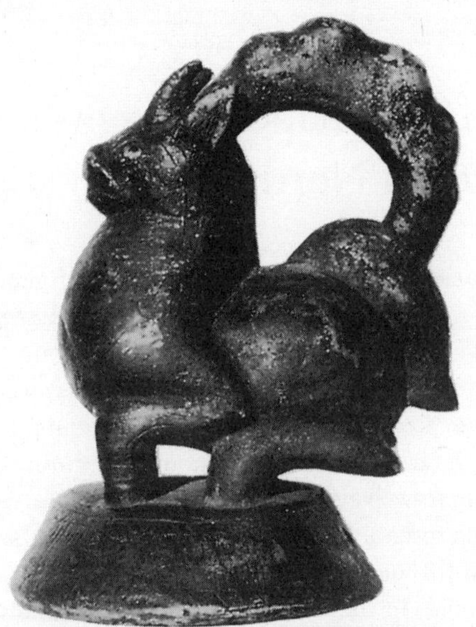

Copy of Group 4 beast weight

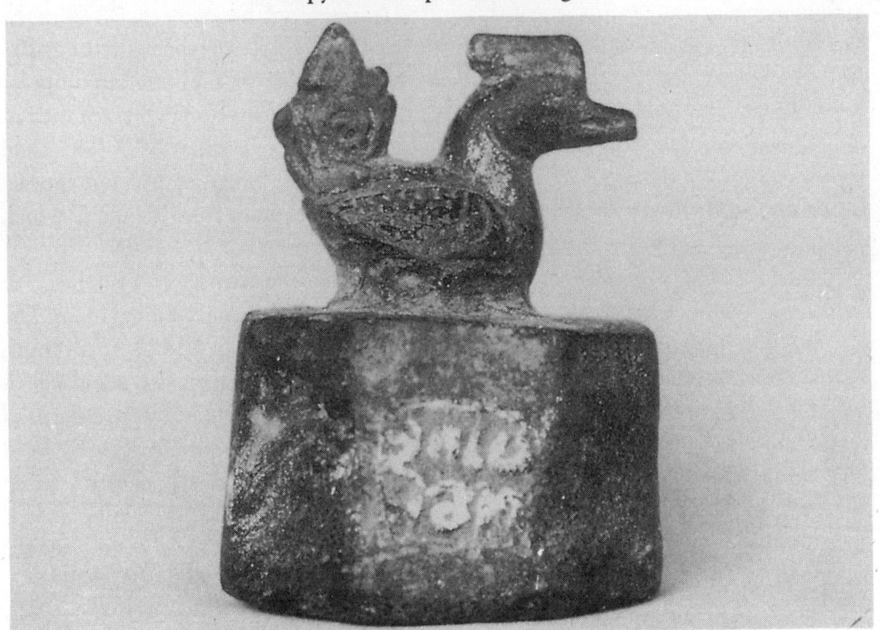

Copy of Group 6 bird weight made in Sagaing at end of last century.

Plate 60: Figurines and Copies of Royal Weights

APPENDIX 3
Animal-Shaped Weights and Figurines of Northern Siam and Laos

The origins and histories of the elephant weights and figurines:
Most, not all, of the figurines, other than the elephant, *to* and *hamsa* weights, are those of the 'Chinese' 12 year cycle[1] which is the same as the oriental solar zodiac. Sets of these animals were known to be made in large numbers by the Chinese for the Mongolian market. They were used in China between about the 3rd century BC and 2nd century AD, though rarely. They became abundant during the period of the 7th to 10th century AD when twelve standing human figures, each bearing the head of one of the animals of the zodiac, were placed in tombs. From this time onwards year-animal figurines spread throughout south-east Asia.[2]

Every one of the animals in the 'Chinese' 12 year cycle is also mentioned in the *Jatakas* (written about the 3rd century BC), along with several other animals found as figurines, e.g. the tortoise, frog, fish, squirrel (mongoose?) and wolf (fox?). With the exception of the pig (boar) all the identifiable figurine animals together with the griffin (*to*), the worm and the deer also are found in the list of Chinese circumpolar constellations (lunar zodiac).[3] Finally, all of the figurine year-animals, except the rat, also appear on the Yunnan bronzes of the 3rd to 1st century BC.[4] The snail representation on the figurines is not found on any of the foregoing lists but appears on the large drums of the Karen people of Burma who originate from Yunnan,[5] which is one of the oldest known locations for these drums.

Before the Thai founded their ancient kingdoms towards the end of the 13th century AD the weights in use were those of the Khmer.[6] The Thai adopted them. The first mention of the *tamlung* as a weight occurs in an inscription of 1183 AD in south Siam.[7] In 1250 AD the northern kingdom of Lan-Na had established standards of weights and measures. In 1391 AD the 'Siamese' obtained the Chinese standards of weights and measures.[8] In the south of Siam the word *baht* had come into use by 1444 AD[9] while before 1691 AD most of the weights were known by the names[10] which they bore in 1908 AD.[11] In 1460 AD the *Annals of Chiengmai* used the Chinese word *peng* to mean a particular mass, though today the term is applied to the bronze animal weights in general. In 1558 AD Chiengmai, capital of Lan-Na, fell to the Burmese, who destroyed its standard weights and measures.[12] At this time the *kakim* silver ingot currency of the Lan-Na kingdom was destroyed and the Burmese introduced their own type of flowered ingot currency. They did not change the Lan-Na unit mass since this was still in use in 1691 AD. Burma then included Chiengmai in its own standardisation of weights and measures, which took place throughout its empire between 1551 and 1581 AD.[13] Between 1569 and 1585 AD many Burmese laws and institutions were introduced into north Siam. The symbolism of the weight elephant and the introduction of the elephant sign on the Chiengmai *tok* silver ingots indicate the likelihood that both commenced their existence between 1558 and 1581 AD.

In the second half of the 16th century AD following the Burmese occupation of north Siam the king of Ayudhya in south Siam required merchants to ensure that their weights corresponded with the royal weights. In 1758 AD King Ekat'at of Ayudhya introduced a law standardising the currency, measures and weights of that country. It seems likely that the present 15,0 gram *baht* mass may have been introduced at this time.

In north Siam the weights may have ceased to be issued under official control when the Burmese finally lost control of Chiengmai in the second half of the 18th century AD.[14] However, since many of these weights are brassy in colour, like the most recent of the Burmese weights, it is probable that the elephant weights continued in production, perhaps unofficially. The high proportion of small animal figurines still existing also indicates the relatively modern origin of many of the elephant weights and animal figurines, since these are the most easily lost. Once Burmese controls had gone, then the other animal figurines would have provided an alternative to make up for any shortage of the elephant weights. That they were not intended as weights, however, also seems likely from a comparison of the mass unit percentage frequencies obtained for the unit masses of 12,7 and 16,2 grams.

In 1858 AD Siam introduced a machine-made flat coinage and thus eliminated the need for assayers, bullion and bullion weighing. Natural conservatism may have caused the old practices to continue but gradually diminish with time.

Symbolism of the weight elephant:

The lion is depicted in Hindu and Buddhist literature as the inveterate enemy of the elephant. Though the elephant is huge and mighty, nevertheless it is vanquished by the lion. In China there is a legend which recounts that the Buddha was attacked by a herd of maddened elephants. He extended his hand and each of his fingers became a raging, roaring lion. They intimidated the elephants which then prostrated themselves and begged for mercy.[15] This well known conflict and its result are often used in metaphor with political meanings.[16] So it seems reasonable to think that when King Bayinnaung of Burma in 1558 AD conquered the king of Lan-Na, who was also a *chakravartin*, he may have seen the conflict in terms of the lion-elephant conflict. Being the victor Burma would have been the lion and the conquered Lan-Na, the mighty elephant. This event may have been the origin of the elephant weights of north Siam. The Burmese, now being the rulers and re-introducing their own kind of ingot currency but retaining the former Lan-Na mass unit, could have introduced the elephant both as a symbol of the conquered nation and as a means of differentiating between the mass unit of the Burmese weights and that of the Chiengmai weights. A possible reason for King Bayinnaung's actions in connection with the Lan-Na mass unit may have been that the king of Lan-Na was esteemed so highly by King Bayinnaung that he was made the last of the pantheon of 37 *nats*, the Burmese cult that paralleled Buddhism.[17] See the beast tail symbolism for another possible reason. An alternative or supplementary explanation for the origin of the elephant weights is the following. King Bayinnaung of Burma attacked Ayudhya in 1563 AD because the latter's king refused to give him one of his seven white elephants. Ayudhya was defeated and Burma captured some of the elephants.[18] Such a booty would warrant commemoration. A clue to the authenticity of the elephant weights may be apparent in that the elephant symbolises the number 8.[19] Most elephant-shaped weights which bear a sign stand on octagonal bases. Though it is possible that the elephant weights may have existed before the Burmese conquests[20], whatever the original symbolism, that given here would still have been appropriate after the conquest.

Locations of occurrence:[20]

In 1884 AD the animal weights in use in Chiengmai were in the forms of elephants and geese.[21] In 1910 sets of Group 2(u) bird weights, Group 2(r) beast weights and elephant weights, together with some varieties of solar zodiac animals, were common in northern Laos, being "very frequent" in places, especially horse shapes and dog shapes. Possibly related to these occurrences are the occasional appearances of the elephant and horse shapes as small signs on the edges of some of the Chiengmai silver ingots known as *tok*. Rarely found figurines in northern Laos were the rat, the cow, bull or buffalo, the goat or ram and the monkey. All animal weights were scarce in southern Laos.[22] From Cambodia fowl 'weights' have been reported[23] as they have from Thailand but these appear to be closer to the Chinese solar zodiac figurines than to the style of the Group 6 bird weights. Cock representations similar to these figurines appear often on coins and statues in the southern part of South-east Asia and in Indonesia. In the 1960's, in north-east Thailand and northern Laos, the lion-like beast weights were thought to be the most abundant while bird weights, supposedly, were most common in the north-west of Thailand.[24] Other animal weights reported were in the forms of elephants, duck-like birds, peacocks (Groups 6 and 7) hastalingas (Groups 3 and 5), while the Chinese solar zodiac (?) animal figurines occurred in small sizes, though only occasionally, and apparently rarely, as weights. The elephant weights were thought to be more abundant in north-western Thailand than elsewhere, especially in the Ping valley, and were common in northern Laos. From the foregoing and other evidence there is little doubt that the origin of the elephant weights was in north-west Siam, the former Lan-Na, and not northern Laos,[25] the former Lan Chang.

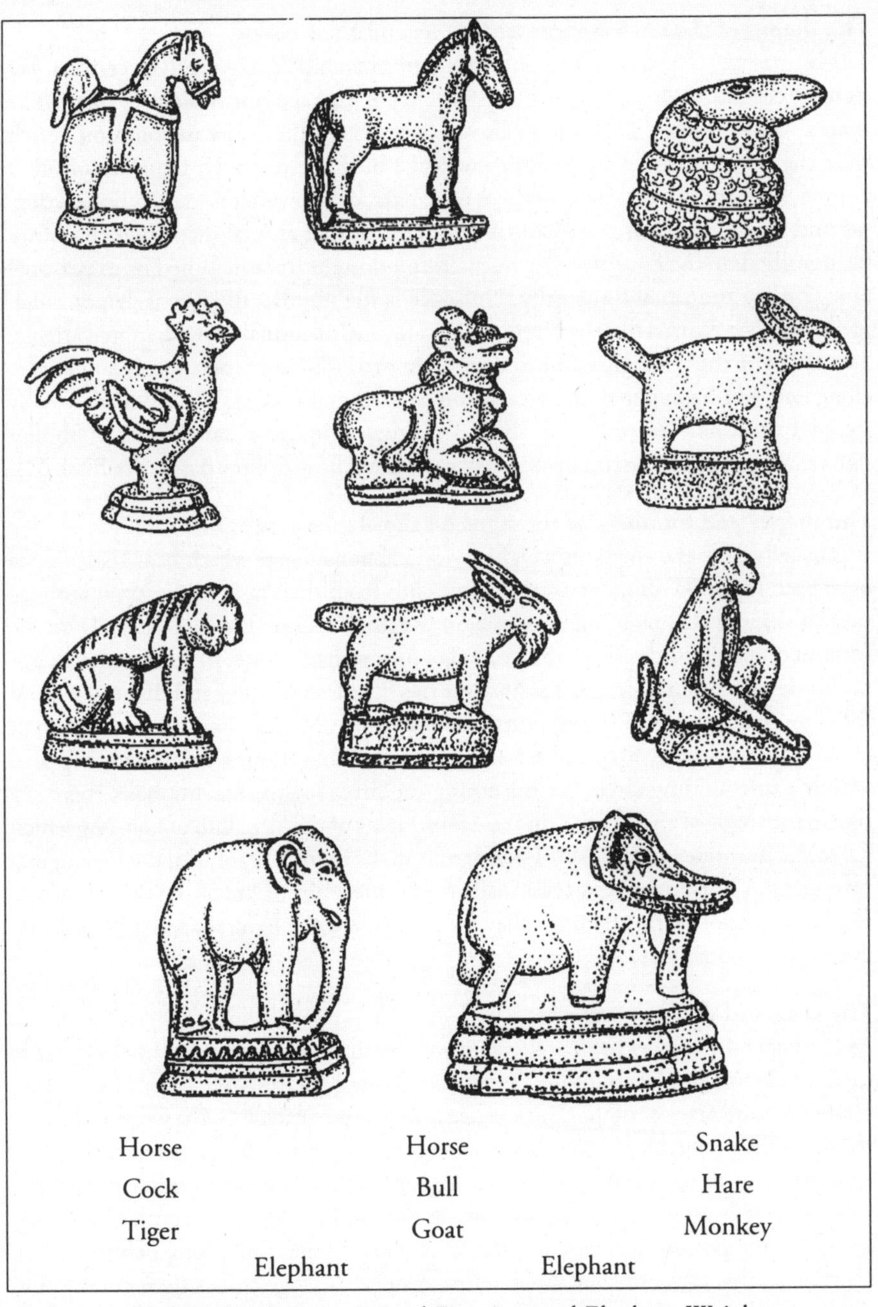

PLATE 61: Siamese Animal Figurines and Elephant Weights

The shapes of the animal representations and their bases:
The bronze weights most commonly found in Thailand during the 20th century AD were of the Burmese lion- and duck-like shapes and the north Siam elephant-like shapes. Other animal shapes were also in use, though they were uncommon. Such were the horse,[26] the snake and the cock-like bird. More rarely, figurines of other animals were found, e.g. rat, buffalo or cow, hare, dragon, goat or ram, monkey, dog, pig and several which were unidentifiable. Those other animal shapes which cannot be identified on the figurines may be malformations or have stemmed from personal fancy or just too small to identify. Not only is the number of animal shapes much greater than in Burma but the shape of any one kind of animal is much more variable. The bases of these weights and figurines are predominantly octagonal, sometimes elongated, but heptagonal, hexagonal and oval bases do occur. The sides are usually stepped or almost vertical and often vertically/obliquely striated. Sets of similar figurines were kept by astrologers. They sometimes have hollow bases (See Plate 61).

The shapes and numbers of the signs on the elephant weights:
It is usually only the elephant weights with octagonal bases which bear signs. Signs have been reported on other animal shapes but from their appearance their authenticity is suspect. The signs on the elephant weights occur on the bases, usually on the front or on the left side. More than one sign on a weight is rare. In general shape, but not in detail, they are similar to those on the Burmese weights and dissimilar from those on the bullet ingots of Ayudhya and Sukhotai. Mostly they are in the form of radiating stars with 4, 5, 6, 7 or 8 narrow or 'petal' shaped rays, sometimes enclosed within a surrounding circle.[27] Concerning the silver ingots, exceptionally there is a sign in the shape of an elephant on the *kakim* ingots of the kingdom of Lan-Na, which fell to the Burmese in 1558 AD. There is one on the margin of some of the Chiengmai *tok* ingots. A horse shape is occasionally found there also. One of the three kinds of *lat* (*lad hoi*) ingot of the Mekong valley always bears an elephant-shaped sign. Another does so occasionally.

The usages of the weights:
As in Burma the smaller bronze elephant weights appear to have been used primarily for the weighing of silver bullion. They were also used as a form of money, since before 1940 it was the practice of the Chinese rice merchants to exchange the weights for rice, copper being a valuable commodity in this copper-short region. Apparently since the establishment of opium plantations at the beginning of the last century the animal weights have been used to weigh opium among other things. However, opium was grown by the peasant farmers from India to Yunnan and Siam long before the last century, so this usage may be much older. From this usage comes their commonly-used modern name, 'opium weights'.

Material:
The weights and figurines are made of a bronze-like material which, like the more recent Burmese animal weights, is often brassy in colour due to a high zinc content. The shortages of copper in the past may explain why about 98% of the number of weights available for study weigh less than 150 grams (10 *kyats* in Burmese terms). In the 1950s, however, there may have been rather more weights up to 300 grams. The larger weights might be fewer because they would have been easier to collect if they were required for making cannon or used for payment instead of money. In times of shortage two or three sets of small weights could have been made from one large weight.

The mass ranges of the weights and figurines:
The weights and figurines occurred in two series. The Burmese weights weighed up to 30 kg. The elephant weights and possibly the figurines, were found in supposedly seven-member sets, weighing from ⅓ *baht* to 20 *bahts* (5 grams to 300 grams).[28]

The accuracy of the masses of the weights and figurines:
Many of the Siamese weights and figurines have been adjusted. Some have had solder added, for example between the legs, in order to increase the mass. Others have had copper wire wrapped around them. Yet others have had metal removed from their bases in order to lighten them. Such adjustments are rare on weights found in Burma. Early French travellers accused Siamese merchants of having two sets of similarly shaped weights, a heavier one used when buying goods and a lighter one for selling the same article. This would be possible not only because the weights would be likely to vary from their means by up to +/- 10%, as with the Burmese weights, but also because two unit masses may have been in operation, which differed by 15% to 20%.

Reference to the table 'Masses and Mass Scales of the Siamese Elephant Weights and Figurines' makes it plain that if the figurines were intended to be weights on one of the known mass scales of the region, unitary, binary or decimal, then they were considerably less accurate than the Burmese weights. Only if an exceptionally large number of members were included in the scale, many with unusual multiples of the unit mass, could the accuracy be improved to within the +/- 10% determined for the Burmese weights. It will be shown later that the figurines are probably relatively recent for the greater part, i.e. the unit mass, and hence the accuracy, would be relatively little affected by the age of the figurines. On the other hand the elephant weights do mainly fall within the +/- 10% range.

The mass units and mass scales:
The table 'Comparison of Changes in the Mass of the Siamese *Baht* and Burmese *Kyat*' sets out what is known of the variations of the unit mass in central and southern Siam

as determined mainly by the masses of the bullet (*pod duang*) silver ingots. (Prior to 1782 AD the *pod duang*, were made to about the same accuracy as the weights of the Burmese empires.) The commercial mass unit and scale in Siam were wholly Chinese while the Indian-originated precious metal mass scale of the central and southern Siamese since before 1637 AD[29] until modern times has been as follows:

80 *baht* = 1 *chang (kati)*	½ *baht*
8 *baht* = 1 double *tamlung*	¼ *baht* = 1 *salung*
4 *baht* = 1 *tamlung (tael)*	⅛ *baht* = 1 *fuang*
2 *baht* = ½ *tamlung*	1/16 *baht* = 1 *pai-song*
1 *baht (tical)*	1/32 *baht* = 1 *pai-nung*

The scale determined for the large base-metal weights *(kud)*, probably datable to 14th century AD Lan-Na or Sukhotai, is 2, 3, 4, 5, 6, 7, 8 and 10 *baht*, no 9 *baht* nor any weights of greater mass being known.[30] A unit mass of 12 or 13 grams was used.

Turning to the *kakim* silver ingot currency masses[31] and using a unit mass of 12,3 grams appropriate to the age and location of the ingots (post 1250 AD to 1558 AD, northern Siam), then the mass scale is a decimal one of 10: 5: 1: ½: ¼. If an equivalent unit mass of 15,4 grams were used then the scale would be a binary one of 8, 4, 1, 0,4, 0,2. The latter scale under these conditions would be anomalous and therefore seems unlikely.

Considering the other main silver ingot currency of northern Siam, the Chiengmai *tok*, the average unit mass of those specimens known to the writers is either 16,2 grams or 13,0 grams. The respective binary and decimal mass scales, neither being well defined, would then be as follows:

8, 5?, 4, 2, 1 ¼, 1, ½, ¼, or

10, 6?, 5, 2,5, 1 ½, 1 ¼, ½, ⅜.

The larger masses of the silver alloy and copper *lat* (*lad hoi*) currencies of Laos and northern Siam appear to have a unit mass of about 12,1 grams and occur in the unitary scale 4, 5, 6 ,7, 8, 9, 10.

The following table 'Masses and Mass Scales of the Siamese Elephant Weights and Figurines' displays the information available to the writers, arranged to show the possible mass scales and associated mass units. The probable mass units and scales are a unit mass of 12,7 grams, yielding a unitary/decimal scale or a unit mass of 16,2 grams, yielding a unitary/binary scale. The former is more likely for the following reasons:

(a) a 16,2 gram unit is similar to the modern Burmese *kyat* mass of 16,3 grams and much greater than the 15 gram mass of the modern Siamese *baht.* Since these weights and figurines were used only in Siam, the mass of 16,2 grams is unlikely.

(b) in Chiengmai in 1615 AD the mass unit was between 11,9 and 12,6 grams while in 1826 AD the mass unit was reported as 11,8 grams.[33]

Turning to the elephant weights, from the 175 weight-masses available, the average unit mass, irrespective of any age variation, was determined to be 13,1 grams. The mass scale was as follows, allowing for ± 10% variation from the unit mass.

60, 40, 25, 20, 10, 5, 2,5, 1, 0,5, 0,25, 0,125

Only one specimen was available for each of the four heaviest members. Ignoring these the scale is decimal above one *baht* and binary below. Apart from the 2,5 member the scale from 20 *baht* and below is similar to that of the Burmese weights, the Group 6 bird weights and the marble weights.

The table lists 29 members of the two mass scales examined, allowing for an accuracy of ±10%. Among the reasons for this large number of members may be that after the departure of the Burmese the figurines could have been made to overcome a shortage of weights, or acceptable weights, to weigh masses such as the Chinese *liang,* the Laotian *lad hoi* ingots, the *pod duang* of south and central Siam and possibly to match the Burmese weights. Alternatively the weights may have been made to different mass units and mass scales at different times, e.g. the 16,2 gram unit mass after 1826 AD, or there may have been two different sets of mass units and mass scales made at the same time. Most probably, however, the figurines were never intended to be weights.

Table 24: Masses and Mass Scales of the "Siamese" Elephant Weights and Figurines

Mass Scales		**Weight Masses**			**No. of Weights as percentage of total weights**
		Calculated	**Actual (Means)**		
Assumed unit masses			**(grams)**	**(grams)**	
12,7	16,2	12,7	16,2	—	—
60	50	762	810	762	0,15
	40		648	—	—
40	30	508	486	501	0,29
30	25	381	405	—	—
25	20	317	324	309,2	0,29
20	15	252	243	268	0,15
15	12,5	190	202	—	—
12,5	10	159	162	157,1	0,59
10	8	127	129,6	125,8	7,47
6,5	5	82,6	81	82,7	1,46
5	4	63,5	64,8	63,5	3,76
4,5	3,5	57,1	56,7	56,6	0,15
4	3	50,8	48,6	50,1	1,46
3,5	2,75	44,5	44,6	44,1	1,32
3	2,5	38,1	40,5	37,9	7,47
2,5	2	31,8	32,4	32,4	11,13
2	1,5	25,4	24,3	25,1	2,34
1,75	1,4	22,2	22,7	22,7	0,29
1,5	1,25	19	20,3	19,2	3,66
1,25	1	15,9	16,2	15,5	9,68
1	0,75	12,7	12,2	12,7	22,84
0,85	0,66	10,8	10,7	10,7	0,59
0,75	0,6	9,5	9,7	9,7	2,93
0,66	0,5	8,4	8,1	8,1	3,37
0,5	0,4	6,4	6,5	6,5	4,98
0,33	0,26	4,2	4,2	4,1	0,59
0,25	0,2	3,2	3,2	3,3	1,32
0,165	0,13	2,1	2,1	2,1	1,03
0,125	0,1	1,6	1,6	1,7	0,73

The mass scales are multiples or fractions of the unit masses (grams). The calculated weight masses (grams) are those which would be obtained if the weights were made accurately to the assumed unit mass. For each calculated weight mass, values of 110% and 90% were derived. These limits were chosen because 97% of the Burmese weights fall within them. The actual weight masses were sorted into classes according to which 110% and 90% limits they fell within. the numbers of weight masses within each class were counted and expressed as percentages of the total number of weight masses available for study, i.e. 683.[11]

Notes:

[1] Shulman, p. 49. Kurz, p. 78.

[2] Fontein, p. 514.

[3] Shulman, p. 49.

[4] From the writers' study of the animals appearing as figurines and incised decoration in Rawson, 1983.

[5] Pirazzoli-T'Serstevens, p. 129.

[6] de Campos, pp. 122, 123.

[7] Le May, p. 61, referring to George Coedes, *Recueil des Inscriptions du Siam*, vols. 1 and 2.

[8] Anonymous, *Some Notes upon the Development of the Commerce of Siam*, p. 81.

[9] Le May, p. 61.

[10] de Campos, p. 134, quoting J. Mandelslo, *Voyages and Travels into the East Indies.* 1669. p.329.

[11] Le May, p. 57.

[12] Gardner, quoting the *Annals of Chiengmai*, 1962.

[13] U Kalar.

[14] Hall, p. 297.

[15] Ball, p. 67.

[16] Iyer, pp. 10, 49, 50, 64.

[17] Htin Aung, p. 102.

[18] Iyer, pp. 46, 47.

[19] Liebert, p. 255.

[20] Much of the information shown in this and the next three sections is based on Gardner, 1962 and 1968.

[21] Braun, p. 30, quoting Ehlers.

[22] Sale, pp. 103, 105, 106.

[23] Forein de Rochesnard, pp. 34, 35.

[24] Gardner, 1962, p. 2.

[25] Temple, 1928, p. 11. The elephant weights are frequently assumed to have originated from Lan Chang which formerly occupied the northern part of present day Laos, perhaps because the name of the country translates into English as 'country of the elephants'. Perhaps too the origin has been deduced, believing the weights to be connected with the three main types of Mekong valley currency ingots, the *lats (lad hoi)*, one of which always bears an elephant sign. The Mekong River, separates north-east Siam and north-east Burma from Laos.

If the set of signs on the weights (Braun, p. 61. Forien de Rochesnard, p. 118) are compared with those on the *kakim* ingots of Lan-Na (Kneedler, plates XIII, XVI; Cresswell, plates II, III), it will be seen that both sets are only various forms of star *(chakra?)* with different numbers of rays. Of those star signs published, 23 on the elephant weights, 30 on the *kakim* ingots, the respective proportions are four-rayed, 54: 43, six-rayed 8: 10, and eight-rayed, 12: 20. The northern Siam *tok* ingots rarely bear signs and even more rarely star signs. Of the three types of Mekong valley ingots, one rarely bears any sign, one bears almost always only the six-rayed star among its three signs and the other bears none to four signs, one of which is usually the four-rayed star and no other kind of star. Since the larger *kakim* ingots usually bear one of at least eight identifiable place names, it is possible to state that all the 13 kinds of four-rayed star signs which are found on them, except for two which are unknown, are associated with the name Chiengmai.

On the evidence presented here and in the text, the elephant weights originated in Chiengmai, capital of the former Lan-Na region, and not Laos, the former Lan Chang region.

[26] These frequently show the raised tail base of the Yunnanese horse.

[27] Braun, pp. 61, 62. Forien de Rochesnard, p. 118.

[28] Gardner, 1968, p. 1.

[29] Le May, p. 58.

[30] Le May, 1932, pp. 17, 18.

[31] Cresswell, p. 13. Le May, p. 57. Le Lacheur, p. 354 ff. Kneedler, p. 9.

[32] Braun, Forien de Rochesnard.

[33] Temple, 1928, p. 11, quoting from McLeod's and Richardson's journals in *Mission from Moulmein to the Frontiers of China.* 1826. Le May, p. 22. The variation in mass depends upon the mass assigned to the Ayudhyan *baht* of that time, i.e. between 14 and 15 grams.

APPENDIX 4
Malayan Gambar

Models:
In Malaya, particularly in the central states of Pahang, Perak and Selangor, infrequently are found animal models (*gambar*) crudely fabricated from lead and tin. (See Plate 62) The animals represented are two kinds of cock, the 'crocodile', the elephant, the tortoise, a fish, a grasshopper and a mythical lion similar to the Burmese *to* of about 1800 AD. The cock and the crocodile are the most common, the fish, elephant and grasshopper, the least.[1]

Usage:
Their usage is a matter of uncertainty. Some may have been used to represent sacrificial offerings in magical ceremonies associated with tin mining. Others may have been used as offerings to sultans.[2] One form of the cock, the *timma*, and the crocodile shape may have seen some use as currencies.[3] They may have been used as weights in the scales of balances, in particular to weigh tin ingots.[4] Several of the animal shapes have rings, usually on their backs, or holes through their noses by which, presumably, they were intended to be suspended. These may have been used as the movable counterpoises on those weighing beams known as steelyards, which long have been common throughout the Far East since ancient times.[5]

Masses:
The masses of the animal models are known to vary from 25 grams, perhaps less, to about 1710 grams and perhaps more. Using a unit mass of 11,3 grams and assuming a variation of ±10%, these masses can be arranged in a single crude scale running 150, 100, 45 or 40, 15, 10, 8, 3, 2, 1 or 1¼. However, there are too few weighings available to place reliance upon any unit mass or scale. Referring to the solid tin ingots used as currency with decreasing frequency from before the 15th century AD until the early 19th century AD, these were cast in crudely graduated masses which, in Perak, fell on an unusually-formed decimal scale of 10, 5, 2½, 1¼, 1, ½, where the *kati* (variable[5] but say about 600 grams) was the unit mass.[7] Using the *bidor* (*viss*) of 1500 grams, the scale becomes a binary one, i.e. 4, 2, 1, ½, ?, ¼, which is suggestive of an Indian origin. Neither scale bears a close relation to that of the *gambar*, nor do the unit masses. To use the *gambar* to weigh ingots on a balance would be distinctly inconvenient and require many sets of such weights throughout the country, of which few, if any, remain. To use the *gambar* as counterpoises would have been consistent with the practices of the time.[8]

Beliefs:

The original inhabitants of Malaya were animist. From about the 6th century AD, Hinduism, particularly Sivaism, entered the country and from the late 14th century AD this was replaced by Islam. During all this time Chinese immigrated and brought with them their own beliefs. Elements of all these beliefs persist[9] and some are apparent in the inferred symbolism of the *gambar.*

Symbolism:

The cock is symbolic of Skanda, the war god son of Siva.[10] It is also an averter of evil in India,[11] and similarly of good fortune in China. In the case of the *timma* cock style, the one to five rings below the animal representation may signify the sun, the cock being, symbolically, a solar bird. On the Malayan tin mines a cock was commonly sacrificed to the earth-spirit when taking over a plot of land, especially before commencing excavations. Later on during the life of the mine, portions of fowls were included among those items offered to propitiate the mine spirits. Fowls were also released to fly away carrying exorcised evil spirits with them.[12]

The grasshopper has somewhat similar symbolic characteristics to the cock. It also is symbolic of Skanda. In Malaya several small creatures were believed to bring tin ore to the mine while others such as the grasshoppers and some birds, were believed to proclaim the location of tin ore by their stridulations and chirpings.[13] In China the grasshopper also signifies good fortune, an abundance of good things.[14]

The crocodile *gambar* sometimes bears a horn on the nose which clearly indicates that it represents a particular form of the Hindu *makara.* This creature also symbolises an abundance of good things. For example, it is supposed to have pearls within its mouth. It also symbolises water, a necessity for those engaged in alluvial mining[15] in order to wash away the lighter quartz grains from the heavier grains of cassiterite (tin oxide). The crocodile and elephant were, like some other large animals, credited with supernatural powers and having human spirits. These could be invoked to drive out spirits of lesser power and for other purposes. In addition, the crocodile was believed to represent the water-god manifestation of the Lord of Beasts, Siva.[16]

The fish, in both India and China, is yet another animal which symbolises an abundance of good things, notably possessions and children, wealth and fertility. It is also an averter of evil and symbolic of water.[17] The tortoise (turtle) has similar symbolisms to that of the fish, and, also like it, is an avatar of Vishnu.[18] According to Hindu mythology the tortoise supports on its back an elephant which, in turn, supports the world. Both animals symbolise longevity and strength. Live tortoises were sometimes placed under the pillars of ancient buildings in China to ensure enduring support.[15] This matter is relevant because, during the mining process, large excavations were made into unconsolidated alluvium or incoherent decomposing granite resulting in the risk of collapsing sides.

Use in magic:

The mine-wizards, *pawangs*, of Malaya, were required to remove the evil spirits from the area to be mined before allowing the felling of any trees or commencing the excavation of the large pits forming the mine. In due course they were required to rid the excavation itself of such spirits, to induce the tin ore to show itself, or to become plentiful, or to cause minerals such as pyrites to change into tin ore. At the same time the benevolent spirits had to be propitiated. The *pawang* needed the help of the animal spirits but the large, and supernaturally powerful animals mentioned previously, could not be allowed in the mining area, not only because of the damage they might do to man, his equipment and the pit, but because their presence, or even their names, might displease the mine spirits. However, models of animals such as those mentioned above, but called by other names, could be used and in fact, are known to have been made in dough for the purpose of removing evil spirits from sick people, the spirits being believed to enter the animal representations[20] which could then be cast away.

Origins and History

The motifs for the animals, at least in part, appear to have been adopted in Malaya during the Hindu period because stone effigies of cocks in Java are referred to this period.[21] The *gambar* themselves may have first been made at about the end of this period (late 14th century AD) because, with the Chinese loss of control of Central Asia, more Chinese merchants had turned to the south-east Asian routes while Chinese miners had commenced working the Burmese lead and gem mines and possibly the Malayan tin deposits.

It may have been at about this time or earlier, during the greatest period of Chinese maritime trade in the 11th and 12th centuries AD, that the Chinese-Malayan Far East Commercial Mass Scale came into use (*picul*, *kati*, *tael*, *candareen*, etc.).

In the 14th century AD, though Chinese mariners along the length of the Malayan coast mentioned the trade in tin blocks, at that time shaped like bushel (basket-shaped?) measures, they made no mention of the *gambar*. They referred especially to the importance of Selangor's tin production. It is from Selangor comes one of the most frequently found *gambar*, the crocodile. It is to Selangor that there is a reference to the crocodile *gambar* being used as a currency. But no crocodile *gambar* are mentioned by the mariners. Seemingly either they were not being made, or were too rare, or were not allowed to be seen by foreigners. Yet in neighbouring Trengannu and elsewhere, the mariners were allowed to see the "gods" which the Malayans had carved from wood.[22] It has been suggested that the concept of the *gambar* was imported from, or re-stimulated by, Burma prior to 1678 AD, due to the activities of the Mon traders from Pegu.[23] If true, this suggests both the place and time (15th century) of their origin. The earliest known record of the currency type of cock, the *timma*, is dated 1775 AD.[24] The other kinds of *gambar* appear to have been mentioned only near the

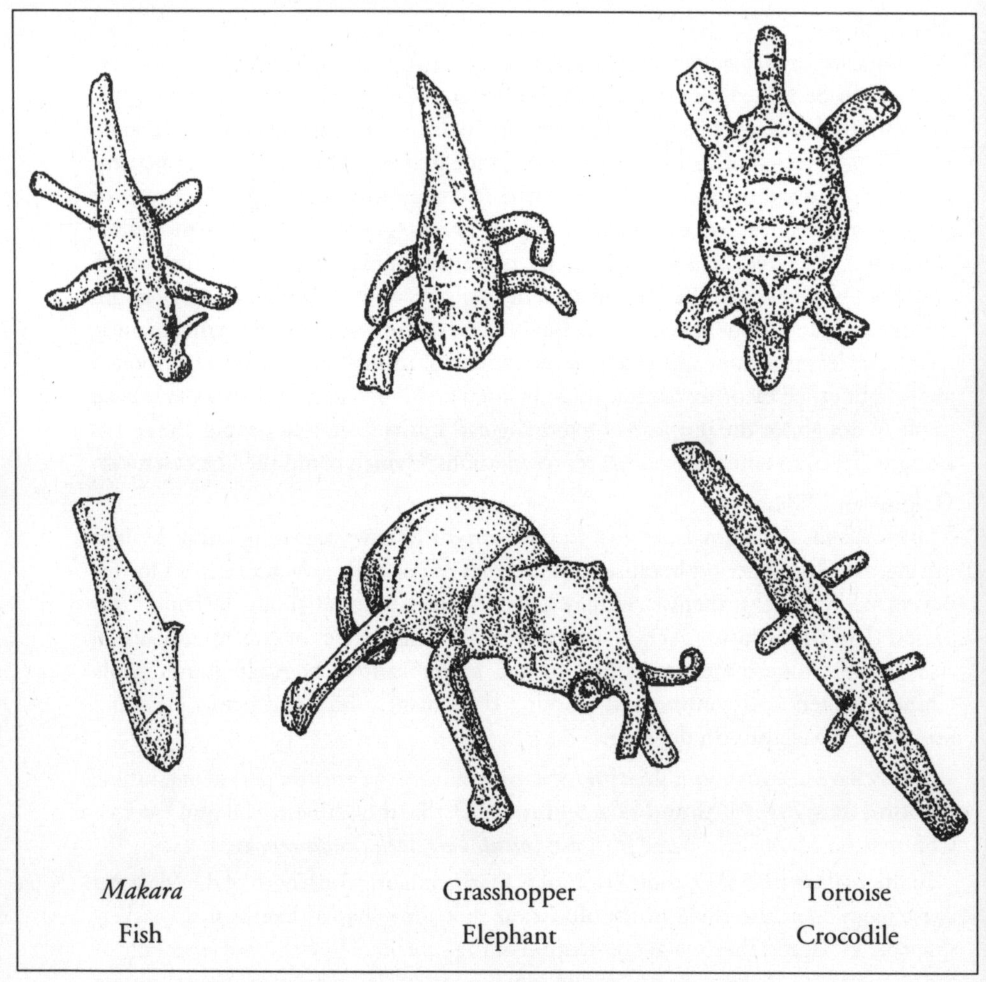

Plate 62: Malayan *Gambar*

end of the 19th century AD. The style of the mythical lion *gambar* suggests Burmese influence of the 19th century AD. During this same century the animal shapes were made by any "bellows-smith or magician" and this may have been the case in earlier centuries also.

Conclusions:
In conclusion the symbolism suggests that the *gambar,* in general, were required for one or both of firstly, the powers and properties attributed to the living animal and/or secondly, their ability to act as scapegoats for the removal of malevolent spirits. More specifically, they were required to secure good fortune, especially abundant water, tin ore and the financial success of the mining operation. They were needed also to avoid misfortune especially from collapsing sides, floods and overall, harm to the miners. If they were ever used as weights they could have been used only as counterpoises. Possibly one of the cock forms and the crocodile could have been used as currencies. In concept and material they may have originated from the metal animal-shaped weights of neighbouring Ramannadesa of the 15th century, though the same motifs in wood could be older.

Notes

1 Pridmore, 1970, p. 440. Shaw and Ali, 1970, provide illustrations as do Temple, 1914 and Quiggins. See also Wootton.
2 Pridmore, 1970, p. 491.
3 Temple, 1913, pp. 41, 46.
4 Temple, 1913, pp. 45, 46.
5 Yule and Burnell, p. 298, see "Datchin". Pridmore, 1971, p. 204. Annandale, pp. 195-200.
6 Prinsep, pp. 115-121.
7 Shaw and Ali, 1970, pp. 3, 4.
8 Brooks, p. 17.
9 Skeat, 1910 pp. 345-372.
10 Liebert, pp. 276-277.
11 Cooper, p. 38.
12 Skeat, 1900, pp. 43, 65, 143, 203, 270.
13 Skeat, 1900, pp. 263, 266.
14 Cooper, p. 76.
15 Iyer, pp. 75-77.
16 Skeat, 1900, pp. 89, 150, 151, 282, 435 – 6.
17 Cooper, p. 68. Eberhard, p. 106. Williams, p. 185.
18 Larousse, p. 362. Iyer, p. 77.
19 Williams, pp. 404-405. Eberhard, p. 44. Cooper, pp. 61, 175.
20 Skeat, 1900, pp. 262 – 269, 432. A Chinese belief also; de Lacouperie, 1888, p. 483.
21 Evans, p. 87.
22 Rockhill, pp. 114 – 121.
23 Temple, 1913, pp. 118, 120.
24 Shaw and Ali, 1970, p. 9, quoting the 1775 AD edition of Steven's *Guide to the East India Trade.*

BIBLIOGRAPHY

Abbreviations:

BRS	*Burma Research Society*, 50th Anniversary Volume, Publication No. 2. Rangoon, 1960.
EB	*Encyclopedia Britannica*. London, 1969.
ER	*Encyclopedia of Religion*. New York, 1987.
ERE	*Encyclopedia of Religion and Ethics*. Edinburgh, 1910. Reprint 1958.
ESEA	*Early South East Asia*. R.B. Smith and W. Watson, eds. Kuala Lumpur, 1979.
EWA	*Encyclopedia of World Art*. London.
HS1, HS2	*Hakluyt Society*, Series 1 and 2. London.
PDF	*Percival David Foundation of Chinese Art*. London.

J — Journal　　v — volume　　cols — columns
Soc — Society　　c — colloquy

Allchin, F. 'Manipur', *EB*. v. 14, pp. 790, 791.

Allchin, B. and Allchin, R. *The Birth of Indian Civilisation*, London, 1968.

Ammann, O. 'Malaya', *EWA*. v. 9, cols. 407-410. London, 1960.

Anderson, F. J. *Riches of the Earth*. New York, 1981.

Annandale, N. 'Weighing Apparatus from the Southern Shan States', *Memoirs Asiatic Society Bengal*. Calcutta, 1917. v. 5, pp. 195-205.

Anonymous. *Coins in Thailand*. Department of Fine Arts. National Museum, Bangkok, 1973.

Anonymous. *Mandalay Palace*. Directorate of Archaeological Survey. Rangoon, 1933? Reprint 1963.

Anonymous. 'Some Notes upon the Development of the Commerce of Siam', *J. Siam Soc.* Bangkok, 1926. pp. 78-102.

Anonymous. 'Weights of Historical Importance', *Myawaddy Magazine*. Rangoon, Nov. 1971. v. 20, pt. 1.

Argan, G.C. 'Emblems and Insignia', *EWA*. v. 4, 1961. pp. 710-712.

Armstrong, E.A. *The Folklore of Birds*. London, 1958.

Aung Thaw. 'Pyu Coins', *Amyatha Yinchenmo Sazang*. Rangoon, 1962. No. 1, pp. 105-112.

Aung Thaw. *Report on the Excavations at Beikthano*. Archaeological Survey Department, Ministry of Union Culture. Rangoon, 1968.

Aung-Thwin, M. 'Kingship in Southeast Asia', *ER*. v. 8, pp. 333-336.

Austin, O.L. and H.S. Zim, eds. *Birds of the World*. London, 1962.

Bacon, E. 'Nomads', *EB*. v. 16, pp. 558, 559.

Backus, C. *The Nanchao Kingdom and T'ang China's Southwestern Frontier.* Cambridge, 1981.

Bagchi, P.C. *India and China.* Philosophical Library. New York, 1951.

Ball, M.K. *Decorative Motifs of Oriental Art.* New York, 1969.

Ball, V. *Tavernier's Travels in India,* 2 v. London, 1889

Barbosa, D. (M. Dames, ed.) 'The Book of Duarte Barbosa', *HS2.* 1918, 1921. v. 48, 49.

Basham, A. L. *The Wonder that was India: a Survey of the Culture of the Indian Sub-Continent before the coming of the Muslims.* London 1959.

Basham, A.L. 'Hinduism', *EB.* v. 11, pp. 507-512.

Battisti, E. 'Zoomorphic and Plant Representation', *EWA.* 1967. v.14, cols. 936-965.

Beale, S. *Travels of Fah-hsien and Sung-yun, Buddhist Pilgrims from China to India,* London, 1869.

Bezacier, L. 'Vietnam', *EWA.* 1967. v. 14, cols. 766-777.

'Vietnamese Art', *EWA.* 1967. v. 14, cols. 777-791.

Birch, S. 'Sycee Silver', *Numismatic Chronicle.* New Series, Texas, 1864. Reprint 1968. v. 4.

Bird, G. W. *Wandering in Burma.* Bournemouth, England, 1897.

Bittel, K. 'Hittite Art', *EWA.* 1963. v. 7, cols. 559-575.

Blawatsky, V. 'Greco-Bosporan and Scythian Art', *EWA.* 1962. v. 6, cols 845-862.

Bock, *C. Temples and Elephants.* Singapore, 1883. Reprint 1986.

Bowman, S., M. Cowell and J. Cribb. 'Two Thousand Years of Coinage in China', *J. of the Historical Metallurgical Society.* London, 1989. v. 23, no. 1, pp. 25-30.

Bowrey, T. (R.C. Temple, ed.) 'A Geographical Account of the Countries Round the Bay of Bengal', 1669-1679', *HS2.* 1903. v. 12.

Boxer, C. R., ed. 'South China in the Sixteenth Century', *HS2.* 1953. v. 106 including:

'The Report of Galeote Pereira'

'The Treatise of Fr. Gaspar da Cruz, OP'

'The Relation of Fr. Martin de Rada, OESA'

Bradley, N. *The Old Burma Road.* London, 1945.

Braun, R. and I. *Opium Weights.* Landau, 1983. p. 240.

Bronson, B. 'The Late Pre-history and Early History of Central Thailand', *ESEA.* 1979. pp. 315-316.

Brooks, T. *An Authentick Account of the Weights, Measures and exchanges, etc. made use of and paid at the Several Ports in the East Indies.* London, 1752.

Brown, L.H., E.K. Urban and K. Newman. *The Birds of Africa.* London, 1982. v. 1.

Bunker, E.C. (N. Barnard, ed.) *The Tien Culture and Some Aspects of its Relationship to the Dong-S'on Culture in Early Chinese Art and its Possible Influence in the Pacific Basin.* New York, 1967. v. 2 .

Bunker, E.C., C.B. Chatwin and A.R. Farkas. *Animal Style Art from East to West.* The Asia Soc. New York, 1970.

Burnay, J. and G. Coedes. 'The Origins of the Sukhodaya Script', *J. Siam Soc.* Bangkok, 1927. v. 21, pt. 2, pp. 87-102.

Burrow, T. 'India-Pakistan: Ancient History', *EB.* v. 12, pp. 136-140.

Bussagli, M. 'Farther India', *EWA.* 1963. v. 7, cols. 913-930.

'Steppe Culture', *EWA.* 1967. v. 13, cols. 375-407.

'Aspects of Artistic Metalwork in Asia', *EWA.* 1964. v. 9, pp. 812-820.

'Monstrous and Imaginary Subjects', *EWA.* 1965. v. 10, pp. 266-269.

Chinese Bronzes. London, 1969.

Cameron, G.G. 'Media', *EB.* v. 15, pp. 68, 69.

Campbell, B. and E. Lack. *A Dictionary of Birds.* England, 1985.

Campbell, J. *Masks of God: Oriental Mythology.* London, 1970.

Carlson, R. 'Transportation', *EB.* v. 22, p. 174.

Carter, D. *The Symbol of the Beast.* New York, 1957.

Chaplin, D. *Mythological Bonds between East and West.* Copenhagen, 1938.

Chhibber, H.L. *The Mineral Resources of Burma.* London, 1934.

Childers, R.C. *Pali Dictionary.* London, 1875.

Christie, A. *Chinese Mythology.* London, 1974.

Cochrane, W.W. *The Shans.* Rangoon, 1915.

Codrington, C.C.S. *Ceylon coins and currency.* Colombo, 1924. Reprint 1975.

Coedes, G. 'Burmese Art', *EWA.* 1960. v. 2, cols. 742-751.

'Burma', *EWA.* 1960. v. 2, cols. 733-742.

The Indianised States of Southeast Asia. Honolulu, 1964. trans. 1968.

Colquhoun, A,.R. *Amongst the Shans.* London, 1885.

Coole, A.B. *Coins in China's History.* Kansas, 1965.

Coolidge, U.P. 'Chinese Sculpture', *EB.* v. 5, pp. 651-655.

Coomaraswamy, A.K. *History of Indian and Indonesian Art.* New York, 1965.

and B. Rowland. 'Indian Art', *EB.* v. 12, pp. 93-103.

Cooper, J.C. *Illustrated Encyclopedia of Traditional Symbols.* London, 1978.

Cotterell, A. China, *A Concise Cultural History.* London, 1988.

Coudert, A. 'Horns', *ER.* v. 6, p. 462.

Cowell, E.B., ed. *The Jataka or Stories of the Buddha's Former Births.* London, 1895. Reprint 1981. vols. 1-6.

Crawfurd, J. *Journal of an Embassy . . . to the Courts of Siam and Cochin China.* London. 1828.

Journal of an Embassy from the Governor-General of India to the Court of Ava, 2 v. London, 1834.

Creel, H.G. *The Birth of China,* London, 1936.

Cresswell, O.D. *'Early Coinage of South East Asia',* Numismatics International. Texas, 1974.

Cuisinier, J. 'Asia, South-Indochina', *EWA.* 1959. v. 1, cols. 848-855.

d'Alviella, G. *The Migration of Symbols.* London, 1894.

Davidson, J.H.C.S. 'Urban Genesis in Vietnam: a Comment', *ESEA.* pp. 304-314.

Davies, C. 'Chinese Jade', *Oriental Ceramic Soc.* London, 1975.

Davies, C.C. *An Historical Atlas of the Indian Peninsula.* Oxford, 1959.

Davies, H.R. *Yunnan, the Link Between India and the Yangtze,* Cambridge, 1909.

de Campos, J. 'The Origin of the *Tical*', *J. Siam Soc.* Bangkok, 1941. v. 33, pt. 2, pp. 119-135.

Decourdemanche, J. *Traité des monnaies, mesures et poids anciens et modernes de l'Inde et de la Chine.* Paris, 1913.

de Groot J. 'Buddhism in China', *ERE.* v. 3, pp. 552-556.

de Lacouperie, T. 'The Cradle of the Shan Race' in A.R. Colquhoun, 1885.

'Babylonia and China', *Archaeological Tract* 7703 g. 18. British Museum, London. No date.

de Lacouperie, T. *The Metallic Cowries of Ancient China (600 BC)* J. Roy. Asiatic Soc. London 1888 v. XX pp. 428 – 439.

de la Loubere, S. *The Kingdom of Siam,* Singapore, 1693. Reprint, 1986.

Denwood, P. 'Nomad Arts and Tibet', *PDF.* 1977. c. 7, pp. 218-233.

de Rochesnard. See J. Forien de Rochesnard.

Desai, W. *History of the British Residency in Burma, 1826-1840.* Rangoon, 1939.

Des Michels, A. *Dialogues Cochinchinois.* Paris, 1871.

Dobby, E.G. and B.B. Fall. 'Vietnam, History', *EB.* v. 23, pp. 1-5.

—— and D.K. Wyatt. 'Cambodia: Archaeology, Architecture and Art', *EB.* v. 4, pp. 678-681.

Dohanian, D.K. 'Indian Architecture', *EB.* v. 12, pp. 86-93.

Duche, J. *The Great Trade Routes.* London, 1969.

Dunan,, M., ed. *Larousse Encyclopedia of Ancient Medieval History.* London, 1964. Reprint 1981.

Dunnigan, A. 'Swans', *ER.* v. 14, pp. 188-189.

Duroiselle, C. The Ari of Burma and Tantric Buddhism. *Annual Report Archeological Survey of India,* 1915 – 1916, pp. 79 – 93, plates LXVIII – LIII.

Eberhard, W. (G. Campbell, trans.) *A Dictionary of Chinese Symbols.* London, 1986.

Einzig, P. *Primitive Money.* London, 1966.

Eliade, M. 'Shamanism', *EB.* v. 20, p. 342.

Shamanism, Archaic Techniques of Ecstasy. USA, 1972.

'Shamanism, An Overview', *ER.* v. 13, pp 202-208.

Elisseeff, V. 'Asiatic Protohistory', *EWA.* 1960. v. 2, cols. 1-38.

Elliott, W. *Coins of Southern India.* 1884. Reprint India, 1970.

Emmerick, R. 'Buddhism in Central Asia', *ER.* v. 2, pp. 400-420.

Evans, I.H.N. 'The Cock in Malayan Art and Its Relation to Currency', *J. Fed. Malay States Museums.* Kuala Lumpur, 1927. v. 12, pp. 87-90.

Fagg, W. 'Metalwork', *EWA.* 1960. v. 9, cols. 790-827.

Ferrand, G. 'Les Poids, Mesures et Monnaies des Mers du Sud', *J. Asiatique.* Paris, 1920. pp. 5-307. Most relevant to our work and not listed separately in our bibliography are the extracts from Antonio Nunes, 1554, pp. 27-52; Sparr de Homberg, ca. 1680, pp. 93-140 and the tables of metric equivalence of the ancient weights of India and Portugal by Jose Gomes Goes, ca. 1920.

Ferrars, M. and B. *Burma.* London, 1901.

Fillozat, J. 'Chronology, Hindu', *EB.* v. 5, pp. 721-722.

Flegg, G. *Numbers, Their History and Meaning.* London, 1983.

Floris, P. (W. Moreland, ed.) 'Peter Floris, His Voyage to the East Indies in the *Globe*, 1611-1615', *HS2.* 1934. v. 74.

Fontein, J. 'China', *EWA.* 1960. v. 3, cols. 393-466.

'Chinese Art', *EWA.* 1960. v. 3, cols. 466-577.

Forbes, C.J. *British Burma and its People.* London, 1878.

Forien de Rochesnard, J. *Les Poids d'Asie.* Paris, 1975.

Frampton, J. *The Travels of Marco Polo,* London, 1929.

Fraser-Lu. 'Burmese "Opium" Weights', *Arts of Asia.* Hong Kong, 1982. v. 1, pp. 73-81.

Frazer, J.G. *The Golden Bough.* London, 1983.

Fredericke, Caesar (T. Hickocke, trans.) *The Voyage and Travell of M.C. Fredericke to the East India and Beyond the Indies,* In *Purchas his Pilgrimage.* London, 1626.

Fry, R. *Chinese Art.* England, 1925.

Fryer, J. (W. Crooke, ed.) 'A New Account of East India and Persia, 1672-1681', *HS2.* 1879. v. 19, 20, 21.

Funke F.W. 'Monstrous and Imaginary', *EWA.* 1965. v. 10, cols. 250-265.

Furnivall, J.S. 'Europeans in Burma: The Early Portuguese', *BRA.* pp. 61-66.

Fugle-Meyer, H. *Chinese Bridges,* Shanghai, 1937.

Gairola, C.K. '*Hamsas* in Indian Art', *Arts of Asia.* Hong Kong, 1982. v. 12, no. 6.

Gait, E.A. A *History of Assam,* Govt. Printing, Calcutta, 1906.

Gardner, D. *Odd and Curious Money of South East Asia: Money Weights.* Colorado, n.d.

Money Weights of South East Asia. Colorado, 1968.

Gear, D. *The Early Coins and Bird Weights of Lower Burma.* Unpublished, 1984.

Giles, H.A. *A Chinese Dictionary.* Chen-wen, 1892. Reprint 1912.

Gill, W. *The River of Golden Sand.* The Narrative of a Journey through China and Eastern Tibet to Burmah, 2 v. London, 1883.

Gimbutas, M. 'Archaeology, Post Paleolithic Period in the Territories of the USSR,' *EB.* v. 2, pp. 268-270.

Glover, I.C. 'The Late Prehistoric Period in Indonesia', *ESEA.* pp. 167-184.

Goetz, H. 'Pala-Sena Schools', *EWA.* 1966. v. 11, pp. 22-37.

Goodrich, L.C. *A Short History of the Chinese People.* London, 1957.

'China: History, Middle Period', *EB.* v. 5, pp. 578-581.

Gorman, J.A. 'Horse', *EB.* v. 11, p. 705, plate 2.

Grancsay, S.V. 'Metalwork, Decorative', *EB.* v. 15, pp. 245-251.

Gray, L.H 'Cock', *ERE.* v. 3, pp. 694-698.

Griaznov, M.P. and E.A. Golomshtok. 'The Pazyryk Burial of Altai', *Archeological Institute of America.* 1933. v. 37.

Griswold, A.B. 'Siamese Art', *EWA.* 1967. v. 13, pp. 10-20.

Groslier, G. *Recherches sur les Cambodgiens.* Libraire Maritime et Coloniale. Paris, 1921.

Guehler, U. 'Further Studies of Old Thai Coins', *J. Siam Soc.* Bangkok, 1944. v. 35, pt. 2, pp. 147-172.

Guirand, F., ed. *New Larousse Encyclopedia of Mythology.* London, 1969. Reprint 1984.

Gutman, P. The Ancient Coinage of Southeast Asia, *J. Siam Soc.* Bangkok, 1978, v. 66. pp. 8 – 21.

Gutman, R. (D.G. Marr and A.C. Milner, eds.) 'Symbolism of Kingship in Arakan' in *Southeast Asia in the 9th to 14th Centuries.* Australian National University. Canberra, 1986.

Gyllensvard, B. 'Metalwork', *EWA.* 1964. v. 9, pp. 819-820.

Hakluyt, R. (J. Beeching, ed.) *Voyages and Discoveries.* Penguin Classic. London, 1972. Reprint 1985.

Hall, D.G.E. *Early English Intercourse with Burma, 1587-1743.* London, 1928.

'Studies in Dutch Relations with Arakan', *BRS.* pp. 67-97.

'The Dagh Register of Batavia and Dutch Trade with Burma in the 17th Century', *BRS.* pp. 99-116.

A History of South East Asia. London, 1968.

Hallade, M. 'Indo-Iranian Art', *EWA.* 1963. v. 8, cols. 1-17.

and R. Heine-Geldern. 'Indonesian Cultures', *EWA.* 1963. v. 8, cols. 41-90.

and L. Lanciotti. 'Myth and Fable, The Orient. China,' *EWA.* 1965. v. 10, cols. 487-492.

Hambis, L. 'Asia, Central ', *EWA.* 1959. v. 1, cols. 815-838.

Hamilton, C.H. 'Gautama Buddha', *EB.* v. 10, p. 36.

'Buddhism', *EB.* v. 4, pp. 354-361.

Hansford, S.H. 'Jade and Other Hard Stone Carvings', *EB.* v. 12, p. 841.

Harrison, B. *South East Asia.* London, 1966.

Harvey, G.E. *History of Burma.* London, 1925. Reprint 1967.

and Than Htun. 'Burma: History', *EB.* v. 4, pp. 440-441.

Heine-Geldern, R. See Hallade M. and R. Heine-Geldern.

Heissig, W. 'Buddhism in Central Asia', *ER.* v. 2, pp. 401-404.

'Buddhism in Mongolia', *ER.* v. 2, pp. 404-405.

Hendrickx-Baudot, M. 'The Weight System in the Harappa Culture', *Orientalia Lovaniensia Periodica.* Belgium. v. 3, pp. 5-34.

Herrmann, F. 'Symbolism and Allegory', *EWA.* 1967. v. 13, cols. 791-840.

Higgins, F. *The Chinese Numismatic Riddle.* New York Numismatic Club, 1910 (Reprint).

Hopkins, T.J. and A. Hiltebeitel. 'Indus Valley Religion', *ER.* v. 7, pp. 215-223.

Hori, A. 'A Consideration of the Ancient Near Eastern Systems of Weight', *Orient.* 1986. v. 22, pp. 16-36.

Htin Aung, M. *Burmese Folk Tales.* London, 1948.

A History of Burma. London, 1967.

Folk Elements in Burmese Buddhism. Rangoon, 1975.

Hudson, G.F. *Europe and China: A Survey of their Relations from Earliest Times to 1800.* London, 1931.

Hunter, W.H. *A Concise Account of the Kingdom of Pegu.* Calcutta, 1785.

Huntingford, G.W. 'The Periplus of the Erythraean Sea', *HS2.* 1980. v. 151.

Hutton, J.H. 'Naga', *EB.* v. 15, p. 1147.

Irwin, J. 'Asokan Pillars: A Reassessment of the Evidence', *Burlington Magazine.* London, 1973, 1974, 1975, 1976.

Iwata, S. 'On the Standard Deviation of the Weights of the Indus Civilisation', *Bulletin of the Soc. Near Eastern Studies in Japan.* 1974. v. 27, no. 2, pp. 13-36.

'Change of Mass Standard in Ancient Mesopotamia', *Bulletin of the Soc. Near Eastern Studies in Japan.* 1982. v. 25, no. 1, pp. 1-16.

Iyer, K.B. *Animals in Indian Sculpture.* Bombay, 1977.

Jacobi, H. 'Chakravartin', *ERE.* v. 3, pp. 336-337.

Jairozbhoy, R.A. *Foreign Influence in Ancient India.* London, 1963.

Janse, O. 'L'Empire des Steppes et les Relations entre L'Europe et L'Extreme Orient dans l}Antiquité, *Revue des Art Asiatiques.* Paris, 1935. tome 9.

Jenyns, R.S. and W. Watson. *Chinese Art.* London, 1963.

Jervis, T.B. *Weights, Measures and Coins.* Bombay, 1834.

Jettmar, K. *The Art of the Steppes.* London, 1967.

Jobes, G. *Dictionary of Mythology, Folklore and Symbols.* New York, 1962.

Jordanus, Fr. (Y. Yule, ed.) 'The Wonders of the East (ca. 1330)', *HS.* 1863. v. 31.

Judson, A. *The Judson Burmese-English Dictionary.* Rangoon, 1852. Reprint 1921.

Kann, E. *The Currencies of China.* Shanghai, 1926.

Kar, E. *Indian Metal Sculpture.* London, 1952.

Keely, H. and C.Price. *The City of the Dagger.* London, 1972.

Kelly, P. *The Universal Cambist.* London, 1835. 2nd ed.

Kelson, K.R. 'Deer', *EB.* v. 7, p. 166.

Kerr, R. 'The Tiger Motif: Cultural Transference in the Steppes', *PDF.* 1977. c. 7, pp. 74-87.

Keyes, C.F. 'Southeast Asian Religions, Mainland Cultures', *ER.* v. 13, pp. 512-520.

Khin Myo Chit. *A Wonderland of Burmese Legends.* Bangkok, 1984.

Kisch, B. *Scales and Weights: A Historical Outline.* London, 1965.

Kneedler, W.H. 'The Coins of North Siam', *J. Siam Soc.* Bangkok, 1937. v. 29, pp. 1-11.

Kosambi, D.D. *The Culture and Civilisation of Ancient India in Historical Outline.* New Delhi, 1970.

Krader, L. 'Hsi-hsia', *EB.* v. 11, p. 800.

Kramrisch, S. *Indische Kunst.* Germany, 1956.

Kuhnel, E. 'Persian Art', *EB.* v. 17, pp. 647-649.

Kulke, H. (D.G.Marr and A.C. Milner, eds.) 'The Early and Imperial Kingdom in Southeast Asian History' in *Southeast Asia in the 9th to 14th Centuries.* Australian National University. Canberra. 1986. pp. 1-22.

Kurz, O. et al. 'Astronomy and Astrology', *EWA.* 1960. v. 2, cols. 39-88.

Kyaw Hlaing. 'Preserve and Care for our Antiquities', *Hanthawaddy Newspaper.* (in Burmese) Mandalay, 22 November 1971.

Kyaw Tun. 'Standard Burmese Weights', *Hanthawaddy Newspaper.* Mandalay, 24 May 1971.

Lanciotti, L. and F. Herrmann. 'Symbolism and Allegory, China and Japan', *EWA.* 1967. v. 13, cols. 832-834.

Larousse. See F. Guirand.

Laughlin, J.L. *A New Exposition of Money, Credit and Prices.* Shanghai, 1931. v. 1, pp. 273-290.

Lee, S.E. *A History of Far Eastern Art.* London, 1964.

Lehman, F.K. 'Burmese Religion', *ER.* v. 2, pp. 575-580.

Lehner, E. *Symbols, Signs and Signets.* New York, 1969.

Le Lacheur, J. 'The Bronze and Silver Bars of the Mekong Valley', *Numismatics International Bulletin.* Texas, 1975.

Lemasurier, J.M.M. 'Iron, Archaeology', *EB.* v. 12, pp. 514-516.

Le May, R. *Asian Arcady,* London, 1926.

Le May, R. 'The Coinage of Siam', *J. Siam Soc.* Bangkok, 1932.

Lessing, F.D. *Ritual and Symbol. Collected Essays on Lamaism and Chinese Symbolism.* Taipei, 1976.

Lester, R.C. *Theravada Buddhism in Southeast Asia.* Michigan, 1973.

Levinson, J.R. 'China, History', *EB.* v. 5, pp. 570-598.

L'Hermitte, G.L. 'Mesures Laotiennes et Siamoises', *Revue de Métrologie Practique.* Paris, 1932. No. 2, pp. 4-6.

Liebenthal, W. 'The Ancient Burma Road — A Legend?', *J. Greater India Soc. India,* 1956. v. XV, no. 1, pp. 1-17.

Lieberman, V.B. *Burmese Administrative Cycles.* Princeton, 1984.

Liebert, G. *Iconographic Dictionary of the Indian Religions.* Leiden, 1978.

Lincoln, B. 'Indo-European Religions, An Overview', *ER.* v. 7, pp. 198-204.

Lion-Goldschmidt, D. and J. Moreau-Gobard. *Chinese Art.* London, 1960.

Litvinskii, B.A. 'Prehistoric Religions, The Eurasian Steppes and Inner Asia', *ER.* v. 11, pp. 516-522.

Liu, Xinru. *Ancient India and Ancient China.* Delhi, 1988.

Loehr, M. 'Ordos', *EWA.* 1965. v. 10, cols. 774-782.

'China: Prehistory and Archaeology', *EB.* v. 5, pp. 570-574.

Lokha Byuha Kyan. See Po Latt.

Lowry, J. *Burmese Art.* Victoria and Albert Museum. London, 1974.

Luce, G.H. and Pe Maung Tin. *The Glass Palace Chronicle of the Kings of Burma.* London, 1923.

Luce, G.H. 'Burma Down to the Fall of Pagan', *BRS* pp. 385-403.

'Economic Life of the Early Burman', *BRS.* pp. 323-375.

'The Ancient Pyu', *BRS.* pp. 307-322.

'The Tan (97-132 AD) and the Ngai-Lao', *BRS.* pp. 201-238.

'The Smaller Temples of Pagan', *BRS.* pp. 179-186.

Luce, G.H. Note on the Peoples of Burma in the 12th – 13th Century AD. *JBRS* Rangoon, 1959, v. XLII, pt. i. pp. 52 – 74,

'Phases of Old Burma', *School of Oriental and African Studies.* 1985. v. 1.

Lyons, E.B. 'Dvaravati, a Consideration of its Formative Period', *ESEA.* pp. 352-359.

MacCulloch, J.A. 'Horns', *ERE.* v. 6, pp. 791-796.

Mackenzie, C. 'The Chu Tradition of Wood-carving', *PDF.* c. 14, pp. 82-102.

Mackenzie, D. *The Migration of Symbols.* London, 1926.

Mahoney, W.K. 'Chakravartin', *ER.* v. 3, pp. 5-7.

Ma Huan. (J. Mills, trans.) *The Overall Survey of the Ocean's Shores, 1433.* Cambridge, 1970.

Maity, S.K. *Early Indian Coins and Currency Systems.* New Delhi, 1970.

Major, R.H., ed. 'India in the Fifteenth Century', *HS1.* 1857. v. 22 including:
'Narrative of the Voyage of Abd-er-Razzak, AD 1442'
'The Travels of Nicolo Conti in the East in the Early Part of the Fifteenth Century'
'The Travels of Athanasius Nikitin'
'The Journey of Hieronimo di Santo Stefano'

Malcom, H. *Travels in South East Asia.* Boston, USA, 1839.

Manrique F. S. (C. Luard and H. Hosten, eds.) 'Travels of Fray Sebastian Manrique, 1629-1643 AD)', *HS2.* 1927. v. 59, 61.

Matthews, L.H. 'Muntjac', *EB.* v. 15, p. 997.

McWhirter, N.D., ed. *Guinness Book of Records.* London, 1985. 31st ed.

Meserve, R. 'Inner Asian Religions', *ER.* v. 7, 238-247.

Milburn, W. *Oriental Commerce,* 2 v. London, 1813. New ed. 1 v. 1825.

Mills, J., trans. Ma Huan's *The Overall Survey of the Ocean's Shores.* Hakluyt Soc. London, 1970.

Milne, L. and W.W. Cochrane. *The Shans at Home.* London, 1910.

Mitchiner, M. *The Origins of Indian Coinage.* London, 1973.

Mollat, H. 'Die standardformen der tiergewichte Birmas', *Baessler Archiv. Neue Folge.* W. Germany, 1984. Band 32, pp. 405-440.

Moortgat, A. 'Mesopotamia', *EWA.* 1964. v. 9, cols. 736-790.

Morland, W. H., ed., 'Relations of Golconda in the Early 17th Century', *HS2.* 1931. v. 66.

Morrison, J.R. *A Chinese Commercial Guide.* Canton, 1834. pp. 62-69.

Mouhot, H. *Travels in the Central Parts of Indo-China, Cambodia and Laos, 1858 – 60,* Singapore, 1864. Reprint 1989.

Mundy, P. (R.C. Temple, ed.) 'The Travels of Peter Mundy in Europe and Asia 1608-1667', *HS2.* v. 45.

Murthy, K.K. *Mythical Animals in Indian Art.* New Delhi, 1985.

Myo Chit. See Khin Myo Chit.

Narain, A.K. 'Saka', *EB.* v. 19, p. 928.

National Weights and Measures Office (De Guojia Jiliang Zangju). *Weights and Measures in China Through the Ages* (*Zhong Guo Gu Du Liang Heng Tu Ti*). Peking, 1981 (in Chinese).

Needham, J. and Wang Ling. *Science and Civilisation in China.* London. v. 4, Physics and physical technology, pts. land 2, 1962.

Neugebauer, O.E. 'Mathematics, History of', *EB.* v. 14, pp. 1101-1104.

Noetling, F. 'Birmanisches Maass und Gewicht', *Zeitschrift fur Ethnologie. Berlin,* 1896. v. 6, pp. 40-46.

Noss, J.B. 'Animism', *EB.* v. 1, pp. 984-985.

O'Connor, S. *Mandalay and Other Cities of Burma,* London, 1907.

O'Flaherty, W.D. 'Horses', *ER.* v. 6, pp. 463-468.

O'Neill, H.B. *Companion to Chinese History.* Oxford, 1987.

O'Neill, J. P. *Along the Ancient Silk Routes: Central Asian Art from the West.* Berlin State Museums, Metropolitan Museum of Art. New York, 1982.

Overmeyer, D.L. 'Chinese Religion: An Overview', *ER.* v. 3, pp. 257-285.

Page, A.J. 'Pegu District', *Burma Gazetteer*. Rangoon, 1917. Reprint 1963. v. A.

Panish, C.K. 'The Coins of North Cambodia', *Amerzican Numismatic Soc.* New York, 1975. Notes 20, pp. 161-175.

Parker, E.H. *Burma with Special Reference to her Relations with China.* Rangoon, 1893.

Peale, S.E. Note on the Old Burmese Route over Patkai via Nangyang, *J. As. Soc. Bengal,* 1879 v. 48, pt. 11. 1879.

Perkins, A. 'Griffin', *EB.* v. 10, p. 924.

Petrie, W.M.F. *Ancient Weights and Measures (in Glass Stamps and Weights).* England, 1926. Reprint 1974.

Measures and Weights. London, 1934.

Phayre, A.P. *Coins of Aracan, of Pegu and of Burma.* International Numismata Orientalia. London, 1882.

Phayre, A. *History of Burma,* London, 1883.

Philips, C.H. and H.M. Vinacke. 'Asia: Social-Economic and Cultural History', *EB.* v. 2., p. 600.

Piggott, S. 'Archaeology: Indian Subcontinent', *EB.* v. 2, pp. 258-261.

Pirazzoli-T'Serstevens, M. 'The Bronze Drums of Shizhai Shan, their Social and Ritual Significance', *ESEA.* pp. 125-136.

Pires, R. (A. Cortesao, ed. and trans.) 'The Suma Oriental of Tome Pires, 1512-1515', *HS2.* 1944. v. 89, 90.

Po Latt. *Lokha Ryuha Kyan.* Ministry of Culture, Burma. Rangoon, 1958 (in Burmese).

Polo, M. (T. Wright, ed., W. Marsden, trans.) *The Travels of Marco Polo.* London, 1904.

Pope, A.U. and P. Ackerman. *A Survey of Persian Art.* London, 1965.

Praz, M. 'Emblems and Insignia', *EWA.* 1961. v. 4, cols. 710-734.

'Demonology', *EWA.* v. 4, cols. 306-335.

Pridmore, F. 'The Native Coinages of the Malay Peninsula', *Spinks Numismatic Circular.* London, 1970, 971, 1972.

Prinsep, J. (E. Thomas, ed.) *Essays on Indian Antiquities.* India, 1858. Reprint 1971. v. 1, 2.

Pyrard, F. (A. Grey and H. Bell, eds.) 'Voyage to the East Indies, the Maldives, the Moluccas and Brazil', *HS1.* 1887. v. 76, 77.

Quaritch-Wales, B. *Prehistory and Religion in South East Asia,* London, 1957.

Quiggin, A.H. *A Survey of Primitive Money.* London, 1949. Reprint 1978.

Rapson, E.J. *Ancient India* 1914. Reprint Chicago, 1974.

Rawson, J. *The Chinese Bronzes of Yunnan.* London and Beijing, 1983. Foreword.

Chinese Ornament. British Museum Pub. London, 1984.

Rice, T.T. *Ancient Arts of Central Asia.* London, 1965.

'Scythians', *EB.* v. 20, pp. 116-119.

Ridgway, W. *The Origin of Currency and Weight Standards.* Cambridge, 1892. Reprint New York, 1976.

Robinson, M. and L. Shaw. *Coins and Banknotes of Burma.* Manchester, 1980.

Robinson, M. *The Lead and Tin Coins of Pegu and Tenasserim.* Cheshire, 1986.

Rock, J.F. *The Ancient Na-Khi Kingdom of Southwest China,* 2 v. Harvard, 1949.

Rockhill, W.W. Notes on the Relations and Trade of China with the Eastern Archipelago and the Coast of the Indian Ocean during the Fourteenth Century. *Toung Pao,* v. XV and XVI. Leiden.

Rosati, F.P. 'Coins and Medals', *EWA.* 1960. v. 3, cols. 700-748.

Rowland, B. 'Buddhist Primitive Schools', *EWA.* 1960. v. 2, p. 703-722.

Rubruck, William of. (P. Jackson and D. Morgan, eds.) 'The Mission of Friar William of Rubruck', *HS2.* 1990. v. 173.

Sahai, S. 'Medium of Exhange in Ancient Cambodia', *J. Numismatic Soc. India.* Benares, 1971. v. 33, pt. 1, pp. 90-104.

Sainson, C. *Histoire Particulière du Nan-Tchao.* Ecoles des Langues Orientales Vivantes. Paris, 1904.

Sale, G. 'Poids Birmans et Laotiens', *L'Art Decoratif.* Paris, 1901. v. 12, pp. 103-105.

Salim Ali. *Book of Indian Birds.* Natural History Soc. Bombay, 1955.

Salmony, A. 'Antler and Tongue', *Artibus Asiae.* Switzerland, 1954.

Sangermano, Rev. *A Description of the Burmese Empire.* Transl. by W. Tandy. Rome, 1833.

Savill, S. 'The Orient', *Pears Encyclopedia of Myths and Legends.* London, 1977.

Scerrato, U. 'Sassanian Art', *EWA.* 1966. v. 12, cols. 702-730.

Schmidt, H.J. 'Textiles, Embroidery and Lace, China, India and Iran', *EWA.* 1967. v. 14, cols. 17-23.

Schroeder, A. 'Annam', *Études Numismatiques.* Paris, 1905.

Scott, J.G. (Shwe Yoe). *A Handbook for Commercial and Political Information.* London, 1906. *The Burman, His Life and Notions.* London, 1910.

Scott, J.G. 'Buddhism in Burma and Assam', *ERE.* v. 3, pp. 37-43.

Semeka-Pankratov, E, The meaning of the term "makara", *Semiotica,* 49 – 3/4 pp. 191 – 242, 1984.

Shabad, T. 'Sinkiang', *EB.* v. 20, p. 568.

'Tsing-hai', *EB.* v. 22, pp. 293, 294.

Shaw W. and M.K.H. Ali. *Tin Hat and Animal Money.* Museum Department. Kuala Lumpur, 1970.

Coins of North Malaya. Museum Department. Kuala Lumpur, 1971.

Shorto, H. 'The 32 Myos in the Medieval Mon Kingdom', *School of Oriental and African Studies.* 1963. No. 26, 572-591.

'The *Stupa* as Buddhist Icon', *PDF.* 1971. c. 2, pp. 75-81.

Shulman, S. *Encyclopedia of Astrology.* London, 1976.

Shwe Yoe. *The Burman, His Life and Nations.* London, 1910.

Sickman, L. and A. Soper. *The Art and Architecture of China* London, 1956.

Sigler, P. *Sycee Silver.* American Numismatic Soc. New York ,1943.

Siikala, A. 'Shamanism, Siberia and Inner Asia', *ER.* London, 1956. v. 13, pp. 208-215.

Skeat, W.W. *Malay Magic,* London, 1900. Reprint 1984.

Skeat, W.W. 'The Malay Peninsula', *ERE.*, v. 8, pp. 345-372.

Smith, B.W. '60 Centuries of Copper', *Copper Development Association Publication 69.* London, 1965.

Smith, R.B. 'South East Asia in the First Millenium AD', *ESEA*. pp. 443-456. and W. Watson. 'Protohistory and the Early Historical Kingdoms', *ESEA*, pp. 257-261.

Smythies, B.E. *Birds of Burma*. Rangoon, 1940.

Snellgrove, D.L. 'Tibetan Art', *EB*., v. 21, pp. 1116-1119.

Soper, A. 'South Chinese Influence on the Buddhist Art of the Six Dynasties Period', *Museum of Far Eastern Antiquities Bulletin 32*. Stockholm, 1960. pp. 47-112.

South, M. ed. *Mythical and Fabulous Creatures*. London, 1987.

Spiro, M.C. *Burmese Supernaturalism*. New York, 1967.

Stamp, L.D. 'Madras', *EB*. v. 14, pp. 56-57.

Stevenson, H.N.C. 'Caste', *EB*. v. 5, pp. 24-33.

Swearer, D.K. 'Buddhism in Southeast Asia', *ER*. v. 2, 385-388.

Symes, M. *An Account of an Embassy to the Kingdom of Ava*, London, 1800.

Taw Sein Ko. *Burmese Sketches*. Rangoon, 1913.

Temple, R.C. 'Currency and Coinage among the Burmese', *Indian Antiquary*. Bombay. 1897, 4 articles; 1898, 9 articles; 1919, 15 articles; 1927, 1 article; 1928, 5 articles.

'The Development of Currency in the Far East', *Asiatic Quarterly Review*. Bombay, 1899. pp. 299-317.

'Burma', *ERE*. v. 3, pp. 17-44.

'The Obsolete Tin Currency and Money of the Federated Malay States', *Indian Antiquary*. Bombay, 1914. v. 43. esp. pp. 92 – 98.

Temple, R.C. On Certain Specimens of Former Currency in Burma, *Indian Antiquary*, Bombay, 1931 v. 60, pp. 70 – 77.

Than Tun. (P Strachan, ed.) *Essays on the History and Buddhism of Burma*. Scotland, 1988.

Thiri Pyanchi and Tha Myat. 'Hintha Bronze Weights', *Yin Kyay Hmu Magazine*. Rangoon, 1959. v. 3, pt. 3.

Thomas, E. 'On Ancient Indian Weights', *Marsden's Numismata Orientalia*. London, 1874. pt. 1.

Thomas, P. *Epics, Myths and Legends of India*. Bombay, 1980.

Toynbee, A. and E.D. Myers. *Historical Atlas and Gazetteer*. London, 1959. Reprint 1962. v. 11, Study of History.

Tsugio, Mikami. *Ceramic Art of the World*. Japan, 1981. v. 13, Liao, China and Yuan Dynasties.

U Kalar. *Maha Yazawungi* (*Great Chronicle of 1724 AD*). Burma Research Soc. Rangoon, 1960 (in Burmese).

Van Linschoten, J. (A. Burnell and P. Tiele, eds.) The Voyage of John Huyghen Van Linschoten to the East Indies', *HS1*. 1885. v. 70, 71.

Varthema, L. di. (C. Badger, ed., J. Jones, trans.) The Travels of Ludovico di Varthema, AD 1503 to 1508', *HS1*. 1863. v. 32.

Veith, I. 'Medicine and Religion in Eastern Traditions', *ER*. v. 9, pp. 312-319.

Vogel, J.P. *The Goose in Indian Literature and Art*, Holland. 1962.

Von Furer-Haimendorf, C. 'Asia: South-India', *EWA.* 1959. v. 1, cols. 839-848.
'India-Pakistan: Anthropology', *EB.* v. 12, pp. 129-133.

Von Manfred, M. *A Concise Etymological Sanskrit Directionary.* Heidelberg, 1964.

Wagel, S.R. *Finance in China.* Shanghai, 1914.

Waida, M. 'Frogs and Toads', *ER.* v. 5, p. 443.

Walker, B. *The Hindu World.* London, 1968.

Walters, W.T. *Oriental Ceramic Art.* London, 1981.

Warren, C. *The Early Weights and Measures of Mankind.* Committee of Palestine Exploration Fund. London, 1913.

Waterbury, F. *Early Chinese Symbols and Literature.* New York, 1942.

Watson, F. *A Concise History of India,* London, 1974.

Watson, W. *The Genius of China.* London, 1973.
Style in the Arts of China. New York, 1975.
The Art of Dynastic China. London, 1981.
and H.M. Vinacke. 'Asia: Archaeology and Pre-History', *EB.* v. 2, pp. 595-598.

Weins, H. J. *Han Chinese Expansion in South China,* U.S.A. 1967.

Willetts, W.Y. 'Chinese Bronzes', *EB.* v. 5, pp. 625-629.

Wilhelm, H. 'Chinese Mythology', *EB.* v. 5, pp. 642-645.

Williams, C.A.S. *Outlines of Chinese Symbolism and Art Motives.* Shanghai, 1932.

Williamson, A. 'Shwebo District', *Burma Gazetteer.* Rangoon, 1929. Reprint 1963. v. A.

Wilson, H.W. *Documents Illustrative of the Burmese War.* Calcutta, 1827.

Wood, W.A.H. *A History of Siam.* London, 1926.

Wootton, A. Notes on Some Forms of Primitive Currency and Money, Seaby's *Coin and Medal Bulletin,* London, Dec. 1964.

Wrights, E. *The Ancient World.* London, 1979.

Wu Cheng Lo (Ch'eng Li-chun rev.). *History of the Weights and Measures of China.* Shanghai, 1937. (in Chinese).

Wyatt, D.K. *Thailand: A Short History.* London, 1984.

Wylie, T.V. 'Tibet', *EB.* v. 21, pp. 1106-1116.

Yapp, M.E. 'Caravan', *EB.* v. 4, p. 860.

Yetts, W.P. *Symbolism in Chinese Art.* Holland, 1912.

Yule H. (H. Cordier, ed.) *The Book of Ser Marco Polo,* London, 1903.

Yule, H. and A. Burnell. *Hobson-Jobson.* London, 1903.
(H. Cordier, ed.) 'Cathay and the Way Thither', *HS2.* 1915. v. 33, 37, 38, 41, including:'
No. 38 'Preliminary Essay'
No. 33 'Odoric of Pordenone'
No. 37 'Missionary Friars', pp. 1-103.
'Cathay under the Mongols', extracted from *Rashid ud din, 1300-1307.* pp. 107-133.
'Pegalotti's Notices of the Land Route to Cathay, 1330-1340', pp. 138-173
'Marignolli's Recollections of Eastern Travel, 1338-1353', pp. 177-269
No. 41 'Ibn Batuta's Travels in Bengal and China (ca. 1347)
'The Journey of Benedict Goes (ca. 1615) from Agra to Cathay', pp. 169-259
A Narrative of the Mission to the Court of Ava in 1855. Oxford,1968

Zakats, Z. 'Emblems and Insignia', *EWA.* 1961. v. 4, cols. 710-734.
Zhong Guo Gu Du Liang Heng Tu Ti. See National Weights and Measures Office.
Zimmer, H. (J. Campbell, ed.) *The Art Of Indian Asia.* New York, 1955.
Myths and Symbols in Indian Art and Civilization. New York 1962.
Zurcher, E. 'Buddhism in China', *ER.* v. 2, pp. 414-421.
Zwalf, W. *Buddhism, Art and Faith,* London, 1985.

Index

Plate numbers are shown at the end of an entry in bold type.

H

I

J

K

L